Michael Köhler and Wolfgang Fritzsche

Nanotechnology

1807–2007 Knowledge for Generations

Each generation has its unique needs and aspirations. When Charles Wiley first opened his small printing shop in lower Manhattan in 1807, it was a generation of boundless potential searching for an identity. And we were there, helping to define a new American literary tradition. Over half a century later, in the midst of the Second Industrial Revolution, it was a generation focused on building the future. Once again, we were there, supplying the critical scientific, technical, and engineering knowledge that helped frame the world. Throughout the 20th Century, and into the new millennium, nations began to reach out beyond their own borders and a new international community was born. Wiley was there, expanding its operations around the world to enable a global exchange of ideas, opinions, and know-how.

For 200 years, Wiley has been an integral part of each generation's journey, enabling the flow of information and understanding necessary to meet their needs and fulfill their aspirations. Today, bold new technologies are changing the way we live and learn. Wiley will be there, providing you the must-have knowledge you need to imagine new worlds, new possibilities, and new opportunities.

Generations come and go, but you can always count on Wiley to provide you the knowledge you need, when and where you need it!

William J. Pesce
President and Chief Executive Officer

Peter Booth Wiley
Chairman of the Board

Michael Köhler and Wolfgang Fritzsche

Nanotechnology

An Introduction to Nanostructuring Techniques

Second, Completely Revised Edition

WILEY-VCH Verlag GmbH & Co. KGaA

The Editors of this Volume

Prof. Dr. Michael Köhler
Technische Universität Ilmenau
Institut für Physik
Postfach 100 565
98684 Ilmenau

Dr. Wolfgang Fritzsche
Institut für Photonische Technologien
Abteilung für Nanobiophotonik
Postfach 100 239
07702 Jena

Cover
The background of the front cover design shows a fragment of Richard P. Feynman's famous classic talk "There's Plenty of Room at the Bottom" given on December 29, 1959. Reproducted with kind permission of Caltech's *Engineering & Science* magazine (**1960**, *23*, 22–36). Feynman's visionary speech can be read in full length at http://www.zyvex.com/nanotech/feynman.html. The Text has been written on a gold surface by Chad Mirkin's group using Dip-Pen Nanolithography (http://www.chem.northwestern.edu/~mkngrp/dpn.htm); notice that for example an "I" is 60 nm of width. Chapter 4.4 of this book deals with these kinds of techniques.

Library of Congress Card No.: applied for

British Library Cataloguing-in-Publication Data:
A catalogue record for this book is available from the British Library.

Bibliographic information published by the Deutsche Nationalbibliothek
The Deutsche Nationalbibliothek lists this publication in the Deutsche Nationalbibliografie; detailed bibliographic data is available in the Internet at http://dnb.d-nb.de

Printed in the Federal Republic of Germany.
Printed on acid-free paper

Composition Mitterweger & Partner GmbH, Plankstadt
Printing Strauss GmbH, Mörlenbach
Bookbinding Litges & Dopf Buchbinderei GmbH, Heppenheim
Wiley Bicentennial Logo Richard J. Pacifico

ISBN 978-3-527-31871-1

Contents

Nanotechnology. M. Köhler and W. Fritzsche

ISBN: 978-3-527-31871-1

Abbreviations and Acronyms

AES	Auger electron spectroscopy
AFM	atomic force microscopy
ALE	atomic layer epitaxy
ATP	adenosine triphosphate
BSE	back-scattered electron
CBO	coulomb blockade oscillations
CFL	capillary force lithography
CFS	chemical force spectroscopy
CMP	chemical-mechanical polishing
CNT	carbon nanotube
cNW-FET	crossed nanowire field-effect transistor
DGL	diffraction gradient lithography
DLP	diffusion-limited patterning
DNM	double-negative material
DPN	dip-pen nanolithography
DUV	deep ultraviolet
EBD	electron-beam deposition
EBDL	electron-beam deposition lithography
EBIT	electron-beam-induced deposition
EBL	electron-beam lithography
ECR	electron cyclotron resonance etching
EDX	energy-dispersive X-ray spectroscopy
ESCA	electron spectroscopy for chemical analysis (XPS)
EUV	extreme ultraviolet
EUVL	extreme-ultraviolet lithography
FIB	focussed ion beam
FIBL	focussed ion beam lithography
HD	high density
HSQ	hydrogen silsesquioxane
IL	interferometric lithography
ISL	iterative spacer lithography
ITO	indium tin oxide
ITRS	International Technology Roadmap for Semiconductors

Nanotechnology. M. Köhler and W. Fritzsche

ISBN: 978-3-527-31871-1

LB film	Langmuir Blodgett film
LEEB	low-energy electron beam
LEEBDW	low-energy electron-beam direct writing
LEESR	low-energy electron-stimulated reaction
MALDI	matrix-assisted laser desorption/ionization
MBS	multi-beam source
MC	molecule cluster
MHA	mercaptohexanoic acid
MOSFET	metal-oxide-semiconductor field-effect transistor
MTJ	magnetic tunnel junction
MWNT	multi-wall carbon nanotubes
NCA	nanochannel alumina
NEMS	nanoelectromechanical systems
NFL	near-field lithography
NIL	nano imprint lithography
NL	Novolak
ODT	octadecanethiol
OMVPE	organometallic vapor-phase epitaxy
PAAF	porous anodic alumina films
PC	polycarbonate
PDMS	polydimethylsiloxane
PE	polyethylene
PE-CVD	physically enhanced chemical vapor deposition
PEPE	perfluoropolyether
PET	plasma etching
PET	poly(ethylene-terephthalate)
PHOST	poly(hydroxystyrene)
PMMA	poly(methyl methacrylate)
PPP	poly(para-phenylene)
PUA	poly(urethane acrylate)
PXL	proximity X-ray lithography
QCA	quantum-dot cellular automata
QD	quantum dot
QDD	quantum dot devices
RIE	reactive ion etching
RSL	reversed spacer lithography
SE	secondary electron
SED	single-electron devices
SEM	scanning electron microscopy
SERS	surface-enhanced Raman scattering
SET	single-electron tunneling
SIMS	secondary ion mass spectrometry
SOI	silicon on insulator
SOQD	self-organized quantum dots
SPL	surface plasmon lithography

SPM	scanning probe microscopy
SPR	surface plasmon resonance
SPRINT	surface plasmon resonant interference nanolithography technique
SST	solid-state technology
STL	sidewall transfer lithography
STM	scanning tunneling microscopy
TEM	transmission electron microscopy; transmission electron microscope
TF	thermal flow
TMAH	tetramethylammonium hydroxide
TRR	tunnel resonance resistor
TSI	terrascale integration (in electronics: more than 1 trillion transistors per chip)
UHV	ultra-high vacuum
VTD	vapor transport deposition
WDX	wavelength-dispersive X-ray spectroscopy
WE	wet etching
XPS	X-ray photoelectron spectroscopy
XRL	X-ray lithography (Roentgen lithography)
ZPAL	zone-plate array lithography

1
Introduction

1.1
The Way into the Nanoworld

1.1.1
From Micro- to Nanotechniques

Microtechnology has changed our lives dramatically. The most striking impact is apparent in computer technology, which is essential for today's industry, and also for our individual life styles. Apart from microelectronics, microtechnology influences many other areas. The size of typical structures that is accessible is in the sub-micrometer range, which is at the limits of optical resolution and barely visible with a light microscope. This is about 1/1000 smaller than structures resolvable by the naked eye, but still 1000 times larger than an atom. Today's developments are addressing the size range below these dimensions. Because a typical structure size is in the nanometer range, the methods and techniques are defined as nanotechnology.

The consequent extension of the resolution limit of microscopes led to instruments with the capacity to resolve features below the wavelength of light: the field ion microscope, the electron microscope, and finally the family of scanning probe microscopes. Now it is possible to image individual molecules, and even single atoms.

Although chemistry and microtechnology appear to be fundamentally different, they are somehow related. They have mutual interests in the area of properties of materials. Microtechnology is not a simple extrapolation of conventional precise mechanical methods down to smaller dimensions. Chemical methods, such as plasma processes, wet chemical etching and photo resist techniques, are predominant compared with cutting or reshaping processes. However, microtechnology follows physical principles. As in classical chemistry, chemical processes in microtechnology use a relatively high number of similar particles. Individual particles play no dominant role, whether in fabrication methods or in applications.

In nanotechnology, the primary role of classical physical principles is replaced as molecular and atomic dimensions are approached. Physical–technical and chemical aspects influence the fabrication and the use and application of nanotechnical structures on an equal basis. The effects of mesoscopic physics, a field that is influenced by and uses quantum phenomena, complement these aspects. In contrast to classical chemistry, small ensembles or even individual particles can play a decisive role.

Nanotechnology. M. Köhler and W. Fritzsche

ISBN: 978-3-527-31871-1

The nanotechnology literature often focuses on the structure size and differentiates between two basic approaches. The *Top-down* approach tries to enhance the methods from microtechnology to achieve structure sizes in the medium and also lower nanometer range. This approach is based on a physical and microlithographic philosophy, which is in contrast to the other approach, where atomic or molecular units are used to assemble molecular structures, ranging from atomic dimensions up to supramolecular structures in the nanometer range. This *Bottom-up* approach is mainly influenced by chemical principles.

The challenge of modern nanotechnology is the realization of syntheses by the *Top-down* and *Bottom-up* approaches. This task is not driven entirely by the absolute structure dimensions, because today macro- and supramolecules extending up to hundreds of nanometers or even micrometers can already be synthesized or isolated from biological systems. So the overlap of both approaches is not a problem. Both techniques provide specific capabilities that can be implemented by the other. The lithographic techniques (*Top-down*) offer the connection between structure and technical environment. The interface with the surrounding system is given in this approach, but it is not really possible with the chemical (*Bottom-up*) approach. At the same time, the integration of nanostructures into a functional microtechnical environment is realized. On the other hand, chemical technologies provide adjustment of chemical binding strength and preferred orientation of bonds, together with a fine tuning according to the numbers of bound atoms or atomic groups and a classification of the spatial orientation based on the number of bonds and their angles.

Therefore, nanotechnology depends on both classical microtechnology, especially microlithography, and chemistry, in particular interfacial and surface chemistry and supramolecular synthesis. Additional basic methods are molecular biology and biochemistry, because nature has provided, with the existence of large molecules and supramolecular complexes, not only examples, but also interesting technical tools [1][2][3]. In the following sections, microtechnical and molecular basics are discussed, prior to particular methods for the creation of nanostructures, their characterization and application.

1.1.2 Definition of Nanostructures

A clear distinction between nanostructures and microstructures is given here arbitrarily using length measurements. Nanostructures are defined according to their geometrical dimensions. This definition addresses technical dimensions, induced by external shaping processes, with the key feature being that the shaping, the orientation and the positioning is realized relative to an external reference system, such as the geometry of a substrate. Of less importance is whether this process uses geometrical tools, media or other instruments.

A narrow definition of nanostructures is that they include structures with at least two dimensions below 100 nm. An extended definition also includes structures with one dimension below 100 nm and a second dimension below 1 µm. Following on from this definition, ultra thin layers with lateral sub-micrometer structure sizes are also nanostructures.

All spontaneously distributed or spontaneously oriented structures in materials and on surfaces are not incorporated in nanotechnical structures. However, this does not exclude the presence of such structures in nanotechnical setups, as long as their dimensions are in accord with the above-mentioned criteria. Also microstructured ultrathin layers are excluded, because they exhibit only one nanometer dimension. Nanodevices are devices with at least one essential functional component that is a nanostructure. Nanosystems consist of several nanodevices that are of importance to the functioning of the whole system.

1.1.3 Insight into the Nanoworld

The realization that there are small things in the world that are not visible to the naked eye extends back into human history. The development of the natural sciences created an interest in the microworld, in order to enable a better understanding of the world and the processes therein. Therefore, the development of new microscopic imaging methods represents certain milestones in the natural sciences. The microworld was approached by extending the range available for the direct visualization of objects through the enhancement of microscopic resolution.

Access to spatial modifications in the nanoworld is not limited to one direction. Long before instruments were available for the imaging of molecules, an understanding of the spatial arrangements of atoms in molecules and solids, in disperse systems and on surfaces had been developed. The basis for this development was the anticipation of the existence of small building elements, which extended back to Greek philosophers (Leukip and Demokrit: "atomos" – the indivisible = smallest unit). This hypothesis was confirmed by Dalton with the discovery of stoichiometry as a quantitative system in materials: chemical reactions are comprised of fixed ratios of reactant masses. Based on the systematic organization of chemical elements, developed by Döbereiner, Meyer and Mendeleyev, into the Periodic Table of the elements, and supplemented by models of the internal structure of atoms, a new theory of the spatial connection of atoms was created: the theory of chemical bonds. It not only defines the ratios of atoms involved in a reaction, but leads also to rules for the spatial arrangement of atoms or group of atoms. We know today that the immense variety of solid inorganic compounds and organisms is based on this spatial arrangement of chemical bonds. Stoichiometry and geometry describe the chemical aspects of molecules and solids. The stability and the dynamics of chemical changes are determined by the rates of possible reactions that are based on thermodynamics and kinetics. Key contributions to the understanding of the energetic and kinetic foundations came from Clausius, Arrhenius and Eyring.

1.1.4
Intervention into the Nanoworld

The scientific understanding of the molecular world and the application of quantitative methods laid the foundations of modern chemistry. Before the quantification of chemical reactions, there was already an applied area of chemistry, for example in mining or metallurgy. However, it was established through an empirical approach. The understanding of the molecular context and its quantitative description, supplemented by the control of reactions by parameters derived from theoretical work or model calculations, improved dramatically the conditions for manipulations in the molecular world. Measurements and quantitative work established the structure-oriented chemistry.

Synthetic chemistry, with its beginnings usually being attributed to the synthesis of urea by Friedrich Wöhler (1828), provides a molecular–technical approach to the nanoworld. The formulation of binding theories and the development of analytical methods for the elucidation of the spatial arrangements in molecules (e. g., IR spectroscopy, X-ray based structure determination, and NMR spectroscopy) transformed chemistry from a stoichiometric- to a structure-oriented science. Modern synthetic chemistry is a deliberate intervention into the nanoworld, because the arrangement of the bonds and the geometry of the molecules are addressed by the choice of both the reaction and the reaction parameters. In contrast to microtechnology, synthetic chemistry uses a large number of similar particles, which show a statistical distribution with regard to spatial arrangement and orientation. So today's molecular techniques connect a highly defined internal molecular geometry with an uncertainty in the arrangement of the individual particles with respect to an external frame of reference.

Recent decades have witnessed the synthesis of an increasing variety of internal geometries in molecules and solids with small and large, movable and rigid, stabile and high-affinity molecules and building units of solid materials. Apart from the atomic composition, the topology of bonds is of increased interest. A large number of macromolecular compounds have been made, with dimensions between a few nanometers and (in a stretched state) several micrometers. These early steps into the nanoworld were not limited to the molecular techniques. Physical probes with dimensions in the lower nanometer range are also suited to the fabrication and manipulation of nanostructures.

During the last few years, the technologies for the fabrication of integrated circuits have crossed the border into the realm of nanotechnology. The smallest structure elements of microelectronic chip devices made by mass production have become smaller than 100 nm. The gate length of solid-state transistors has reached the mid-nanometer range. The road map (ITRS 2001) demands a gate length of 13 nm by the year 2013 (Semiconductor Roadmap 2001) [4]. Nanosized solid-state devices are meanwhile realized in various research laboratories. The requirements of the large semiconductor industry provide a very powerful impetus for the further development of nanotechnologies. Other fields of device development also demand ever better nanofabrication tools. Thus, the entry into the nanoworld is propelled forwards by strong economic forces beside the purely scientific and general technological interests.

1.2 Building Blocks in Nanotechnology

Nanotechnology utilizes the units provided by nature, which can be assembled and also manipulated based on atomic interactions. Atoms, molecules and solids are therefore the basic building blocks of nanotechnology. However, there is a fundamental difference from the classical definition of a building material used in a conventional technical environment, which also consists of atoms and molecules in solid materials. The smallest unit in technical terms includes an enormous number of similar atoms and molecules, in contrast to the small ensembles of particles – or even individual particles – addressed in nanotechnology. This puts the definition of material into perspective. The properties of a material are determined by the cooperative effect of a huge number of similar particles in a three-dimensional arrangement and by a mixture of only a few types of similar particles (e. g., in an alloy). Many physical properties of materials require a larger ensemble of atoms for a meaningful definition, independent of the amount of material, for example, density, the thermal expansion coefficient, hardness, color, electrical and thermal conductivity.

With solid materials, it is known that the properties of surfaces may differ from the bulk conditions. In the classical case, the number of surface atoms and molecules is small compared with the number of bulk particles. This ratio is inverted in the case of nanoparticles, thin layers and nanotechnical elements. The properties of nanostructures are therefore more closely related to the states of individual molecules, molecules on surfaces or interfaces than to the properties of the bulk material. Also the terminology of classical chemistry is not fully applicable to nanostructures. Key terms, such as diffusion, reactivity, reaction rate, turnover and chemical equilibrium, are only defined for vast numbers of particles. So their use is limited to the case of nanostructures with small numbers of similar particles. Reaction rate is replaced by the probability of a bond change, and diffusive transport by the actual particle velocity and direction.

However, not all definitions from classical physics and chemistry are unimportant at the nanoscale. The consideration of single particles is preferred compared with the integral discussion of particles in solid, liquid or gaseous media. Because the dimensions extend to the molecular scale, the importance of the chemical interactions between particles is greatly enhanced compared with the classical case.

Nanotechnical elements consist of individual particles or groups of particles with different interactions between the atoms (Fig. 1). The following types can be distinguished:

Building block type	Analogy in classical materials
Single atom	–
Group of similar atoms	elemental solid (e. g., metals)
Group of different atoms with similar interactions between adjacent particles	compound solids (e. g., glass or salt crystals)
Single molecule	–
Group of different atoms with different interactions between adjacent particles	molecular solid (e. g., polymer)

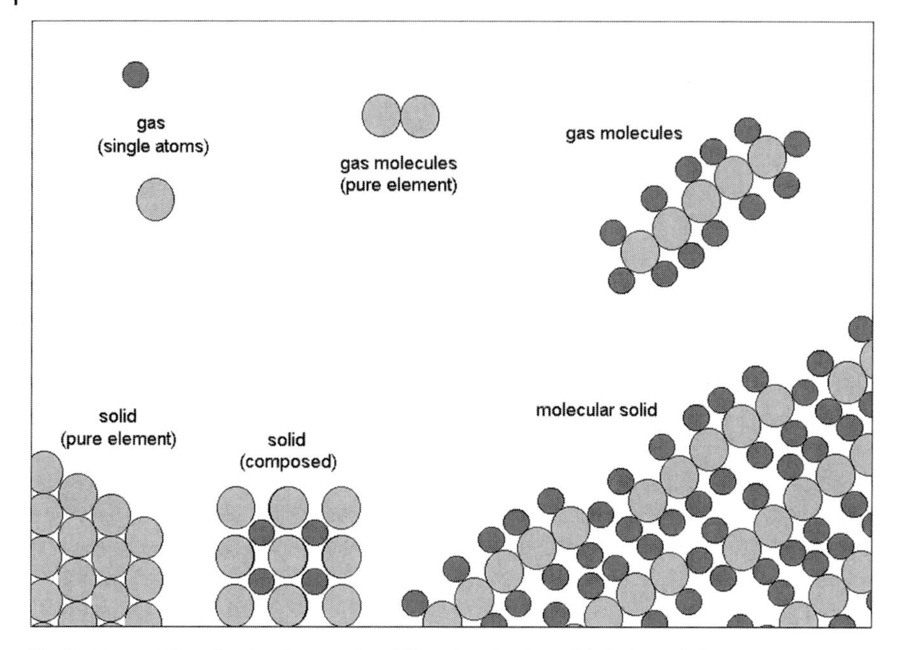

Fig. 1 Composition of molecules, atomic solids and molecular solids (schematics)

The dimensions for individual particles can be quite different. Atoms have diameters of about 0.1 nm; individual coiled macromolecules reach diameters of more than 20 nm. In an extended state, these molecules exhibit lengths of up to several micrometers. In principle, there is no upper size limitation for molecules. Technical applications usually use small molecules with typical dimensions of about 1 nm besides polymers and solids with three-dimensional binding networks. Synthetic molecules, such as linear polymers, exhibit, typically, molar masses of 10 000 to 1 000 000. These values correspond to particle diameters of 2–10 nm in a coiled state in most instances.

Apart from the molecules, both elemental solids and compound solids are essential for nanotechnology. They are, for example, prepared as nanoparticles with dimensions ranging from a few atoms up to diameters of 0.1 μm, corresponding to about 100000000 atoms. Similar values can be found in structural elements of thin atomic or molecular layers, in monomolecular films or stacks of monolayers. A number of one hundred million seems large, but it is still small compared with the number of atoms in standard microtechnological structures.

It is not usually the single atom, but small solids, large individual molecules and small molecular ensembles that are the real building blocks for nanotechnology. The nature of their connection and arrangement determines the constructive potential and functions of the nanotechnical devices and systems. Besides the standard lithographic methods known from microtechnology, a wide range of chemical techniques are applied in nanotechnology, from fields such as synthetic, surface, solid state, colloid and biomolecular and bioorganic chemistry. In addition to the importance of chemical methods in many microlithographical processes, these methods are increasing in influence in the nanometer range to become a key component in addition to the so-called physical techniques for the creation of small structures.

1.3 Interactions and Topology

Shaping and joining of materials to devices, instruments and machines is the prerequisite for functional technical systems. The spatial modification of material surfaces and the three-dimensional arrangement of the components result in a functional structure. This principle applies to both the macroscopic technique and the nanoworld. However, the spatial arrangement and functions at the nanometer scale cannot be described adequately by the classical parameters of mechanics and material sciences. It is not the classical mechanical parameters of solids, but molecular dimensions and individual atomic or molecular interactions (especially the local character of chemical bonds) that determine the arrangement and stability of nanostructures, their flexibility and function.

The properties of a material are controlled by the density of bonds, their spatial distribution and the bond strengths between the particles. For shaping and joining, the processes are determined by the strength and direction of positive interactions between the joining surfaces. In classical technology and usually also in microtechnology, a separation between the bonding forces in the bulk material and the surface forces has some significance. Both internal and external bonds are based on interatomic interactions, the chemical bonds. With the dimensions of nanotechnical objects approaching molecular dimensions, a combined consideration of both internal and external interactions of a material with its environment is needed. Besides the spatial separation of a material, the orientation of the internal and the surface bonds also determines the properties of materials or of material compounds.

Conventional technology uses materials with isotropic properties. Isotropy means that these properties are approximated as being similar in all spatial orientations of the solid. Restrictions are as a result of materials being created in an inhomogeneous process (e. g., wood) or materials transformed by processes inducing a preferred orientation (e. g., shaping). The macroscopic model of ideal isotropy is also not valid for single-crystalline materials such as silicon, gallium arsenide or other typical microelectronic materials. A single-crystalline solid excludes the statistical distribution of interatomic distances and of bond orientations. It includes elementary cells consisting of a few atoms, and a randomly oriented plane results in a density fluctuating with the angle of this plane. In addition, the bond strength between atoms is localized and is determined from its orientation. Such elementary cells create the solid in a periodic arrangement in an identical orientation. So the anisotropy of the particle density and bond strength on the atomic scale is transformed into macroscopic dimensions.

However, non-crystalline materials created by surface deposition processes can also show anisotropy. Almost all thin layers prepared by evaporation or sputtering exhibit anisotropy due to the preferred positioning by an initial nucleation and a limited surface mobility of the particles, which results in grain boundaries and the overall morphology of the layer. Even spin-coated polymer layers have such anisotropic properties, because the shear forces induced by the flow of the thin film lead to a preferred orientation of the chain-like molecules parallel to the substrate plane.

The transition from an almost isotropic to an anisotropic situation is partly based on the downscaling of the dimensions. For example, a material consists of many small crystals, so these statistically distributed crystals appear in total as an isotropic material. A classification of isotropic is justified as long as the individual crystals are much smaller than the smallest dimension of a technical structure created by the material. The dimensions of nanotechnical structures are often the same as or even less than the crystal size. The material properties on the nanometer scale correspond to the properties of the single crystals, so that they possess a high anisotropy even for a material with macroscopic isotropy.

The anisotropy of a monocrystalline material is determined by the anisotropic electron configuration and the electronic interactions between the atoms of the crystal. It is based on the arrangement of the locations of the highest occupation probability of the electrons, especially of the outer electrons responsible for chemical bonds. The length, strength and direction of the bonds as well as the number of bonds per atom in a material therefore determine the integral properties of the material and the spatial dependence of these properties.

The decisive influence of number, direction and strength of interatomic bonds is even stronger for the properties of molecules. Although molecules can have symmetrical axis, outside of such axis practically all properties of the molecule are strongly anisotropic. A material consisting of molecules can exhibit isotropic properties at a macroscopic level, as long as the orientation of the molecules is distributed statistically in all directions. At the nanoscale, anisotropy is observed, especially in the case of monomolecular layers, but also for molecular multilayers, small ensembles of molecules, clusters and individual molecules.

Because nanotechnological objects consist of anisotropic building blocks, it is usually not possible to construct systems where objects of the same type are distributed statistically with respect to their orientation. On the contrary, preferred directions are chosen, and also the connection to other molecules occurs in preferred orientations. So the anisotropic connection network of smaller and larger molecules and small solids leads to a constructive network of objects and connections, with anisotropically distributed stronger and weaker bonds both at the molecular level and in larger modules. These networks of bonds create connection topologies, which cannot be described simply by their spatial distribution. Depending on the character of the bonds between the particles, various complex topologies can interact with each other, depending on the point of view (e. g., conductivity, mechanical hardness, thermal or special chemical stability) of the description of the connection strength.

The discussion of topological connections in three-dimensional objects at the nanometer scale assists with the evaluation of properties, which are only described in an integral manner for classical solids. These properties are essential for the function of nanostructured devices, for processes involving movement, for chemical transformations, and for energy- and signal-transduction. The spatial relationship is of particular importance for the evaluation and exploitation of mesoscopic effects, which are unique for nanosystems, such as single quantum and single particle processes.

1.4
The Microscopic Environment of the Nanoworld

Nanometer structures are abundant in nature and the technology. The general tendency of nature towards the spontaneous creation of structures by non-equilibrium processes leads to the formation of more or less regular structures with nanometer dimensions. Such objects exist in a variety of time scales and exhibit rather dissipated or conserved character. Typical structures can be found in cosmic dust, in the inorganic structures of solidified magma, or in the early seeds of condensing atmospheric water vapor.

In contrast to many inorganic structures, the nanoscopic objects in nanosystems are not spatially independent, whether they are in technical systems or in natural functioning systems. They are always embedded in an environment or at least adjusted to interactions in a larger setting. Nature demonstrates this principle in an impressive manner. The smallest tools of life, the proteins, have dimensions of a few nanometers up to some tens of nanometers. They are usually found in closed compartments, in cells or cell organelles. Often an arrangement into superstructures, as in for example, cell membranes, can be observed. These tools for the lower nanometer range are produced in the cells as biological microsystems, and are usually also used by these cells. The slightly larger functional nanoobjects, such as cell organelles, are also integrated into this microsystem environment. The smallest biological objects with a certain functional autonomy are viruses. With dimensions of several tens of nanometers up to a few hundred nanometers they are smaller than the smallest cells, nevertheless they can connect thousands of individual macromolecules into a highly ordered and complex structure. However, they are not able to live on their own. Only when they (or their subsystems) interact with cells in a more complex nanomachinery are they able to reproduce and to induce biological effects.

This principle of integrating small functional objects into a wider environment is common in technical applications (Fig. 2). It can already be seen in conventional construction schemes, e. g., in the combination and functional connection of several units in the hood of a car. This principle is essential in microtechnology. Electronic solid-state circuits combine individual electronic devices, such as wires, transistors and resistors in a chip. The circuits are arranged on a circuit path, and these paths are assembled into machines. Approaching the nanotechnology range, even more levels of geometrical and functional integration are required, to make the nanoobjects usable and the interface functional. The large distance between the macroworld with typical dimensions of centimeters to meters and the structure sizes of the nanoworld has to be considered. This gap is comparable to the difference between a typical machine and up to near cosmic dimensions (Fig. 3).

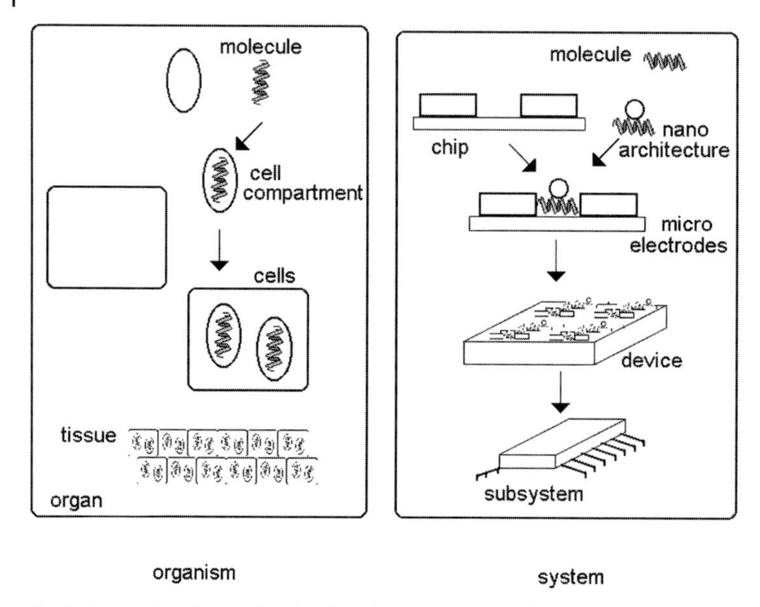

Fig. 2 Integration of natural and technical nanosystems in a functional microstructured environment

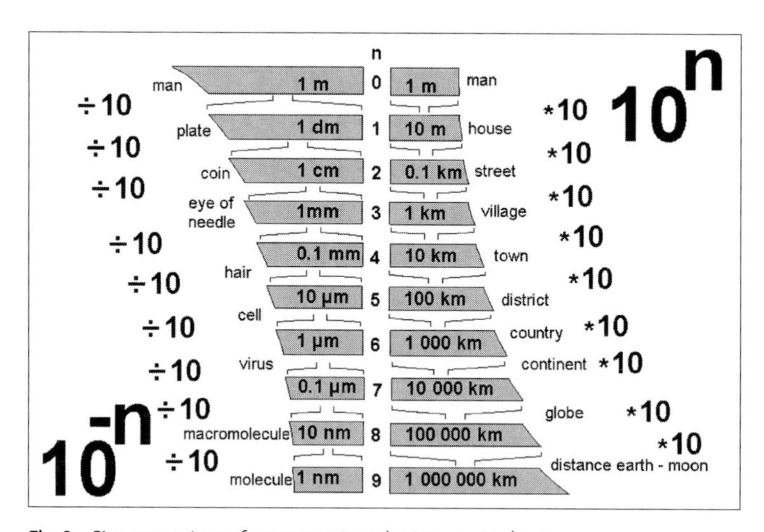

Fig. 3 Size comparison of macroscopic and microscopic objects

The application of microtechnological objects requires the integration of microchips into a macroscopic technical environment. Such an arrangement is needed to realize all interface functions between the micro- and macroworld. The lithographic microstructures are not accessible for robotic systems as individual structures, but only in an ensemble on a chip with the overall dimensions in millimeters. The smallest lateral dimensions of such a structure are in the medium to lower nanometer range, but the contact areas for electrical access of the chip are in the millimeter range. This principle of geometric integration is also utilized in nanotechnology, in this case the microtechnology is used as an additional interface level.

Although selected nanostructures can be produced independent of microtechnology, a functional interfacing of nanosystems requires the interaction with a microsystem as a mediator to the macroscopic world. Therefore, a close connection between nano- and microtechnology is required. Additionally, a variety of methods originally developed for microtechnology were further developed for applications in nanotechnology. So, not only is a geometrical but also a technological integration observed. Nevertheless, apart from the methods established in microtechnology and now also used in nanotechnology (such as thin film techniques), other methods like photolithography and galvanic techniques are typical methods in the micrometer range; and scanning probe techniques, electron beam lithography, molecular films and supramolecular chemistry are specific methods in the nanometer range.

2
Molecular Basics

2.1
Particles and Bonds

2.1.1
Chemical Bonds in Nanotechnology

In addition to the elementary composition, the interactions between the atoms determine the properties of the materials, and therefore, of devices. So knowledge of chemical bonds in a device is essential for its functioning. In classical technology, bond properties are described as collective phenomenon, and general material parameters are utilized for a fairly indirect characterization.

Apart from the properties known from the bulk materials, surface and interface properties in particular exert an increasing influence as structure size decreases. The material properties in microscopic dimensions often differ dramatically from the bulk properties. Besides the dominant role of surfaces and interfaces, the individual bond is also no longer negligible relative to the sum of the bonds in the nanostructure. Often individual bonds or single molecules determine the properties and function of a nanostructure.

In general, all types of positive interactions between particles represent bonds. Interactions between atoms, groups of atoms, ions and molecules can vary widely with respect to their character and their strength. To differentiate, these interactions were divided into classes known as bond types. These classes are well suited for a description of bonds. In contrast to classical synthetic chemistry where strong bonds are important, often the medium and weak bonds are of particular importance in nanotechnology. The importance of weak bonds increases with the increasing size of the aggregates constructed, which is comparable to what happens in nature. While in the field of strong bonds the differentiation of bond types is easy, the area of weak bonds is determined by the parallel existence of several interactions with a wide range of strengths and characters. Molecular geometries are not just described by the topology of covalent bonds. Other types of bonds as well as weak interactions contribute substantially to the establishment and conservation of given geometries, and therefore have to be considered. Thus, the following sections will introduce the key classes of chemical bonds and discuss their importance to nanotechnology.

Nanotechnology. M. Köhler and W. Fritzsche

ISBN: 978-3-527-31871-1

2.1.2
Van der Waals Interactions

All of the shells of atoms interact with each other. When atoms approach each other, the electrons of one atom deform the distribution of the electrons of the other atom. This deformation disturbs the charge distribution in a way such that the sum of the energy of the two approaching atoms is lower than the sum of the atoms initially. This difference in energy determines the strength of the bond. If this effect is not influenced by other bonds (e. g., by the exchange of electrons), the bond energy is fairly low. The Van der Waals bond is a weak bond. At room temperature, the bond between individual atoms can be easily thermally activated and broken.

Van der Waals bonds are nevertheless of particular importance in nanotechnology, because the building units are usually solids and consist of molecules instead of individual atoms. If two or more atoms connected by strong ionic, covalent, coordinative or metal bonds, then the interactions of the electron shells with surfaces and molecules are in conjunction with the Van der Waals bonds. As a result, as the number of atoms in a molecule increases, this molecule is able to bind to a substrate based solely on Van der Waals bonds. One consequence of this effect is the decreasing vapor pressure of homologue compounds with increasing molecular size.

The Van der Waals bond is therefore a basic type of bond, which becomes important due to the cooperative effect of many atoms bound to each other. The mobility of molecules is determined by the size. Another parameter is the partial dissociation of Van der Waals bonds by intramolecular movements. If atoms or groups of atoms are only connected by freely rotating bonds, the rotation of one part of the molecule can thus induce the separation of the respective bond. With fixed bonds, all bonds are distributed in a cooperative manner. Van der Waals bonds play an important role in hydrophobic interactions. They are essential in resist technology and therefore in the whole field of micro lithography. They are also essential for living cells, especially in the creation of the three-dimensional structure of proteins. In cells, hydrophobic interactions are a prerequisite for the composition of lipid bilayer membranes and the inclusion of membranous proteins in these layers. In analogy to such structures in nature, Van der Waals interactions are important in nanotechnology, especially in the field of supramolecular chemistry for the arrangement of complex molecular aggregates based on smaller units.

2.1.3
Dipole–Dipole Interactions

Owing to the differences in electronegativity, molecules consisting of different atoms normally exhibit an inhomogeneous electron distribution. Only when the bonds are symmetrical is this distribution not apparent in the surroundings. Otherwise, an electrical polarity of the molecule is observed. Such molecules, with one or more dipole moments, attract each other. The intensity of polarity determines the strength of the dipole–dipole interaction.

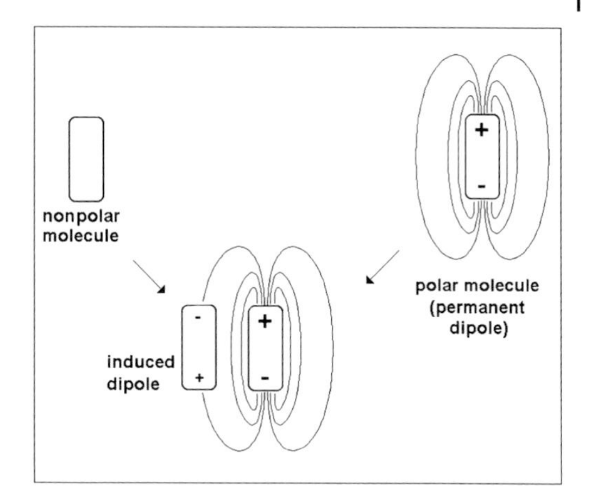

Fig. 4 Bonding through induction of a dipole moment into a nonpolar molecule

Dipole–dipole interactions are also observed in cases when only one half exhibits a permanent dipole moment. Because the electron shell can be deformed by external fields, a molecule with a permanent dipole moment is able to induce a deformation and therefore a polarization resulting in a dipole moment (Fig. 4). The energy gain in such cases is usually lower than the interaction of permanent dipole moments. The bond energies are determined by the tendency of the electron shell to be polarized. If this capability is low (hard shells), only weak dipole moments are induced, the resulting bond is therefore weak. In shells with high capability for polarization (soft shells), significant dipole moments can be induced.

Dipole–dipole interactions are widely distributed. They account for undesired effects in microtechnology, because they are responsible for unspecific interactions. These interactions result in deposition on surfaces or unspecific binding of individual molecules/particles (Fig. 5). In particular, electron rich heavy atoms exhibit readily polarizable electron shells, so that they are sensitive to unspecific adsorption. In

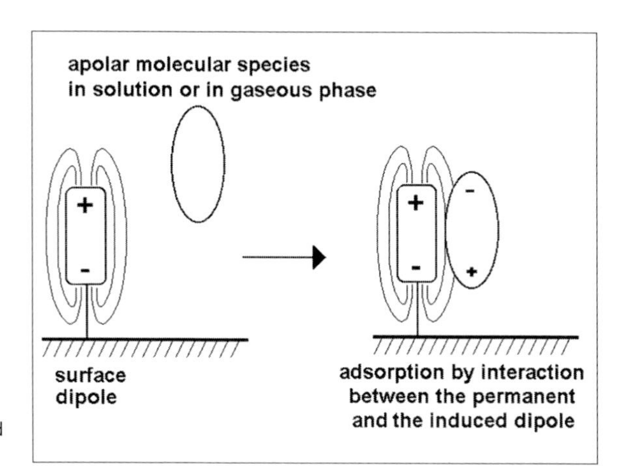

Fig. 5 Surface bonding of non-charged molecules by dipole induction due to the interaction with an immobilized dipole molecule

gas reactors, such as vacuum equipment, unspecific adsorption is minimized by the heating of reactor surfaces and substrates, through thermal activation of desorption. Tighter bound particles on substrates are treated by etching through sputtering, which is not applicable for sensitive substrates, such as in the case of substrates with ultrathin and molecular layers. To counteract these processes of undesired adsorption in liquid phase processes, ultra pure substances and solvents are used.

Coupled dipole bonds are utilized in the three-dimensional folding or arrangement of synthetic macro- or supermolecules. The application of less specific bonds for the design of molecular nanoarchitecture in nanotechnology is still a long way off the level demonstrated in nature.

2.1.4 Ionic Interactions

Where there are large differences in the electronegativities of atoms, a transfer of one or more electrons from the less to the more electronegative interacting partner is observed. The resulting bond is not determined by the binding electrons, but by the interactions of the ions created by the electron transfer. The strength of this bond is comparable to a covalent bond; it is therefore a strong chemical interaction.

Pure electrostatic interactions between ionized atoms, as in the case of salts, are of less interest in nanotechnology. In contrast, molecular ions and also polyions are of particular interest. Macromolecules often exhibit a multitude of similar functional groups. If these groups are ionizable and can be readily ionized (e. g., as a result of dissociation processes), this effect results in polyionic macromolecules. They can interact with small ions of opposite charge, but also with similarly charged polyionic partners, resulting in the creation and stabilization of multiple ultrathin layers or complex molecular aggregates.

Surface charges, electrostatic repulsion and electrostatic bonds are essential for the manipulation of macromolecules, supermolecular aggregates, micelles and nanoparticles in the liquid phase. Nanoheterogeneous systems can be created, stabilized or collapsed by adjustment or compensation of surface charges.

2.1.5 Metal Bonds

The creation of strong chemical bonds by exchange of binding electrons can also take place without asymmetric distribution of the electron density. If the exchange occurs only in one direction, a single covalent bond is created (cf. Section 2.1.6). If the exchange takes place in several spatial directions and is furthermore combined with a high mobility of the binding electrons, a so-called metal bond is created.

Through the simultaneous existence of bonds in various spatial directions the metal bond is present in a three-dimensional network of equal bonds. Clusters are created where a limited number of atoms are involved. For large numbers of atoms, an ex-

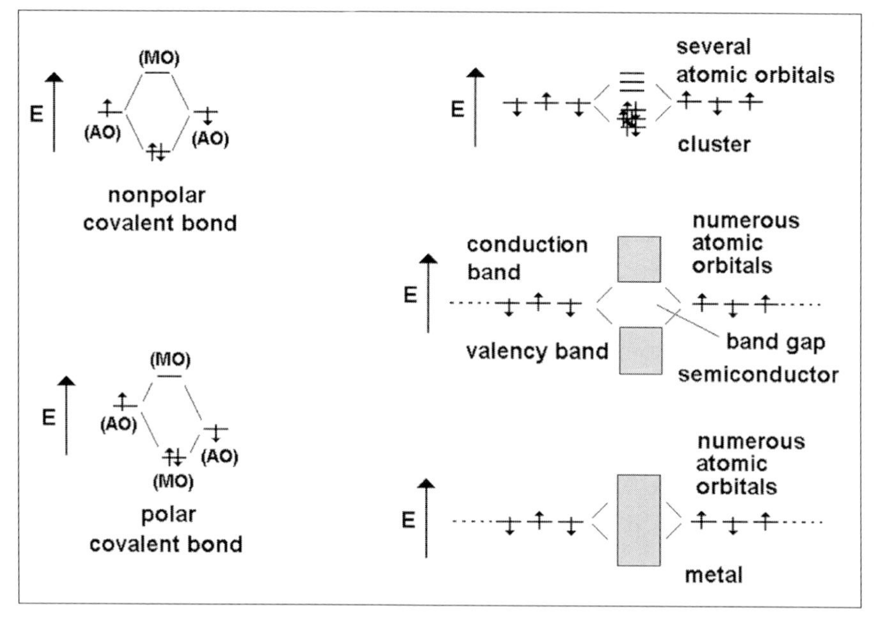

Fig. 6 Comparison of the energy levels of molecules (left) and metal nanoparticles or extended solids (right)

tended binding network leads to a three-dimensional solid. Owing to the high mobility of the binding electrons, this solid is electrically conductive (Fig. 6).

The metal bond is of special interest in micro- and nanotechnology due to the broad application of metals and semiconductors as electrical or electronic materials. Additionally, metal bonds facilitate the adhesion and both electrical and thermal conductivity at interfaces between different metals and inside alloys. Completeness or discontinuity of metal bonds in the range of molecular dimensions inside ultrathin systems determine the nanotechnological functions, such as tunneling barriers realized by local limitations of the electron mobility or the arrangement of ultrathin magnetic layers for magnetoresistive sensors leading to a change in magnetic properties at constant electrical conductivity.

2.1.6 Covalent Bonds

Strong bonds occur in the interaction of two atoms with unpaired electrons, resulting in doubly-occupied binding orbitals (Fig. 6). While the density distribution of electrons does not differ significantly from the density distribution of the free atoms, the differences in the electronegativity of the binding atoms results in polarity for covalent bonds. In contrast to the typically extended solids in the case of the metal bond, in some cases the covalent bond can lead to particles consisting of only two atoms,

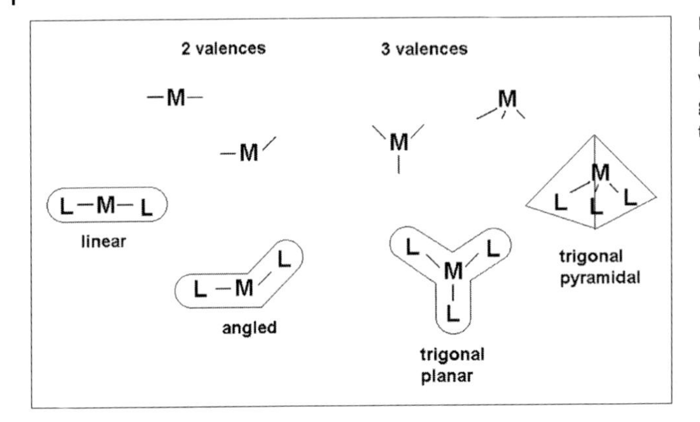

Fig. 7 Relationship between atomic valences and molecular geometry for two- and three-valent atoms

e.g., oxygen or nitrogen found in the air. Covalent bonds can also affect just several or a high number of atoms. So the results can be linear, disk-shaped, globular molecules or solids extended in three dimensions.

The fixed rules for the electron density distribution are of importance in nanotechnology, theses rules being based on the number of possible bonds per atom, the number of non-binding outer electrons, and the angle between the bonds. They apply for all bond types with electron exchange as essential distribution to the bond, such as polar and apolar atomic bonds, coordinative bonds and hydrogen bonds. These bonds are directed.

The geometry of the bonds around an atom is influenced by its valence. Bivalent atoms create linear or bent structures and trivalent atoms result in trigonal-planar or trigonal-pyramidal geometries (Fig. 7). Regular geometries around atoms with four valences are planar square or tetragonal, which are deformed in the case of asymmetric substitutions. Square pyramids are typical for five valences, and octahedral or trigonal pyramids for six valences (Fig. 8).

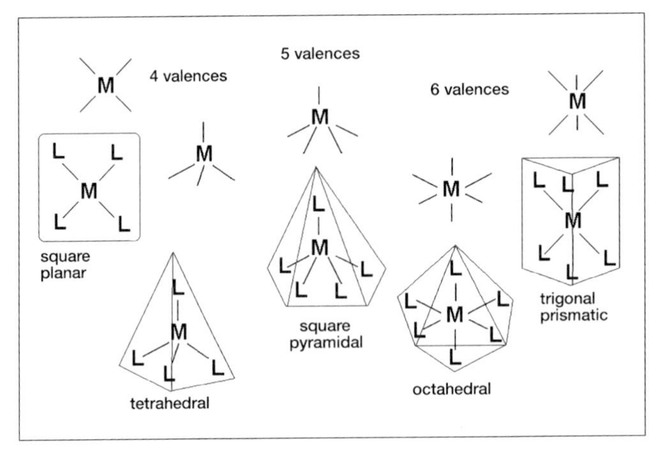

Fig. 8 Relationship between atomic valences and molecular geometry for four-, five- and six-valent atoms

The three parameters “valence”, “polarity” and “direction” create a complex set of rules for the architectural arrangements based on covalent bonds. The orientation and arrangements of bonds determine not only the topology of bonds, but also the mobility of atoms and groups of atoms relative to each other. So the sum of bonds affects how the bond topology determines a certain molecular geometry or allows degrees of freedom for spontaneous activated intramolecular mobility, and how the external pressure affects the mechanical relaxation of molecules. Also, without intramolecular bridges the free rotation of bonds could be limited due to double bonds.

The creation of molecular nanostructures relies on the degrees of freedom of individual bonds on the one hand, and the rigidity (limitation of mobility) of certain parts of the molecules on the other hand. Hence double bonds, bridged structures and multiple ring systems of covalent units are important motifs for the molecular architecture in molecular nanotechnology.

2.1.7 Coordinative Bonds

Bonds are created by the provision of an electron pair by one of the binding partners (the ligand) for a binding interaction. A prerequisite is the existence of double unoccupied orbitals at the other binding partner, so that a doubly occupied binding orbital can be created. According to the acid–base concept of Lewis, electron donors are denoted as Lewis bases, electron acceptors as Lewis acids.

The central atom in such a coordinative bond is usually the respective acid; the ligands are the Lewis bases. Such coordinative bonds are typically found with metal atoms and metal ions, which always exhibit unoccupied orbitals. Thus metal atoms or ions in solution usually exist in coordinated interactions. The metal central ion or atom (“central particle”) is surrounded by a sphere of several ligands and creates a so-called “complex”; therefore these bonds are also denoted as complex bonds.

The stability of complex bonds lies between the strength of the weaker dipole–dipole interactions and of covalent bonds, thereby covering a wide range. Coordinative bonds are therefore particularly well suited to the realization of adjustable binding strengths and thus to adjustable lifetimes of molecules. This is of great importance for construction in supramolecular architecture. Nature also uses this principle of finely tunable binding strengths of complex bonds, e. g., in the Co- or Fe-complexes of the heme groups of enzymes.

Similar to the covalent bonds, the coordinative bonds are also coupled to the spatial orientation of the binding orbitals (Figs. 7 and 8). Because the central particles are usually involved in two or more bonds, their orbitals determine the geometries of the complex compounds. Two- or multiple-valent ligands often bind on one and the same central particle. When multiple binding ligands interact with several central particles, complexes with several cores are created. Such compounds are promising units for supramolecular architectures and therefore of special interest in molecular nanotechnology.

Beside anions and small molecules, ring-shaped molecules, extended molecules and parts of macromolecules can act also as ligands. Covalent and coordinative bonds are then both responsible for the resulting molecular geometries. Because the central particle and often also the ligands are ions, coordinative bound architectures in addition to exhibiting complex and covalent bonds also display ionic and dipole–dipole interactions, representing a complex structure.

2.1.8 Hydrogen Bridge Bonds

The hydrogen bond is a specific case of a polar covalent interaction. It is based on hydrogen atoms, which create interactions between two atoms of fairly strong electronegative elements. In this way one of the atoms is relatively strongly bound as a covalent binding partner, and the second significantly weaker. A classic case of hydrogen bonds occurs in water, where they are responsible for the disproportionately high transition points of water.

Hydrogen bonds are observed when the bond of a different atom to hydrogen is so polar that the separation of the hydrogen atom almost certainly occurs. So oxygen and nitrogen, and to a certain degree sulfur also, are the preferred binding atoms for hydrogen bonds.

The individual hydrogen bond is of relatively low energy, distributing only a weak contribution to the overall energy. In addition, it is easily cleaved. However, several hydrogen bonds between two molecules can stabilize the created aggregate significantly by inducing a cooperative binding.

Hydrogen bonds lead to less specific adsorption processes; therefore they belong to the class of bonds responsible for disturbances at surface modifications or on layer deposition. In contrast to Van der Waals bonds and dipole–dipole interactions, hydrogen bonds are localized and oriented, so that they contribute significantly to specific interactions. In this respect, they are similar to coordinative bonds. So hydrogen bonds play an important role in both the supramolecular chemistry and the supermolecular synthesis of biomolecules.

2.1.9 Polyvalent Bonds

Nanotechnology is based on the creation and dissociation of connections due to interactions between atoms or molecules. Reduced dimensions result in a lower relative precision for external tools, so the accuracy of manipulations has to be realized by the specificity of chemical bonds instead of by external means. A fine-tuned reactivity is required, which is not possible with the limitations of the individual bonds from of the classes mentioned earlier.

Instead, through the differentiation of a few types of discrete individual bonds, chemical reactivity and specific stability can also be achieved with a digital binding principle, characterized by the arrangement and number of bonds determining specificity and stability. The energy of the individual bond has to be sufficiently small, so that it does not result in a stable final binding and can be dissociated if needed.

Van der Waals bonds fulfill the requirement of weak interaction energies, but they do not exhibit positional specificity. So they are not ideal for digital binding, and participate only as background bonds. The requirements of both low binding energy and positional specificity are met by many coordinated interactions as well as by the hydrogen bridge bond. These two bonds therefore play a central role in the realization of molecular and supramolecular architecture in living systems. Additionally, the arrangement of a synthetic nanoarchitecture depends on these bonds. In such systems, the strength of an individual bond matters less than the number, position and relative mobility of binding groups, which determine the geometry and stability of larger molecular architectures.

While individual weak bonds are easily broken, a cooperative effect occurs in the case of coupled bonds, when several bonds only dissociate together. This phenomenon is well known from the melting behavior of double-stranded DNA. The thermally induced separation of the two strands connected by hydrogen bridge bonds requires increased temperatures with increased strand length and a higher density of hydrogen bridges (GC/AT ratio). Over a length of about 40 bases, the melting temperature does not increase further, pointing to an independent movement of strand sections above a critical length.

The mobility of molecular groups determines the size of cooperative effective sections in larger molecule, which are able to bind externally in a polyvalent manner [1]. The cooperative sections can be extended by the inclusion of rigid groups, such as conjugated double bonds, bridges based on dipole–dipole interactions or coordinated interactions. This is a prerequisite for stabile polyvalent interactions between large molecules based on multiple weak bonds. Natural molecular architectures demonstrate the synergetic use of different bond types. So the binding pockets of enzymes or antibodies often utilize a complex system of hydrogen bridge bonds, dipole–dipole interactions, Coulomb and Van der Waals interactions.

The strength of polyvalent bonds consisting of one type of individual bond is determined by the strength of the individual bond and the number of bonds connected by the rigidity of the molecule. So building units with a high rigidity and a compatibility with the liquid phase are of specific interest in nanotechnology. Linear aliphatic polymers do not fulfill these requirements without the introduction of groups for additional rigidity, in contrast to biological macromolecules, such as double-stranded DNA or proteins. In synthetic chemistry, rigid and connected macrocycles are appropriate candidates. Other interesting materials are substituted metal clusters, nanotubes and other nanoparticles. They provide an extremely high rigidity based on the strong bonds between the atoms of the cluster or the particle, resulting in a coupling of surface bonds as regards mobility. In such cases, the interactions of groups of weak individual bonds represent polyvalent bonds.

Polyvalent bonds, which are strongly coupled weak bonds, provide a base for nanoarchitectures. While the activation barrier for the establishment and the dissociation of individual bonds is low, the simultaneous activation of a group of coupled weak

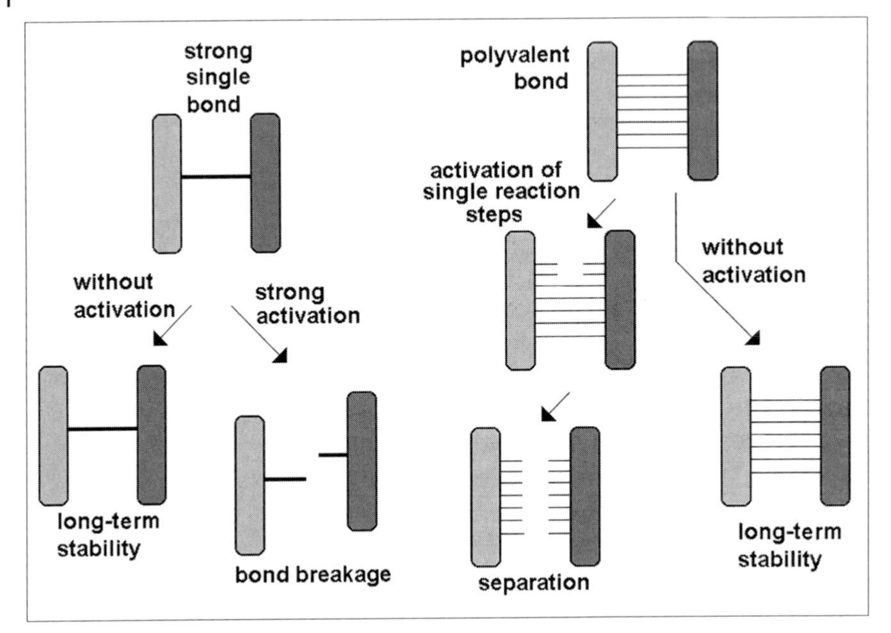

Fig. 9 Influence of strong individual bonds and polyvalent bonds on the stability of molecular complexes: in contrast to individual bonds, polyvalent bonds can be separated by a successive moderate activation of weak bonds

bonds is extremely unlikely. So after creation aggregates are stabile in the long term [1]. Only under extreme conditions or as a result of factors that assist in the successive opening of the weak individual bonds (e.g., the catalytic effect of an enzyme), can polyvalent bonds be reversed (Fig. 9).

The synthetic challenge for molecular nanoarchitecture is to avoid the creation of complex three-dimensional polymeric networks by spontaneous aggregation, but to control the aggregation so that in every step individual units are assembled at defined positions. A prerequisite is a high efficiency in the coupling reactions combined with a low probability for competing reactions.

2.2
Chemical Structure

2.2.1
Binding Topologies

A relationship between the internal geometry of molecules and the coordinates of the external reference system is essential for nanotechnology. In the following, the basics of chemical structure are discussed from the viewpoint of spatial determination. Often, terms such as "structure" and "molecule" are not sufficient to describe all of the geometric and dynamic aspects of the interactions of molecules with nanotechnological structured surfaces and the construction of supramolecular architectures on the surfaces of solids. The external and internal mobility of molecules as well as the full effect of strong and weak interactions have to be included.

The term "molecule" will be discussed in the gas phase. Here, all atoms with the same (averaged) directional components of translation form a molecule. The joint movement is based on interatomic binding forces. The proximity alone is not sufficient as a parameter, because there are conformations in molecules with rotating bonds where atoms are in a close proximity but are without a direct strong bond. Also, the absolute strength of a bond is not sufficient for a description. There are molecules in the gas phase held together by hydrogen bonds, e. g., acetic acid, which exists in the gas phase as a dimer. The criterion of common translation vectors can be transferred to the liquid phase. However, it is not applicable when the translation of the particles is hindered, e. g., in solidified matrices.

For this reason, the relative (instead of the absolute) strength of binding topologies will be used for the characterization of a particle. A binding topology includes a linear arrangement or a network of bonds, which in its entirety is more stabile than all other bonds through atoms in its proximity. It is independent of the type of bonds, and also weaker interactions such as hydrogen bridge bonds or cohesive forces are included. This approach allows the general discussion of single molecules, micelles and nanoparticles.

An estimation of the strength of the binding topologies requires the discussion of a particle and the environment as one system. Strong particles exhibit stronger individual internal bonds compared with weaker external ones. In this sense, a small alkane (such as ethane in the condensed phase) represents a very strong unit. Transformation into the gas phase is easy, in contrast to the significantly higher temperatures required for breakage. The transformation of long-chain molecules of polyethylene, which have the same covalent bonds as in ethane, into a mobile phase requires strong thermal activation (melting) or the substitution of the solid-state interactions between the molecular chains by interactions between dissolved molecules and solvent molecules (solvation). Mechanical forces lead to the breakage of the covalent bonds, but not to an extraction of a molecule as a unit from the solid. The sum of the weak interactions with the environment is stronger than an individual intramolecular bond in the topology. The movable macromolecule is a relatively weak unit in the binding topology. Molecules with a large number of internal stabilizing interactions represent a stronger unit than the unfolded molecule. It is not the strength of a covalent bond network alone, but the sum and the arrangement of all intramolecular interactions that determine the binding topology of a particle.

Nanotechnology utilizes different levels of internal stabilization of particles to realize durable devices with strong bond structures. The different technological steps use units with a wide range of strengths. The stability criterion is the lifetime. For a successful device all components must be functionally preserved over the whole lifetime, which is typically in the range of years. When creating nanoarchitectures, intermediate units have to be stabile only for the given process step, which can be in the second or even millisecond range. Single units used in the technology frequently possess the character of reactive intermediates with even shorter lifetimes.

Particles with shorter lifetimes include molecular aggregates with weaker bonds, such as Van der Waals or hydrogen bridge bonds, as the interactions connecting the subunits. Typical examples are microemulsions and micelles. Also, coordinated compounds are relatively unstable aggregates as in the case of the high exchange rates of ligands.

The geometry of rigid molecules is determined completely by the binding topology. This applies to solids with dense three-dimensional binding networks, but also to molecules consisting of two atoms, small linear molecules with multiple atoms such as carbon dioxide, simply bent molecules such as water, and highly symmetrical molecules such as benzene. Various conformations of one and the same molecule represent different geometries at the same binding topology. With an increase in the number of free rotating bonds, the number of possible geometries of particles with the same binding topology also increases.

The internal mobility of particles represents a challenge to nanoconstruction. Chemical stability does not imply spatial stability. Mobility required in coupling steps could be incompatible with certain functions or with subsequent steps in the synthesis. The restriction of degrees of freedom of mobility is an essential instrument for molecular nanotechnology. On the other hand, internal mobility of particles is also an important instrument, because many chemical and physical functions require mechanical flexibility. Functional nanoarchitectures call for balanced and not maximal mobility.

2.2.2 Building Blocks of Covalent Architecture

An ideal approach to molecular nanostructures that have covalent bonds utilizes presynthesized units (which are easily prepared and manipulated in a homogenous mobile phase) and their coupling to substrate surfaces. This general principle corresponds to the classical mechanical construction approach, which builds complex units from prefabricated building blocks, or to traditional solid phase synthesis, which builds chain molecules by subsequent coupling of molecular groups.

Molecules consisting of covalent bonds can be grouped according to formal constructive properties. Even complex binding topologies have their roots in a few basic types of units (Figs. 10 and 11):

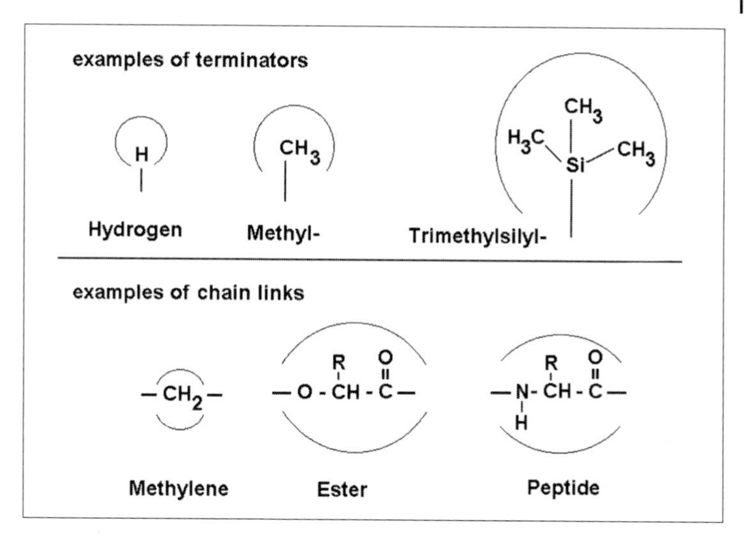

Fig. 10 Examples of molecular building blocks with a terminator function (top) and a chain link function (bottom)

- single binding elements ("terminators"), e.g., alkyl or trimethylsilyl groups
- double binding elements ("chain elements"), e.g., alkenes, simple amino acids
- three-fold binding elements ("branches"), e.g., substituted amino acids
- four-fold and higher branched elements

There is a large number of multiple branched elements. They can usually be traced back to a combination of units from the above-mentioned first three classes. So the three-fold binding phloroglucin (1,3,5-trihydroxybenzene) can be thought of as being assembled from three chain elements (CH) and three branches (COH).

There are three types of chain elements, based on the symmetry of the coupling group (Fig. 12, A–C):

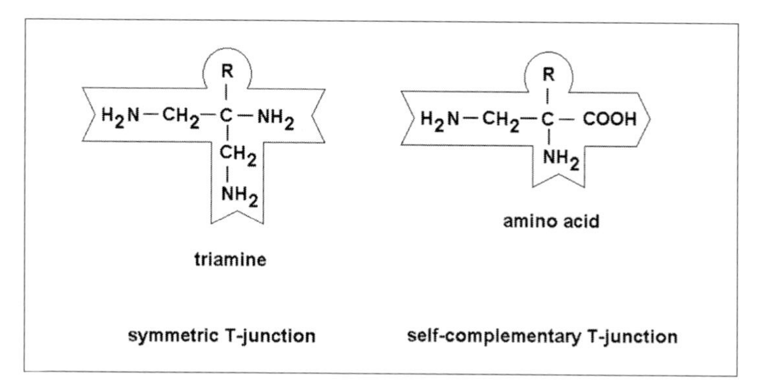

Fig. 11 Examples of molecular building blocks with branch geometry

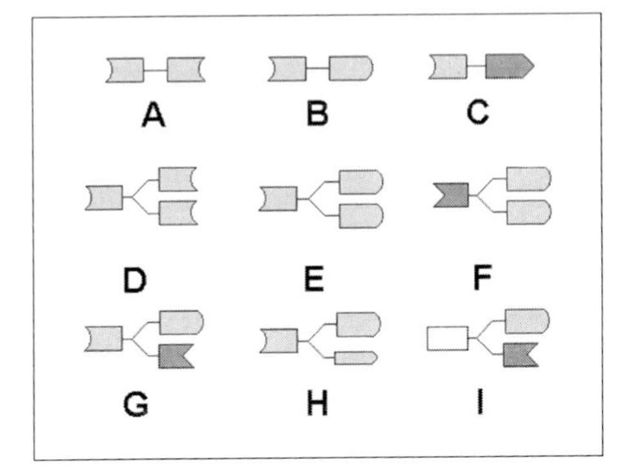

Fig. 12 Variations of connections of coupling functions in molecular chain links (top) and branches (bottom)

- two identical coupling groups, e. g., alkane diol (A)
- two different groups complementary to each other, e. g., amino acids (B)
- two different and not complementary groups, e. g., amino alcohols (C)

The branches can be divided into six basic types (Fig. 12, D–I):

- with three identical coupling groups, e. g., glycerol (D)
- with two identical and a third, complementary, group, e. g., lysine (E)
- with two identical and a third, non-complementary, group, e. g., diamine alcohol (F)
- with three different groups, including two complementary to each other, e. g., tyrosine (G)
- with three different groups, with one complementary to the two other, e. g., hydroxyl alkyl amino acids (H)
- with three different non-complementary groups (I)

Silicon and carbon are well suited as units for the construction of complex three-dimensional architectures due to their four valences. In contrast to carbon, with its stable C–C bonds and a wide variety of chemical methods for preparation and manipulation, in silicon structures Si–C and Si–O bonds prevail. The carbon atom is the center of an elementary tetragon in the sp^3-hybridized state; the same basic geometry is formed by $Si(O)_4$ tetrahedra. Both structures are responsible for highly branched spatial structures. In its sp^2-hybridized state, carbon represents a simple branch leading to planar structures.

The properties of molecular systems combine the fairly design-oriented aspects of the planned architecture on one side with the rather technological aspects on the other. The binding topology is determined by the number of coupling groups per unit, which is influenced by the potential of internal connections. In contrast, the symmetrical properties of the units determine the choice and the order of reactions leading to the architectural arrangements.

2.2.3
Units for a Coordinative Architecture

The scheme for covalent bonds (Section 2.2.2) is also applicable to other types of bonds. The central atom of coordinative compounds typically exhibits ligand numbers of between 4 and 6. When the ligands with additional coupling groups are bound in a stabile manner to the central atom, complex compounds can act as a chain element or branch.

Complexes consisting of monovalent ligands and with a complete saturation of the electron vacancies frequently exhibit only a low stability. Polyvalent ligands stabilize to a significant extent through the distribution of electron pairs from two or more donor groups. Such so-called chelate complexes are well suited as units for supramolecular architecture.

The geometries of molecular groups based on complex compounds are determined by the symmetries of the electron shells of the central atom, which are determined by the atomic number and the degree of ionization of the central atom. In general, metals positioned further left and low in the Periodic Table create coordination spheres of higher numbers than metals from the top right. An additional point affecting the geometries of complex architectures is that the overall number of coupling groups of a coordinative compound is related to the ratio of the number of coupling groups per ligand to the valence of the ligands inside the complex. So a six-fold coordinated central atom and three bivalent ligands with one external coupling group each results in a three-fold coupling complex, which is a simple branch. Changes to the oxidation number of the central atom affect not only the stability of the individual coordinative bond, but often the geometry of the coordination shell also.

Chelate ligands with four or more electron pair donor groups are able to build stable chelate bonds with several central atoms simultaneously, thereby creating stable multiple-core complexes or polymeric complex structures. Another route to multiple-core complexes is the subsequent reaction of ligands with each other (such as additions onto double bonds or condensation) while preserving the coordinative bond. Helical supramolecules known as "helicates" can be constructed through the combination of multiple-valent bridge ligands with multiple central atoms [2].

2.2.4
Building Blocks for Weakly Bound Aggregates

In addition to covalent and coordinative bonds, dipole–dipole interactions, hydrogen bridge bonds and Van der Waals bridges can also lead to the assembly of molecular building blocks for nanotechnology. Normally (except at very low temperatures) a single bond is not sufficient to stabilize a particle consisting of several atoms. So it is preferable that polyvalent bonds are involved, usually exhibiting a mixture of the bond types discussed previously.

The collective effect of weak bonds is enhanced when the participating particles themselves consist of several atoms bound together by strong bonds. Such a building

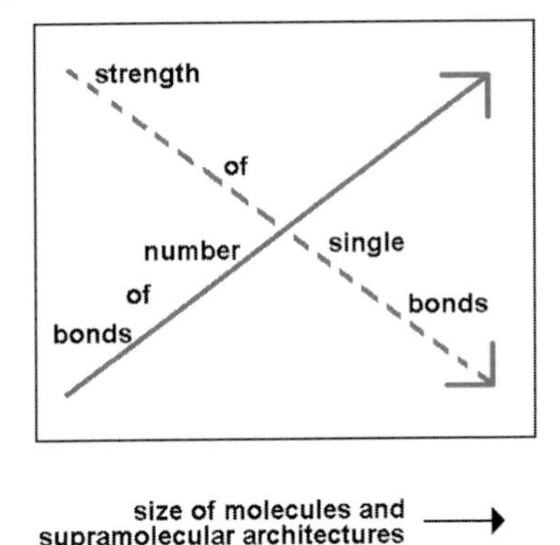

Fig. 13 Principle of construction of complex molecular architectures by complementary changes in number and strength of bonds realized in the system

block demonstrates a general principle of a binding hierarchy: with the increasing size of the molecular aggregates the average strength of the bonds between the particles decreases, and the number of simultaneous bonds rises (Fig. 13).

Many particles in the liquid phase, which influence the reaction but will not actually be involved in the final molecular structure that is created, are bound by weak and shorter-lived interactions. These particles include the solvent molecules, and also other small molecules, which determine the rate and selectivity of the chemical reactions by preferred interactions with certain regions of the reacting molecules and solid surfaces. Typical examples are small amounts of competing solvent or surface-active substances, and also metal ions that stabilize reactive intermediates by short-lived coordinative interactions or by influencing the concentration of free ligand groups. These small molecules therefore act by assisting the important molecules in the creation of molecular nanostructures.

2.2.5 Assembly of Complex Structures through the Internal Hierarchy of Binding Strengths

The general principle of binding strength hierarchy underlying complex molecular structures has already been demonstrated in biology (Fig. 14). Proteins show the strong modularity applied by nature in the nanoworld: the basic structure of the complex molecule is one-dimensional (primary structure) and this structure is created by covalent (strong) bonds.

The secondary structure is two- or three-dimensional and is stabilized by multiple hydrogen bridges as well as dipole–dipole-interactions, applying bonds that are individually weak but sufficient to give significant support to the structure. The tertiary structure is always three-dimensional and is preserved by a number of bonds. At the same time, mobility is induced through the low activation barriers of many weak bonds. The building blocks used for the primary structure are limited in number (20) and chemistry (amino acids). Every block consists of four elemental units that have specific tasks, with three of these being identical in all blocks: a coupling unit (carboxyl group), a second and complementary coupling group (amino group),

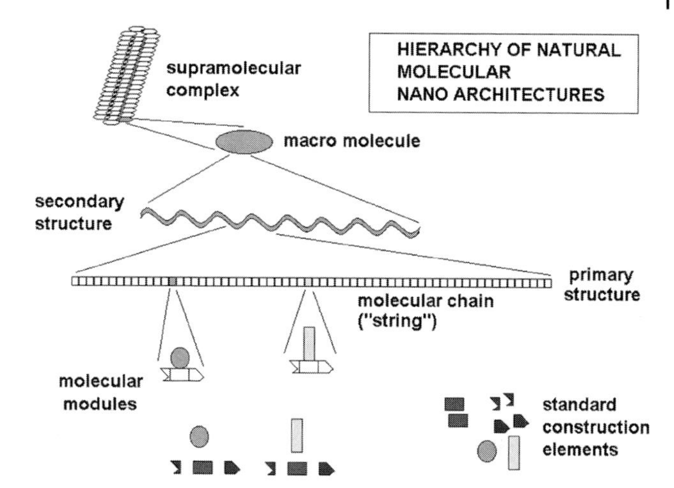

Fig. 14 Principle of construction hierarchy of supermolecular structures in nature (e. g., protein or protein complexes) consisting of simpler building blocks (e. g., atoms or amino acids)

a central connective unit (methine group of central carbon) and a variable unit (bound to the central carbon). These four units are connected by stronger bonds (resistant to hydrolysis) compared with the bonds between the individual building blocks (peptide bonds non-resistant to hydrolysis).

Using this hierarchy and enzymes as highly specific molecular tools, nature has succeeded in the realization of highly complex, three-dimensional structures of thousands of atoms in a highly defined and functionally optimized manner, without the need for macroscopic tools. Only the modular arrangement in combination with control of the binding strength and the resulting intramolecular mobility (and through the application of self-assembly mechanisms) overcomes the optimization problem.

Classical and also supramolecular synthetic chemistry are not yet able to apply this strategy of biomolecular systems to other classes of compounds and new functions. They do, however, use an analogue approach of connecting modules of molecular constructions with decreasing bond strengths through increasing unit size (Section 5.4).

2.2.6
Reaction Probability and Reaction Equilibrium

A chemical reaction occurs when the two reaction partners come close to each other and the sum of the relative movements overcomes the activation barrier. Thus the reaction is closely connected with both the internal and the external movement of the molecules. In an ideal gas, the reaction probability for particles of a particular geometry is based on the spontaneous distribution of energy on the degrees of freedom of mobility. In addition to the number of particles N in a given volume V and the frequency factor f (which includes the geometric properties), the ratio of the mobility energy (RT) to the activation barrier (E_a) determines the reaction rate r (Arrhenius equation). For a reaction that includes only one partner, for example in a decay processes, the following applies:

$$r = dc/dt = -c{\cdot}f{\cdot}e^{-E_a/RT} \text{ with } c = N/(L{\cdot}V) \text{ and } L = 6 \times 10^{23} \text{ mol}^{-1} \qquad (2.1)$$

In classical chemistry, the reaction rate is a result of the sum of the reaction probabilities of the individual molecules. Because of the huge number of particles, even for small concentrations and volumes (e. g., 600 000 000 molecules in 1 μL in a 1 nanomolar solution), the deviations in individual reaction probabilities are negligible relative to the overall reaction rate.

In nanotechnology small ensembles of particles, sometimes even a single particle, do play an important role. A cube of 10 nm, and even a higher concentration of 10 mmol L^{-1}, contains only six particles on average.

The lifetime of a nanotechnical device can be estimated from the ratio of the reaction probability to the tolerable failure rate. If an individual molecule plays a key role and determines the functionality, the effects of the reaction probabilities are of particular importance. Fast decays or other competing reactions should have a lower probability than the planned molecular reaction. This principle leads to a strict avoidance of competing reactions because of the absence of potential reaction partners for such reactions. This rule is softened for individual elements with a function realized by several parallel-processing molecules, thereby correlating with the number of tolerable malfunctions per individual element.

However, nanotechnology has the means to overcome this problem. An important tool to manage reaction probabilities is the control of the spatial movement of particles. While in classical chemistry this movement is based on spontaneous thermal mobility (at least in the vicinity of solid surfaces), nanotechnology applies directed transport processes, such as migration in an electrical field in a liquid phase. Cold plasma, ion or neutral particle beams are well-suited to the directed transport of particles onto solid surfaces. They utilize electrical fields to force charged or secondary neutralized particles in a given direction.

Adsorption and binding equilibria are often observed on microstructured substrates, and also sometimes on nanostructured surfaces. Therefore, forward and reverse reactions have to exhibit a certain probability in a given time range. This process can be described by an equilibrium constant K_{ad} (binding constant), e. g., for the binding of particles onto surfaces:

$$K_{ad} = k_{ad}/k_{des} \qquad (2.2)$$

Classical chemical kinetics based molecular statistics are not applicable to processes involving a small number of particles. This is the case for the situation in a chemical equilibrium. For microstructured binding spots known from biochips with several millions of similar molecules, the effect can even be observed: a chip of 3 μm^2 for an assumed foot print of 0.3 nm^2 per molecule exhibits about 10^7 binding places. With a binding constant of 10^6, on average 9 999 990 places are occupied, but 10 remain free. So a prerequisite for the statistics of large numbers is not given. For a reduced binding area of 300 nm^2, the statistical method fails even for binding constants below 10^3.

Also in the classic case, reactions with conditions close to equilibrium are driven by probability in nanotechnology. In addition to the external electronic, physical and chemical factors, a molecular quantum noise can also be observed and reduces the signal-to-noise ratio. This phenomenon is of particular importance for operations with large libraries of substances, as, for example, in the case of DNA computing [3][4][5][6][7].

Molecular operations in nanotechnology should be considered under the aspect of a certain distance from the equilibrium. So nanotechnology can be compared with biomolecular morphogenesis, where essential processes also take place away from the equilibrium. The kinetic trick in nature is the application of fast reactions with small participating particles (e. g., equilibrium in protonation or the creation of coordinated compounds) to support slow processes away from the equilibrium, such as the construction of supermolecules or subcellular structures. Hence enzymatic processes become important in nanotechnology. In the classical chemical sense, enzymes are catalysts, and the enhanced reactions are usually close to equilibrium. For the same number of similar particles this view is identical to inorganic and non-biogenic organic catalysts. However, for small numbers of particles and the application of a probability approach, the question arises as to whether the enzyme is activated after coming into contact with a substrate molecule. Depending on the activation of the individual catalyst molecule, the substrate is preserved or converted into the product. This behavior is similar to a classical tool. Because of the probability conditions, and the level of small particles, catalysts are tools rather than reaction partners.

3
Microtechnological Foundations

3.1
Planar Technology

Standard methods of mechanical fabrication usually fail for devices with dimensions in the medium and lower micrometer range. Therefore, novel techniques with extremely high precision and resolution have been developed for microtechnology. In contrast to the typical dimensions of 0.3–1 mm for precision mechanics, standard methods of microsystem technology provide dimensions down to 0.25–1 μm [1]–[9]. The precision has also increased, from 1–10 μm down to the sub-micrometer range and extends below 20 nm for the advanced microelectronic circuits.

Apart from the dimensions of the structures created, the means of fabrication has also changed. In general, an increase in the effectiveness of spatial restriction can be observed. Biological systems have the ability to create complex structures from the inside, e. g., from within the volume. On the other hand, technical fabrication usually starts from the outside and acts on the surface. The only exception is the shape of small components of the materials, so that the geometry of these small particles determines the material properties, but not the external geometry. This principle of fabrication of starting from the outside extends from standard mechanics into microtechnology.

To achieve the desired high resolution and precision in microfabrication, one has to accept technological restrictions as regards the choice of materials. For precision mechanics certain preconditions still apply: lathes only work on surfaces with round cross-sections, a planing machine requires an even surface. Even surfaces are optimal for key methods of microtechnology. Such surfaces are not only needed for individual technological steps, but they provide a whole technological platform for an optimal application of nearly all methods. This joint platform also offers maximal compatibility between the different technological steps. Because the even surface is its key parameter it is called planar technology. Planar technology utilizes materials with one or often two parallel level planes. Typical examples are the so-called wafer, glass or silicon disks that are utilized in the mass fabrication of microelectronic chips. Because several modification steps occur on the surface of these materials they are referred to as substrates.

Nanotechnology. M. Köhler and W. Fritzsche

ISBN: 978-3-527-31871-1

Planar technology has the following advantages:

- the substrate plane provides a simple geometric reference
- the substrate plane allows for marks for a lateral positioning
- substrates with a radial symmetry are possible, differences in positioning due to symmetry are minimized
- no inhomogeneity with respect to thermal strain and radiation intensity due to substrate geometry during exposure, deposition and etching processes
- thin layer depositions are homogeneous compared with rough or bent surfaces
- projecting methods (with e.g., light, electron or ion optics) require only a small depth of focus
- in fluid media, rotating substrates induce a constant thickness for the diffusion layer, resulting in homogeneous transport processes for electro or surface chemical processes
- for most inspection techniques, only one plane or a minimal depth of focus has to be measured
- substrates with planar parallel surfaces can be easily partitioned into chips of the same thickness

Planar technology is usually realized as layer technology. This means that the lateral dimensions of the materials on a substrate are some orders of magnitude larger than the thickness. An exception is the substrate itself, which can be etched to relatively large depths as in the case of silicon deep etching. The thicknesses of the layers have to be adapted to the functional and specifically to the microtechnological requirements. This means that for cost-efficient technologies, the layer thickness should not exceed the dimension of the lateral structures by more than an order of magnitude. Lateral dimensions in the lower micrometer range can be realized by standard techniques, as long as the vertical dimensions are in the same range (Fig. 15 a and b). This rule is also valid for nanostructures.

The deposition of various thin layers of different materials is a proven method in microtechnology. The thickness of these layers is usually around 1 μm, but it can vary between below 100 nm and several micrometers. This range is clearly differentiated from the thick layer technique (from several tens to 100 μm) and the ultrathin layers (about 1–10 nm). The latter are of particular importance in nanotechnology.

The term "layer" includes several types (Fig. 16):

A. Continuous layers with homogeneous thickness prepared on a planar substrate with parallel top and bottom planes.
B. Structured layer of homogeneous thickness on a planar substrate, with parallel top and bottom planes, but with additional faces
C. Continuous layer of homogeneous thickness on a topographically structured substrate
D. Structured layer of homogeneous thickness on a topographically structured substrate
E. Continuous planarizing layer on a topographically structured substrate
F. Structured planarizing layer on a topographically structured substrate

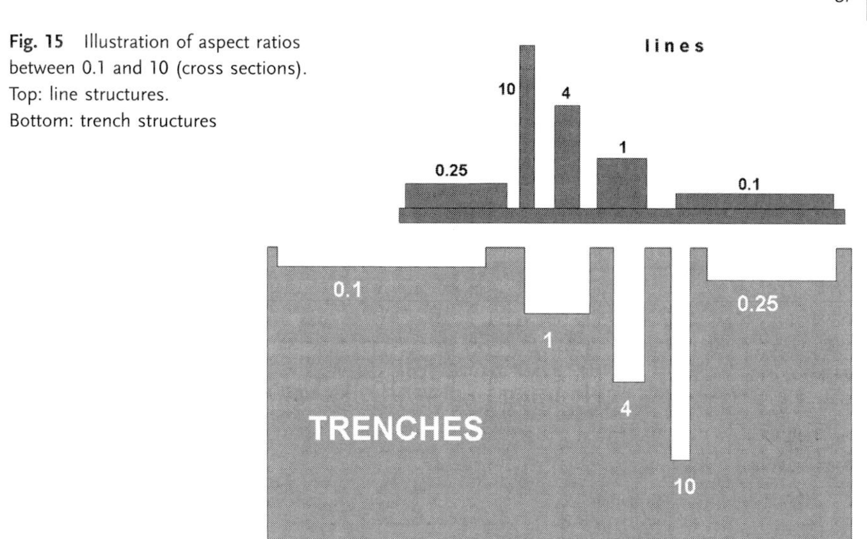

Fig. 15 Illustration of aspect ratios between 0.1 and 10 (cross sections).
Top: line structures.
Bottom: trench structures

According to this definition, structural elements that are not connected to each other can also be regarded as a layer. They can be prepared by fabrication of a continuous layer and subsequent microstructuring steps (subtractive process) or by separate deposition of the individual structures (additive process). The subtractive method begins with the deposition of the functional layer, prior to the creation of an etching mask with windows corresponding to the areas of the functional layer to be removed by the

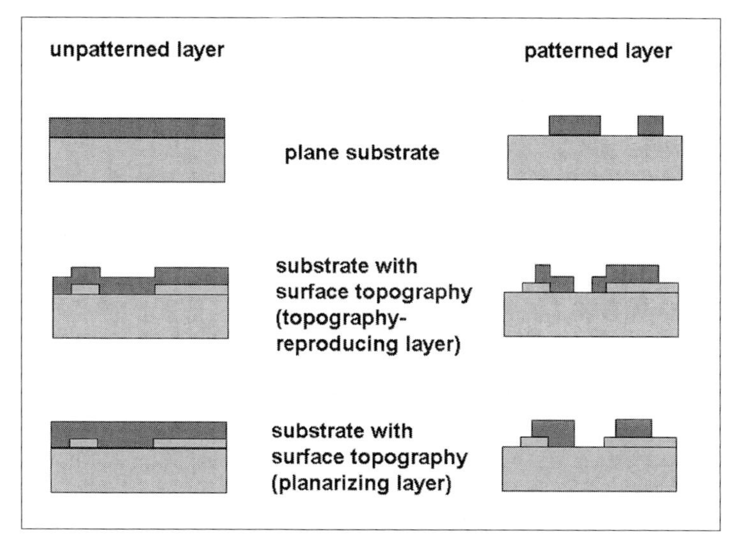

Fig. 16 Variations of closed and structured layers on plane substrates and substrates with a relief (cross sections)

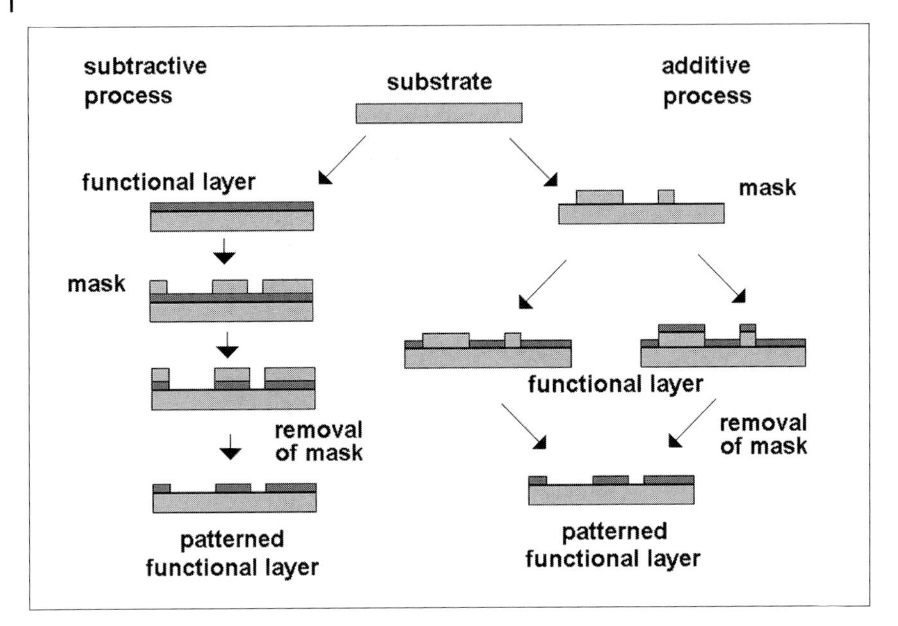

Fig. 17 Subtractive and additive patterning of layers (cross sections)

etching process. The additive method initially applies the mask onto the substrate, prior to deposition of the functional material. So the functional material is limited to the unmasked regions of the surface after removal of the masking material (Fig. 17).

Most devices exhibit more than a single layer. Stacks of layers are used to achieve a combination of materials. A stack can include layers with a wide range of thicknesses. The order of the layers is determined by the function, but is also influenced by technological restrictions with respect to the order or orientation in the stack. Such restrictions include interdiffusion zones in the interface between two layers, mechanical tension gradients induced by growth conditions, a difference between deposition and ambient temperature combined with different thermal extension coefficients of the layers. Other conditions influencing the sequence of the layers are the selective etching behavior of layer materials or the extension of grain boundaries or complex morphological parameters on the interface between two materials.

For technical application, normally nanostructures have to be embedded into a microtechnical surrounding. This creates the challenge of combining thin layers with ultrathin layers in one stack with layer thicknesses stretching over three orders of magnitude. In such cases, the choice of layer arrangement and the method for structuring are of particular importance. In addition to the preparation, this wide range of thicknesses also challenges the characterization of layers and structures (cf. Section 6).

3.2 Preparation of Thin Layers

3.2.1 Condition and Preprocessing of the Substrate Surface

Layer techniques are key to the integration of the materials for microtechnical devices into a system, because the substrate surface should approximate a planar state through the deposition and microfabrication steps. By utilizing layer deposition techniques, such planar surfaces can be realized in a cost-efficient manner.

The conditions on the substrate surface determine the quality of the deposition process. The chemical composition of the surface influences the binding between the substrate and the layer, thereby determining the adhesion of the layer and the chemical resistance of the interface. It establishes the starting conditions for the layer preparation, and the properties of the interface zone determine the functional properties of the layer. The preprocessing and cleaning of the surfaces and the control of the chemical and physical surface properties is an essential prerequisite for microstructuring methods and in particular for nanotechnology [10].

The surface conditions of substrates depend on the materials and the preprocessing steps. Surfaces tend to adsorb gas molecules and particles. Charging effects, which are critical in the case of electrically isolating surfaces, supports this process. Previous vacuum processes enhance the deposition of molecular components, because reduced pressure assists with the transfer of molecules of low vapor pressure into the gas phase. This is an issue with the volatile hydrocarbon components of oils, e. g., in oil-based vacuum pumps. The pressure increase observed moving from vacuum to atmospheric pressure leads to a condensation from the gas phase, resulting in molecular adsorbates and even to continuous adsorbate films on the substrates and the walls of the reaction vessels. The condensation of water from air is avoided in vacuum equipment by the application of dry noble gases for pressure equalization.

Most metals and semiconductors react with air. Any water that is present is more reactive than the oxygen, resulting in fairly continuous layers of oxides and hydroxides. The thickness of these layers ranges from islands and discontinuous monolayers to continuous oxide layers of several micrometers thickness. Some metals create oxide passivation layers in air and in many process media. Typical thicknesses are in the nanometer range. If these layers are complete, further oxidation is prevented.

Beside oxygen and water, hydrogen sulfide also reacts with many metals leading to at least a partial creation of metal sulfides on the surfaces. Water-containing oxide films or adsorbate layers are able to react with the carbon dioxide of the air, leading to carbonates and hydrogen carbonates. This variety of potential byproducts on surfaces is further extended by process steps with reactive components. The application of halogenide ions in cleaning and etching baths or of halogen compounds in plasma processes supports the creation of halogenides on the surface. Also phosphates, sulfites and sulfates can be found, depending on the compounds utilized.

An unspecific etching step under vacuum conditions is a typical approach to achieving the defined surface conditions. The application of chemically inert but accelerated particles (usually noble gas ions) breaks bonds on the surface and transfers the desorption product into the gas phase. It is a universal approach for practically all types of surface contaminations. It has the advantage of being conducted in the same vacuum vessel as the subsequent deposition steps, so the cleaned surface can be further processed without a break in the vacuum and a time delay. However, such a universal removal process eliminates not only contaminations but also parts of the functional layer. Hence it has to be carefully controlled in the case of thin layers, and is not applicable for ultrathin layers.

More gentle cleaning procedures need to differentiate between contaminations and the actual layer material. Therefore, the choice of cleaning procedure requires a certain knowledge of the type of contamination. Standard procedures utilize, for example, aqueous acids and bases as well as oxidizing solutions to convert organic and salt contaminations into water-soluble compounds prior to removal. Instead of the unspecific etching step with energetic noble gas ions, reactive plasma cleaning procedures are also applied. They apply, for example, oxygen for the selective removal of organic contaminants without interactions with non-oxidizable substrate material.

The adhesion of a layer onto the substrate is based on chemical interactions between atoms of the substrate and the layer. Van der Waals interactions alone result in poor adhesion. The process management is simplified when the chemical properties of the substrate, layer and interface are not too different, so that the stability of the interface layer is similar to that of the individual materials. The bonds in the interface layer are usually less stable compared with the bonds inside the layers, which is also the case for strong bonds between substrate and layer. The strength of individual bonds and the density of the bonds influence the layer adhesion. Apart from covalent bonds, surface charges also result in binding interactions. One example is functional groups observed in an ionized state. Although usually compensated for, like charges can lead to local failures.

The controlled introduction of adhesion-promoting materials in the interface or as an intermediate layer (adhesion layer) leads to significantly enhanced binding interactions. Adhesion layers consist of materials able to form strong bonds with both the layer underneath and the top layer, when the lower and top layers themselves interact only weakly. Noble metals bind to other metals by metal bonds. On the other hand, they exhibit a low affinity toward oxygen, resulting in a low adhesion on oxide surfaces. Transition metals, such as titanium and chromium, which are easily passivated, form very stable bonds with oxygen. So they exist in air with oxide surface layers. They remain metallic when they are deposited in high vacuum or under very low partial pressures of oxygen and water, so that in a subsequent step a noble metal can be deposited directly on top of this layer. Because base metals form stable bonds with oxide or hydroxide surfaces, on such surfaces they can act as adhesion promoters for the subsequent metal layers (Fig. 18).

Metal layers often stick poorly to hydrocarbon surfaces, such as to many polymers. Films of polymers usually exhibit good adhesion between each other, as long as a certain mobility of groups of atoms within the context of the molecular chain is ob-

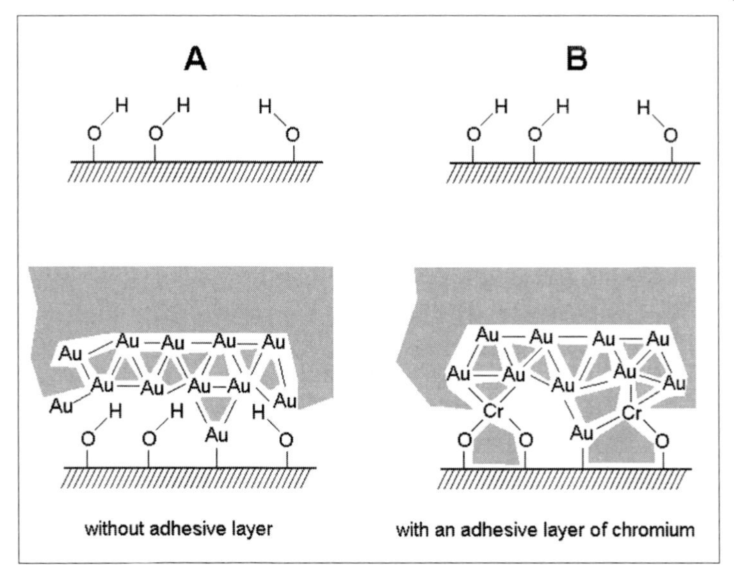

Fig. 18 Mechanism of adhesion layer of transition metals (e. g., chromium). A: Poor adhesion of a gold layer on a substrate with hydroxyl groups without adhesion layer. B: Increased adhesion of a gold layer on a hydroxyl surface due to an adhesion layer of chromium

served. This applies for many linear polymers that are deposited from a liquid phase. Therefore, polymers with polar functional groups are utilized as adhesion promoters for metal layers or inorganic compounds.

3.2.2
Layer Deposition from the Gas Phase

Layer deposition from the gas phase represents an important group of preparation techniques for thin and ultrathin layers. Often it is the only practical way to achieve thin layers of homogeneous thickness.

A general feature of these techniques is the transport of precursors for the layer preparation through the gas phase to the substrate surface, where they are involved in layer growth. In the case of two-dimensional deposition (all techniques without direct-writing deposition), the trajectories of the particles are statistically distributed, as are the points of contacts of the particles. The morphology of the resulting layer depends significantly on the mobility of the layer-building particles on the surface. If the particles have no mobility, layers with a rather unordered arrangement of the individual particles develop. These layers are amorphous. The fixation of the particles and therefore the tendency towards an amorphous layer correlates with the reciprocal of the ratio of the melting temperature of the deposited material and the absolute temperature, and with the decrease in bonds between the deposited atoms. In the case of a low surface mobility, a solid-state structure exhibits only a low order in small dimensions based on particle diameter and the interaction between the nearest neighbors.

An increased mobility of the particles on the surface leads to condensation and crystallization centers as seeds for the layer development. Only in certain cases does the substrate induce the creation of adjacent seeds into monocrystalline layer elements. Every seed usually leads to an individual grain, which is either amorphous, an elementary crystallite, or consists of several other crystallites. The resulting layers are always heterogeneous, including a variety of morphological domains. Depending on the local arrangement of atoms or molecules, one can differentiate between an amorphous, partially crystalline and polycrystalline arrangement. Grain boundaries demonstrate changes in the orientation of crystal planes and anisotropy axis, and lead to increased free volumes. In the case of heteroatom layer composition, chemical differences can also be observed.

Size and density of the grains depends on particle mobility. Higher mobility results in only a few seeds, leading to extended large morphological domains. The lower the mobility, the denser and smaller the structure of the grains, and the wider leads the effects of the starting seed arrangement into the grains. For low mobility, this effect leads to column-like structures in thin layers, with column diameter correlated with mobility. Because the surface mobility of particles correlates with the melting temperature, a direct dependency of the layer morphology on the ratio of deposition to melting temperature is observed.

If the surface mobility of a particle is very low, nearly every contact with a surface atom leads to a bond. So small inhomogeneities and surface roughness is enhanced in the process of further layer deposition. Dendritic structures are observed, and at a certain density sponge-like structure are observed. This process requires low deposition temperatures. So thin layers of nanoporous materials can be prepared (Fig. 19a). Depending on the material and deposition conditions, a wide pore width distribution ranging from atomic dimensions (0.1 nm) up to micrometers is observed. Such layers are for example catalytically active surfaces, chemical absorption layers, or absorption layers for electromagnetic radiation.

An increased surface mobility of the particles results in denser structures. Individual particles come to a halt after contact with several neighbors. For an overall low mobility (i. e., the removal from given positions exhibits a significant activation barrier) the layers become denser, but the thermodynamically most stable positions are usually not reached. Only in the vicinity of atoms is a rather ordered arrangement achieved, on the atomic scale it is not ordered. So the state is amorphous or glassy (Fig. 19b). Grain boundaries and domains are found inside such amorphous layers.

Higher surface mobilities of deposited particles reduces the number of seeds created, and the seeds grow through the subsequent addition of further atoms or molecule in energetically favorable positions and orientations. So regionally ordered arrangements, that means small crystallites, develop. With further layer growth, these crystallites come close to each other, and finally a continuous layer is created. The

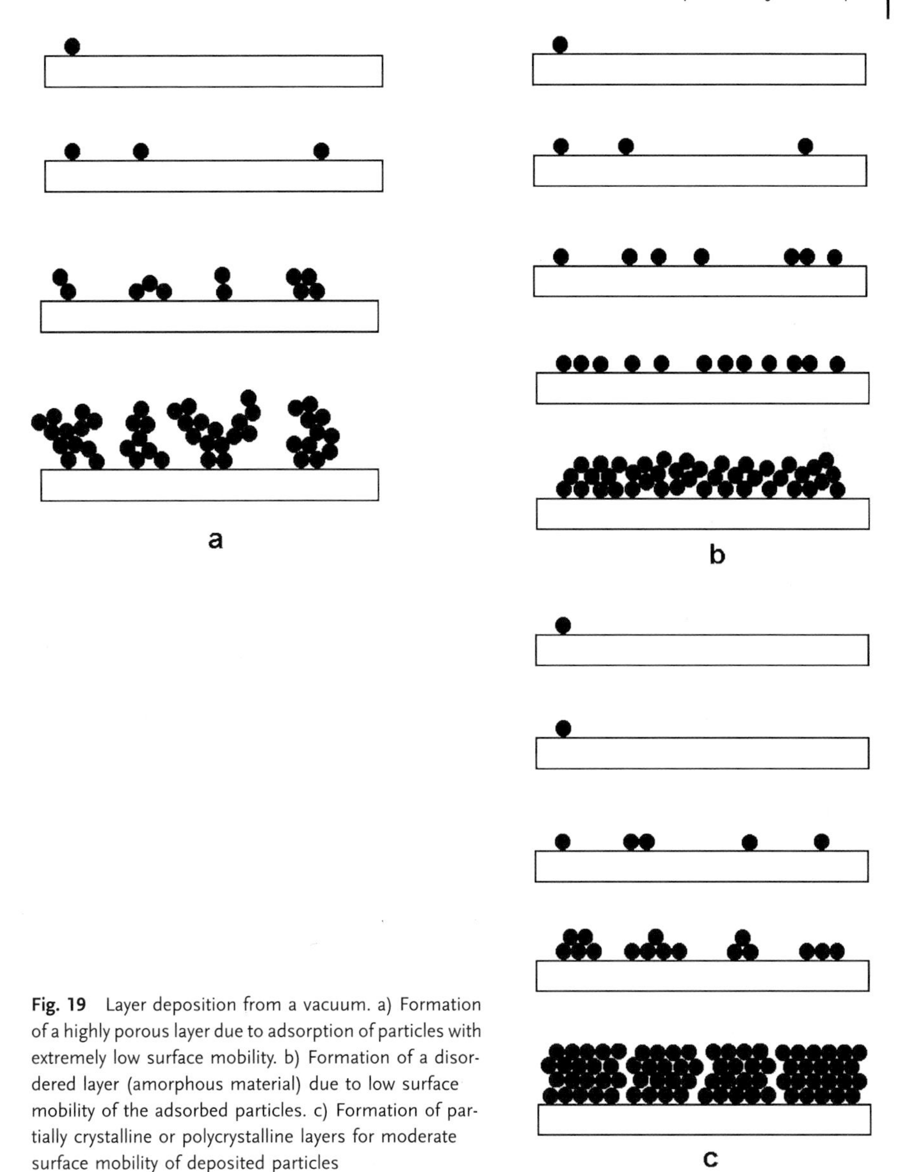

Fig. 19 Layer deposition from a vacuum. a) Formation of a highly porous layer due to adsorption of particles with extremely low surface mobility. b) Formation of a disordered layer (amorphous material) due to low surface mobility of the adsorbed particles. c) Formation of partially crystalline or polycrystalline layers for moderate surface mobility of deposited particles

crystallographic orientation of the individual crystallite and sometimes also the deviation of the exact position of the seeds lead to the development of grain boundaries (Fig. 19c). The result is a polycrystalline layer.

The dimensions of the morphological domains are usually between a few and some hundred nanometers, thereby mirroring the range of interest for nanotechnology. While in microtechnology the dimensions of domains have only an integral influence on material properties, in nanotechnology even the influence of discrete morpholo-

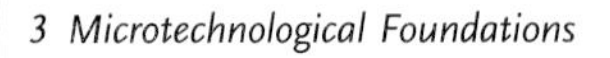

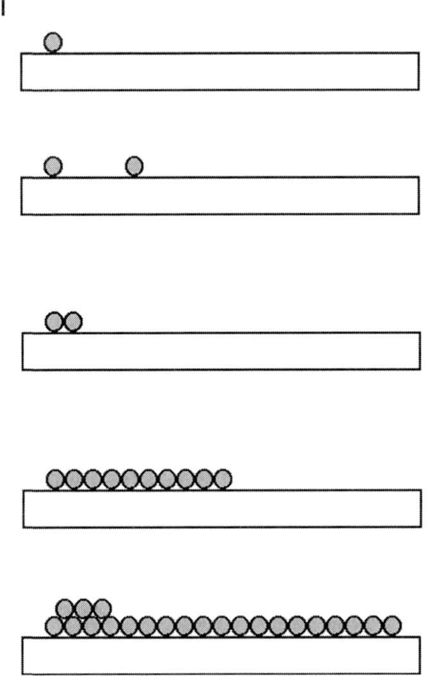

Fig. 20 Creation of single-crystalline layer resulting from the deposition of particles with very high surface mobility

gical domains can play a role. To minimize the influence of individual morphological domains on the function of the devices, nanotechnology is aimed at layers with very small domains and amorphous materials without grains, such as glassy layers or single-crystalline structures.

The fabrication of single-crystalline layers requires epitactical layer growth. Such growth is important for the production of stacks of layer for compound semiconductors [11][12]. Layer arrangements of compound semiconductors are applied in optoelectronics. Epitaxial depositions start from a planar, single-crystalline surface. Seed induction is minimized, the layer growth occurs on the faces of the atomic layers, thus being created plane by plane (Fig. 20). This process requires two preconditions: the lattice distances of both substrate and deposited material should be comparable; and there should be a certain particle mobility to allow for minimum binding on lattice planes but maximum growth at lattice steps.

3.2.3 Evaporation

This deposition technology is based on the evaporation of suitable material from a source by thermal activation (electrical heating, electron beam, lamp heater, etc.) in a vacuum process. Because of the low gas density, the thermally activated gas particles move with virtually no interactions through the reaction vessel and deposit on the walls and substrates placed in the reaction chamber. Movement of the substrates allows for layers of homogeneous thickness despite the inhomogeneous density of the evaporated particles.

The energy of the particles reaching the substrate surface is equivalent to the thermal energy of the source, despite the side processes, for example, wall contact and cooler particles. The higher the vacuum, the nearer the particle energy is to the source temperature, with the following equation describing the relationship between particle speed v, mass m and temperature T (in K) for every translational degree of freedom:

$$kT = mv^2 \tag{3.1}$$

With thermal evaporations, particles of the evaporated material initially contain kinetic energy equivalent to the surface temperature of the material in the evaporator. This energy is reduced on impact with cooler gas particles and the walls of the reactor. Only with pressures low enough to cause the free path lengths of gas particles to be equivalent to the source-substrate distance, the particles emerging from the source are able to carry their energy to the substrate surface.

Typical particle energies for evaporation processes are in the range of a few tens up to some 100 meV, so that the energy input into the substrate is moderate even for high deposition rates. Substrate holders of evaporators are often equipped with heating and cooling systems to control the layer morphology.

The kinetic energy of particles is usually below the binding energy of the surface components. Apart from reactive compounds or substances which desorb easily, the chemical state of the surface is preserved. The relatively low kinetic energy of the particles leads to their low surface mobility, usually resulting in strong structured surfaces with fine grains, and column-like or even fractal structures. A disadvantage of this approach is that the removal of contaminants is missing, so that vacuum evaporation requires thorough cleaning of the surface.

The preservation of sensitive surface states is a huge advantage for nanotechnology. In contrast to other deposition techniques, sensitive substrates can also be processed. Thus evaporation is better suited, such as sputtering or reactive plasma deposition for the deposition of metal films onto a molecular monolayer or an individual macromolecule as well as supermolecular arrangements.

3.2.4 Sputtering

Sputtering utilizes a mechanical activation (instead of a thermal process) to dissociate particles from a source, the target [13]. This approach provides the important advantage that materials which are difficult to evaporate can also be utilized for deposition. Inert ions created in a plasma and accelerated in an electrical field are used for the mechanical activation (Fig. 21). Their kinetic energy must exceed the binding energy of the particles in the target material significantly. To achieve sufficient deposition rates, an excess of 2–3 orders of magnitude is usually applied. A prerequisite for such energies is sufficient free path length for the acceleration of the ions, requiring at least

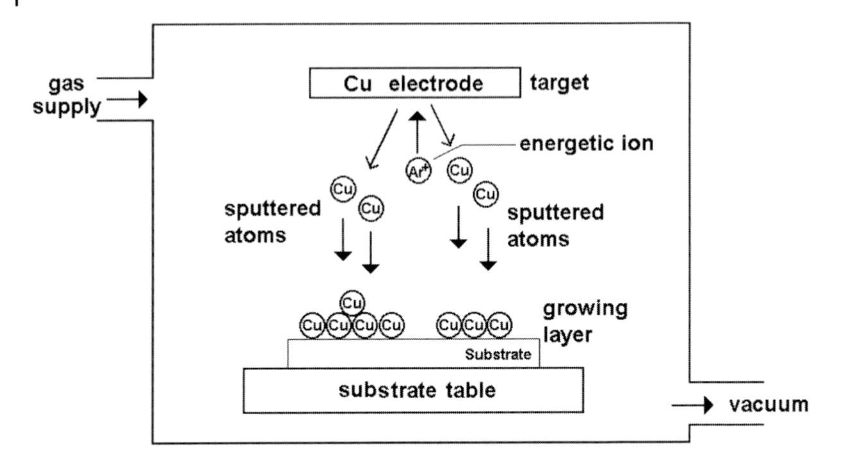

Fig. 21 Scheme of a reactor for sputter deposition

moderate vacuum (mTorr) or even high vacuum (μTorr). Because the particles experience only a few contacts on their way through the gas phase, they retain a significant amount of kinetic energy and exhibit a narrow angular distribution.

The energetic difference between evaporation and sputtering is shown schematically in Fig. 22. The area of the circles and arrows represent the energy. In sputter processes, the kinetic energy of a small particle exceeds that of a similar particle in evaporation (Fig. 22A) by several orders of magnitude (Fig. 22B), so that at impact the sputtered particle exhibits the energy of a particle that is several orders of magnitude heavier than a thermalized one (Fig. 22C).

Thus a particle that usually originates from the target and reaches the substrate surface exhibits sufficient energy to induce chemical reactions on the surface, such as the desorption of adsorbed particles or the activation of chemical bonds between the substrate and layer particle. Sputtered layers stick better than evaporated layers, and have a higher stability towards surface failures. However, they are not appropriate for depositing layers on sensitive surfaces, such as organic films. In this case, a certain gas pressure, a limited temperature of the substrate and a limited energy of the deposited particles are required.

The energy input into the substrate is significant, especially for higher sputter rates. The energy originating from the kinetic energy of the particles is often exceeded by energy due to the plasma. For thermal reasons, the surface mobility of particles is higher than with evaporation processes, and comparable layers are usually more compact but sometimes morphologically rougher. The roughness of sputtered thin layers is normally significantly lower than for evaporation at room temperature. This is the reason for the preference of sputter processes for the fabrication of ultrathin metal layers on organic substrates in nanotechnology.

Layers of various qualities can be realized by the choice of deposition process, vacuum, particle energy and deposition rate. They range from highly compact layers with

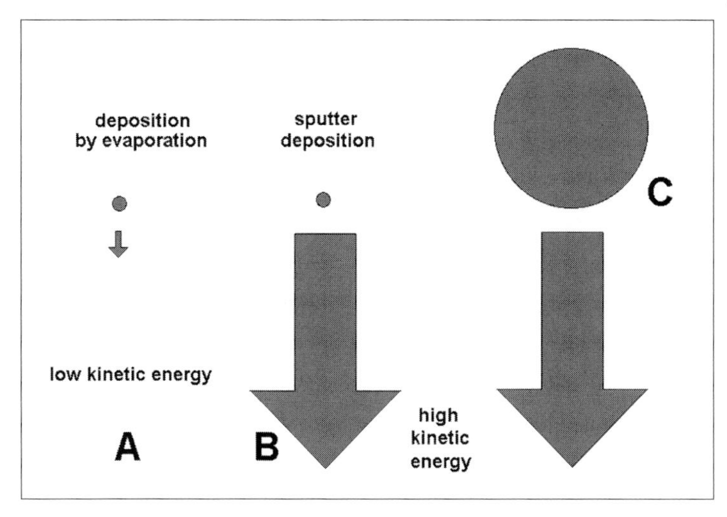

Fig. 22 Scheme comparing the energy impacts of thermalized particles (A) and energetic particles (B) with heavy particles (C) with moderate speed

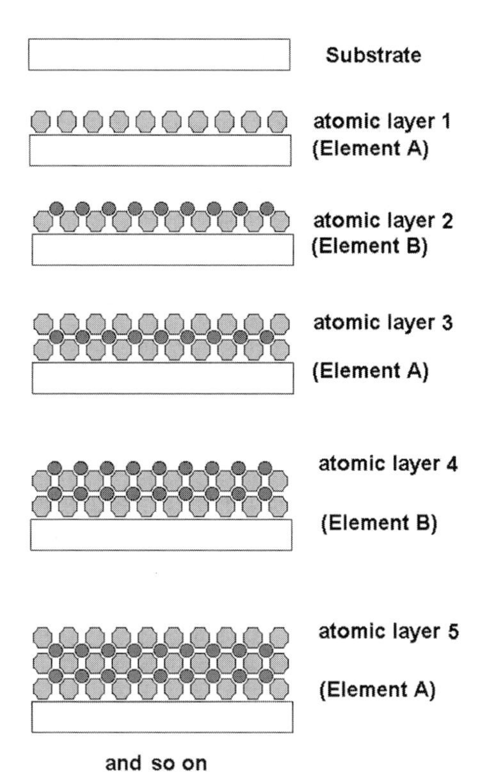

Fig. 23 Layered reactive formation of ultrathin layers by an ALE process (M. Ritala and M. Leskelä 1999) [20]

properties approximating the bulk material to extremely nanodendritic and highly porous material, such as Silbermohr.

The addition of reactive gaseous components during the sputter process also allows a chemical conversion of the target material to take place [14]. Such a "reactive sputter deposition" can be used to prepare metal oxide layers utilizing metal targets. Because of the high energy of the particles in the gas phase, a collision there has a high reaction probability. The same applies for the substrate surface. Under the addition of different amounts of oxygen and nitrogen, metal and semiconductor targets can lead to non-stoichiometric oxides, nitrides, oxynitrides and other compounds. In addition, reactive sputter processes achieve high deposition rates for oxides, because dielectric layers can be prepared from conductive substrates, and the sputter efficiency is higher than for non-reactive sputtering of oxide targets due to the usually lower binding energy of metals.

3.2.5
Chemical Vapor Deposition

Chemical vapor deposition (CVD) is based on the addition of the starting substances for layer deposition in the gaseous phase. The layer composition is determined by the parameters of the activating plasma and, most notably, the types and the proportions of the gases [15][16][17]. The conditions ensure that the condensing products are created on the substrate but not in the gas volume. CVD processes normally exhibit higher pressure compared with sputtering. Particles arriving on the substrate have a broader angular distribution of orientation, and their energy is low. Because reactive components are involved and the layer creation is based on a chemical conversion at the substrate surface, in the presence of functional groups the particles form bonds with the substrate, despite the low kinetic energy of the arriving particles.

Incomplete conversion of the gases into layer material leads to byproducts, which should be gaseous or at least volatile. Thus silane (SiH_4) and ammonia (NH_3) react to give silicon nitride (Si_3N_4 as a layer) and hydrogen.

The preparation of monocrystalline layers on monocrystalline substrates is a key process for many applications, such as in solid-state electronics and optoelectronics [18]. Such epitaxial layers exhibit a high quality and provide optimal conditions for the control of charge transfer in miniaturized devices. The preparation of multi-component epitaxial layer arrangements prepared from mono atomic layers of different materials are denoted as atomic layer epitaxy (ALE) [19]. Thus, the next layer develops only on the complementary previous layer and every layer has a self-limited growth based on the chemical properties of the substrate material. The buildup of a binary material by ALE is shown schematically in Fig. 23. The first layer of substance A is deposited on the substrate. Because this material does not react with itself, the growth stops after the formation of a monolayer. Subsequently, a second material B is deposited on the substrate, which forms a chemical compound with A but not with itself. The third step again includes deposition of A, and so on. The process can be controlled precisely, so that not only is the composition and degree of ordering of individual layers determined, but even the exact

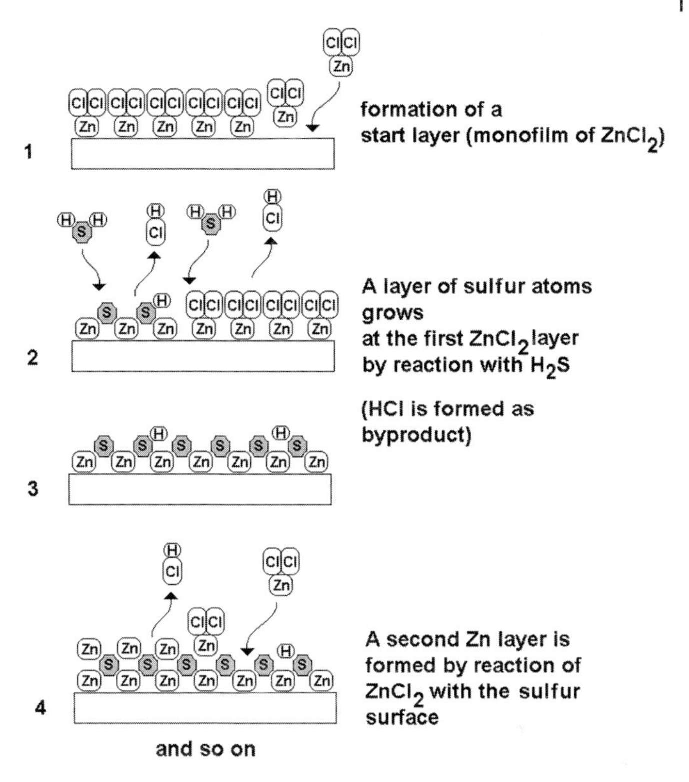

Fig. 24 Generation of zinc sulfide layer by a reactive synthesis based on alternating addition of hydrogen sulfide and zinc chloride using an ALE process (M.Ritala and M. Leskelä 1999)

number of individual atomic layers. As an example, Fig. 24 shows the preparation of an epitaxial zinc sulfide layer through ALE from alternating reactions with $ZnCl_2$ and H_2S [20]. In principle, layers containing just one element can also be fabricated by ALE, when the valences of the deposited atoms required for the subsequent construction are blocked by protecting substituents. In this case, deposition and activation steps occur in an alternating pattern. This approach represents an analogy at the atomic level to solid-phase based organic synthesis, where two excluding complementary binding reactions or protection group chemistry is used. ALE is also suitable when a high precision of layer thickness and/or layer composition is required, without the need for epitaxial layers. This wider application field, with numerous examples of ALE utilizing a wide range of materials, is detailed in Table 1.

Tab. 1 Examples of compound materials that are created by ALE processes (including polycrystalline and amorphous materials) [20]

Metals	Ge, Cu, Mo
Semiconductor	Si
Carbide	SiC
Nitrides	AlN, GaN, InN, SiN_x
Nitrides (metallic)	TiN, TaN, Ta_3N_5, NbN, MoN
Oxides, dielectrics	Al_2O_3, TiO_2, ZrO_2, HfO_2, Ta_2O_5, Nb_2O_5, Y_2O_3, MgO, CeO_2, SiO_2, La_2O_3, $SrTiO_3$, $BaTiO_3$
Oxides (transparent conductors and semiconductors)	In_2O_3, ITO, $In(O_xF_y)$, ZnO, Ga_2O_3, NiO, CoO_x
Oxides (supraconductor)	$YBa_2Cu_3O_{7-x}$
Sulfides	ZnS, ZnSe, ZnTe, CaS, SrS, CaS, SrS, BaS, CdS, CdTe, MnTe, HgTe, La_2S_3, PbS, In_2S_3, $CuGaS_2$
Fluorides	CaF_2, SrF_2, ZnF_2
III/V semiconductors	GaAs, AlAs, AlP, InP, GaP, InAs

3.2.6
Galvanic Deposition

Galvanic deposition can be applied for the creation of continuous thin layers as well as for the direct preparation of structures. While the fabrication of thin layers is of limited importance, galvanic deposition is an important additive structuring technique.

Additive shaping processes working with molecular disperse or ionic phases are the material-constructive pendant to the etching processes. The transport of material from the mobile phase to the surface influences the fabrication of the desired structure, and can be controlled by a directed transport together with a localized deposition onto an unstructured area, or by mask structures directing the deposition into the holes. Mobile phases can be fluids but also gases.

In the case of metals, galvanic deposition presents a simple method for the additive buildup of nanolocal structures in the liquid phase. Galvanic layer deposition includes the supply of the layer materials in the form of salts, and the deposition by electrical current. A prerequisite is a conductive layer on the substrate, which is wired and able to deposit metal ions from solution cathodically prior to conversion into solid metal:

$$M^+ + e^- \rightarrow M_s \tag{3.2}$$

The composition of the galvanic bath, the electrical potential, the temperature and the current density determine the quality of the deposited layer. Compact layers are usually the result of low deposition rates. Because metal ions are provided in the form of complex compounds, the choice of ligand L is essential for the quality of the deposited layer and structure:

$$[ML_n]^m + k\, e^- \rightarrow M + n\, L^{[(m-k)/n]} \tag{3.3}$$

(M = metal, L = ligand, k, n, m = stoichiometric and charge coefficients)

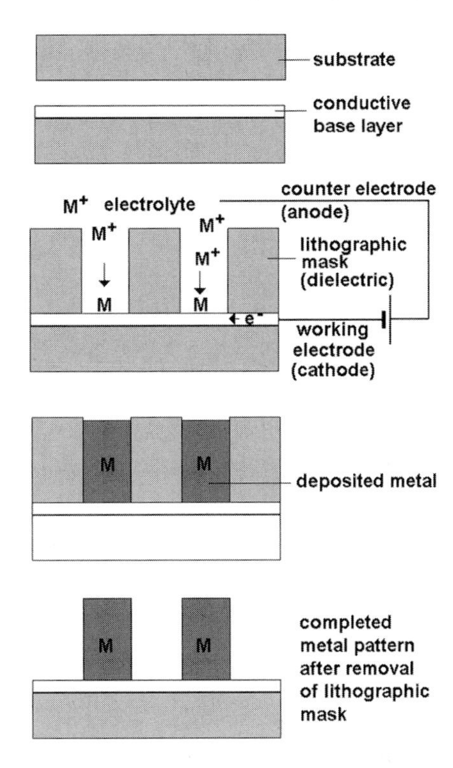

Fig. 25 Additive preparation of microstructured layers by galvanic deposition in the windows of a lithographic mask

Galvanic deposition allows the fabrication of structures with high aspect ratios, if adhesion masks with steep flanks are utilized (Fig. 25). Also, geometrically defined nanostructures can be fabricated for amorphous or microcrystalline layers, and even submicrometer and nanometer structures with high aspect ratios are possible.

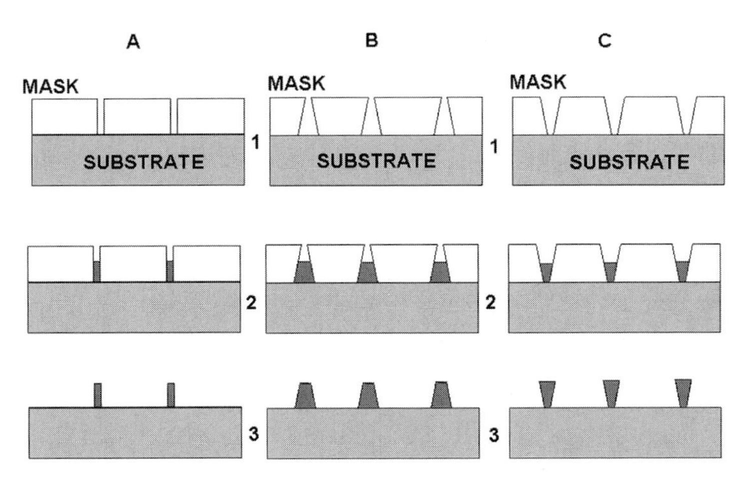

Fig. 26 Additive preparation of nanostructures with different flank angles by galvanic deposition in resist structures with various flank profiles

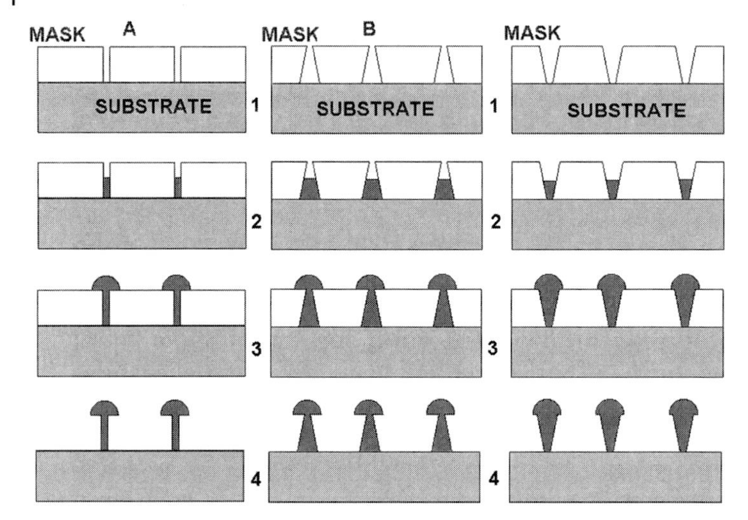

Fig. 27 Preparation of mushroom-like nanostructures with various profiles by galvanic deposition

Nanogalvanic deposition uses mask of non-conductors, and deposits metal from the solution in the windows onto the conductive substrate. Mask materials are polymer resists, preferably PMMA or similar compounds, which are nanostructured by electron, ion or X-ray lithography. Galvanic deposition transfers the mask window as a plastic negative onto the metal structure. Therefore, the flank geometry of the mask window determines both the lateral dimensions and (in connection with the thickness of the deposited material) the three-dimensional shape of the metal structure. The mask can usually be removed later by using an isotropic solvent, leading to, for example, structures with large aspect ratios and structures with undercut edges (Fig. 26). One example is the fabrication of metal columns of 75 nm diameter and 700 nm height (aspect ratio 9.3) by cathodic deposition of nickel in an electronic beam structured array of holes in a resist layer [21]. Complex flank profiles can be prepared by partial galvanic overgrowth of the mask layer surface or the use of multiple layer resist technology (Fig. 27).

3.2.7
Deposition by Spinning (Spin Coating)

For spin-coating, a solution of the layer-building material is placed onto a plane and centrally rotating substrate. Fast rotation of the substrate results in a centrifugal floating of the material, leading to a decrease in resist thickness. Simultaneously, differences (even significant ones) in substrate height will be completely leveled (Fig. 28). Evaporation of at least one solvent component results in a jump in the viscosity of the resist layer. As a result of an overlap of this viscosity change with the fluid-induced

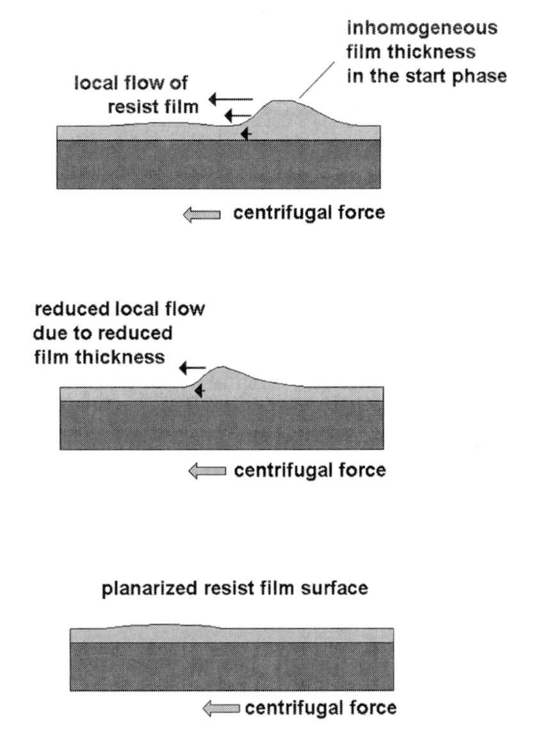

Fig. 28 Compensation of initial thickness inhomogeneities and preparation of homogeneous layers by the effect of centrifugal forces during spin coating

decrease in layer thickness the resist floating stops after a small period of time, so that layers with highly reproducible thicknesses can be prepared.

Spin-coating is a convenient method. It delivers layers of high thickness homogeneity and can extend down below a thickness of 100 nm. Spin-coating planarizes structured surfaces better than the other deposition techniques. There is only a residual relief due to the evaporation of the solvent after solidification of the layer. This shrinkage is approximately proportional to the local resist thickness. Because heating decreases the viscosity of the resist, subsequent tempering steps can reduce the observed relief in the case of densely located structures (Fig. 29). One problem is that topographic failures are higher than the final resist layer, because they lead to long-ranging inhomogeneities in the thickness. For low topographic relief, spin-coating is an interesting deposition technique for nanotechnology.

A precondition for an application of spin-coating is the feeding of the layer-building material as a spinable resist. It requires viscous floating of the resist material. The layer material has to be soluble in a solvent which exhibits an adequate evaporation rate during the spinning process, so that a stabilization of the chosen layer thickness is achieved. It requires the application of a mixture of solvents with different evaporation rates. The solvent with the lower evaporation rate facilitates the layer stabilization, whereas that with the higher rate leads to a homogeneous, failure-free layer. The remaining solvent is usually removed in a subsequent drying step.

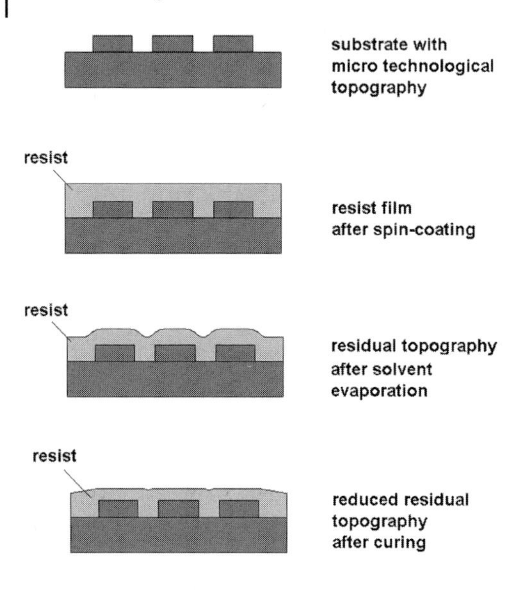

Fig. 29 Appearance of residual relief after planarization of substrate topography due to shrinkage of resist layers during the drying process and subsequent minimization of this relief by a local floating process at elevated temperature (curing)

The requirements of solubility and solvent parameters limit the choice of spinable layer materials significantly. Organic polymers, and also biopolymers, are easily deposited by this method. Inorganic layers are accessible when the atoms of the planned layer are chemically integrated into the resist and the organic components of the layer are later removed, e. g., by thermal treatment or plasma processing. One example of this approach is the so called spin-on glass layer. Normally oxide layers are involved, because the polymers are removed in an atmosphere of oxygen.

An important point for nanotechnological applications is the stretching and orientation of chain-like molecules such as polymers in spin-coating processes (Fig. 30). Although the layers are solidified as glassy or rubber-like, they are not fully isotropic.

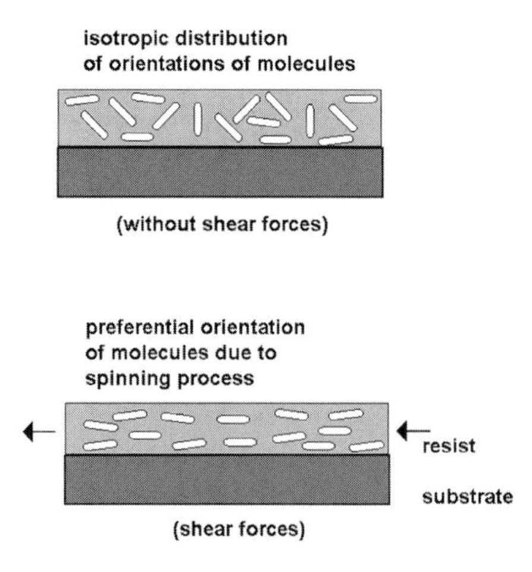

Fig. 30 Alignment of rod- or chain-like molecules within resist layers by shear forces during spin coating

3.2.8
Shadow-mask Deposition Techniques

In micro- and nanostructures, a local deposition of material from the gas phase results instead of a homogeneous layer. One simple approach to this problem is the application of an aperture at a defined distance from the substrate surface. Because the structures are created at the moment of material deposition (and without an etching step) this process represents an additive technique.

Local deposition of directed particles allows the fabrication of structures with very small dimensions. This technique applies a shadow mask which contains the shape of the final structure as a hole, which is positioned in close proximity to the substrate. A material beam with low divergence is required. Because the structure is generated below the mask holes, the technique represents a negative process (Fig. 31).

The structure generated on the substrate can be significantly smaller than the mask opening, as long as the distance between mask and substrate is optimal and the material beam is highly focused. Structures of down to 15 nm diameter can be prepared using micro beads with a diameter of 1.6 μm as distance mounts and mask openings of about 50 nm [22]. Slit-shaped masks yielded lines of similar dimensions.

A special shadow-mask process utilizes a small slit created in the cantilever of a scanning force microscope (Fig. 32, for scanning probe techniques cf. Section 4.4). Using the scanning tip and the laser-based z-axis positioning of the cantilever, the cantilever is aligned over the surface, which is scanned in the xy-direction. A particle beam of low divergence from an evaporation or ionization source is directed onto the cantilever and through the slit onto the substrate surface. The beam diameter has to be less than the lateral cantilever dimensions to shield the beam from the cantilever,

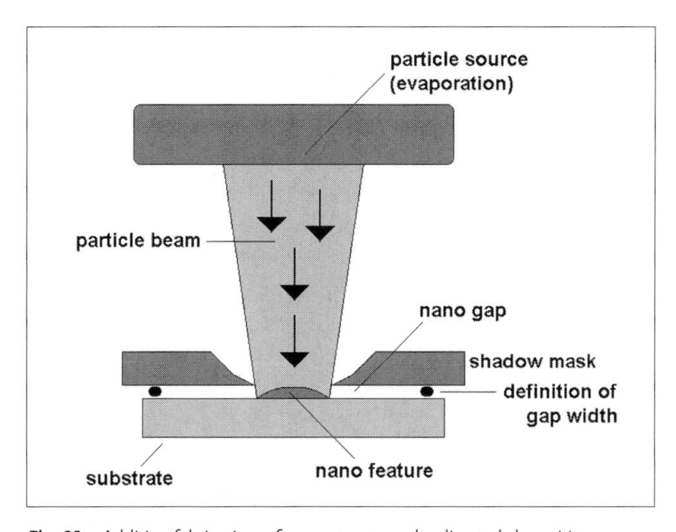

Fig. 31 Additive fabrication of nanostructures by directed deposition through a shadow mask with a small distance to the substrate

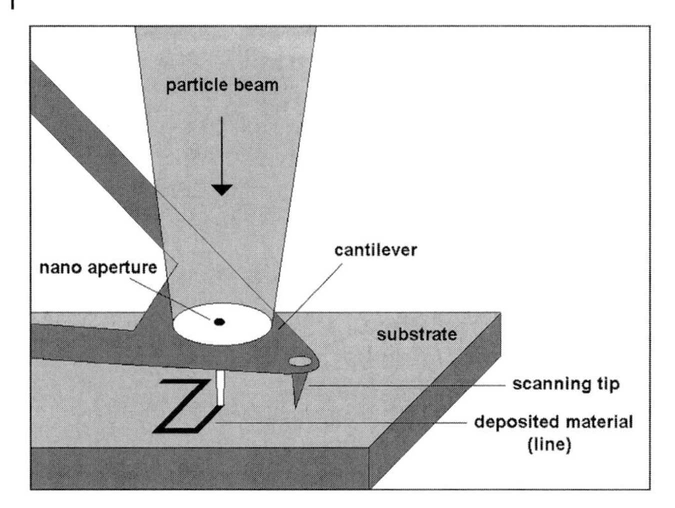

Fig. 32 Direct-writing additive fabrication by deposition based on shadow masks using a scanning probe cantilever with aperture

except for the small part which is shining through the mask opening. The size of the projected structure depends on the aperture of the opening in the cantilever [23]. An additional switchable aperture ("beam blanking") allows any pattern to be created on the substrate. This method is ideal for the preparation of a few small structures, but is problematic for larger areas and longer writing times due to the time requirements and the deposition of material on the cantilever, which means that the cantilever properties are changed.

3.3 Preparation of Ultrathin Inorganic Layers and Surface-bound Nanoparticles

3.3.1 Ultrathin Layers by Vacuum Deposition Processes

While the term thin layers describes thicknesses of about 1 μm, the term ultrathin layers describes thicknesses in the medium and lower nanometer range. It refers in particular to molecular monolayers and stacks of molecular layers, but also describes less defined arrangements.

Ultrathin inorganic layers can be monocrystalline, amorphous-isotropic or polycrystalline and, respectively, amorphous-anisotropic. Inorganic ultrathin layers are prepared by standard vacuum processes known from the production of thin layers (cf. Section 3.2). Preconditions are homogeneous, smooth and clean substrate surfaces as well as a high homogeneity in the deposition process and in film formation.

Epitaxial deposition techniques allow for the preparation of a highly defined single atomic layer on a monocrystalline substrate. Ultrathin layers consisting of several atomic layers normally exhibit atomic steps. These layers, consisting of a few atomic layers, can also be prepared by sputter processes. Although this approach yields no ordered monolayers, the precision of the average layer thickness is in this size range and the fabrication of fairly homogeneous, closed layers with thicknesses down to the sub-nanometer range is possible.

One application of ultrathin inorganic layers is magnetic multilayer systems. The magnetic decoupling of a ferromagnetic layer by ultrathin layers of non-ferromagnetic metals yields materials with a high magnetic field coefficient with respect to electrical conductivity, with an application as magnetoresistive sensors (GMR sensors).

Another preparation technique for ultrathin inorganic layers is the CVD process. In contrast to the techniques described earlier, all of the precursors for a CVD process are provided in the gas phase, requiring reactive precursors. CVD methods are described for a variety of materials in microtechnology. Extra thin layers can be achieved by layer deposition induced by external energy sources (PE-CVD), such as HF or microwave excitation. The rather inactive educts are transformed into highly reactive particles, e.g., by ionization or homolytic bond cleavage with the formation of reactive radicals. So the control of energy application (e.g., for only a short period of time) leads to control of the layer growth.

Thus a change of gas composition in between periods of energy irradiation allows the realization of a stack of different layers. Owing to the influence of the type of electronic excitation and the potential distribution near to the substrate on the layer deposition, the control of the energy input yields defined layer properties perpendicular to the surface. The layer elements with, for example, different electrical isolation or mechanical stress have thicknesses down to the lower nanometer range.

3.3.2
Deposition of Ultrathin Films from the Liquid Phase

The choice of the type of deposition for ultrathin inorganic materials from the liquid phase depends strongly on the material properties. Metal layers can be created by electrochemical methods, such as cathodic, or by electrodeless deposition. A cathodic deposition requires a conductive substrate. An even layer formation is required to realize homogeneous layers. Surface-controlled processes are preferred, due to the occurrence of autoenhancement of inhomogeneities in transport-controlled deposition processes. A high quality of the substrate is important for the creation of a homogeneous layer deposited by cathodic ablation.

Similar requirements are valid for electroless deposited metal layers. The fabrication of such layers in the liquid phase represents an electrochemical process. In contrast to the cathodic deposition, the electrons required for the reduction of the metal ions are provided by a reducing agent and not an external current source. So only a local current flow is observed (Fig. 33).

To achieve the deposition of material on the substrate, spontaneous reactions between metal ions and reducing agent have to be avoided. The composition of the me-

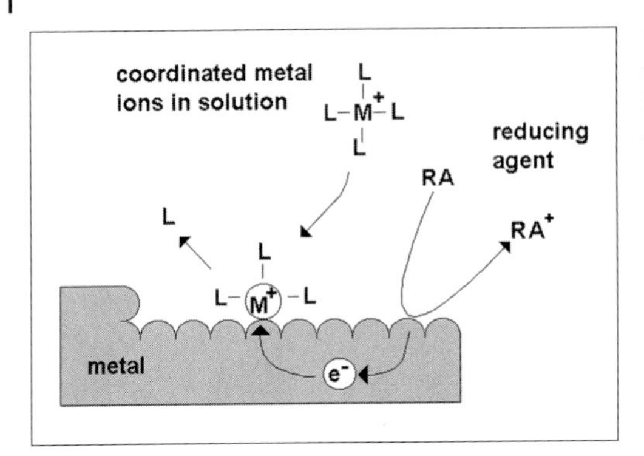

Fig. 33 Principle of electrodeless deposition of metal- or semiconductor layer (M, metal; L, ligand; RA, reducing agent

tallization solution and the deposition protocol have to ensure that the metal seed formation occurs only on the surface or even only on selected elements, such as nanoparticles or individual molecules. These seeds grow to larger clusters or closed thin layers.

3.3.3 In Situ Generation of Ultrathin Inorganic Films by Chemical Surface Modification

Ultrathin inorganic films can be prepared by the chemical transformation of layer or bulk material on a surface by reaction with reaction partners from the gas or liquid phase. A typical example in microtechnology is the surface oxidation of metals and semiconductors. Using increased temperature in air or an oxygen atmosphere, and possibly with controlled injection of water vapor, thin films of silicon oxide are fabricated on monocrystalline silicon. This process is widely applied for the formation of dielectric layers in microelectronics and microsystem technology.

Utilizing high-temperature processes in an atmosphere containing the respective reaction partner, thin surface films of various thicknesses can be formed on metal surfaces. The quality of the surface films is essentially determined by the cleanliness and the composition of the initial surface. The layer thickness is influenced not only by the duration of the process, but also by the mobility of the reaction partners in the formed layer, because for the layer growth either particles from the initial layer material diffuse to the free surface, or components of the gas phase have to diffuse to the interface between the deposited layer and the initial surface.

Typical thicknesses are about 1 μm for isolation layers, but significantly less for capacitive coupling. Of special interest are spontaneously-grown ultrathin oxide films for the fabrication of electronic tunneling contacts, requiring isolation layer thicknesses of about 1–2 nm. They are realized by the deposition of easily oxidizable metals, which are deposited in a vacuum free of both humidity and oxygen, prior to a short incubation period with an increased partial pressure of one of active components such as water or oxygen. After removal of the reactive gas and flushing with inert gas, the formed very thin oxide or oxidhydratic layer is immediately protected and stabilized by

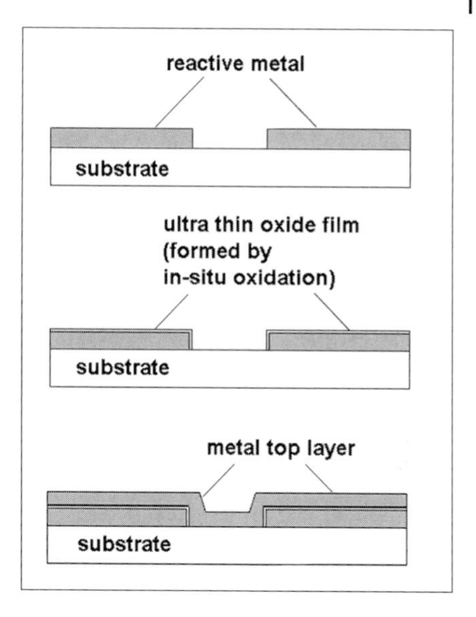

Fig. 34 Generation of isolating or semiconducting layers by spontaneous oxidation of metal layers on the surface on disruption of the vacuum

an additional metal layer (Fig. 34). Niobium and aluminum are examples of metals suitable for the controlled formation of such ultrathin oxide layers.

Thin metal oxide, oxidhydrate, hydroxide or salt films can also be prepared on metal substrates in liquid phase. Comparable to the galvanic deposition, these processes are initiated by an external potential, but in this case in an anodic polarization. The strength and duration of the polarization is determined by the electrical potential and, in particular, the composition of the electrolyte. Oxide and hydroxide films are formed from the aqueous phase or at least with the addition of water to non-aqueous electrolytes. The preparation of salt-like layers requires the addition of the respective anions and occasionally should be carried out under water-free conditions.

3.3.4
In Situ Formation of Ultrathin Inorganic Layers on Heteroorganic Materials

Ultrathin inorganic layers are prepared by the transformation of a surface film of organic material, when this film contains materials that readily form stable inorganic films. This is often the case for metals and semiconductors with stable oxides. One example is the formation of very thin oxide films on treating silicon-containing polymers with an oxygen plasma or with ion beam radiation in the presence of oxygen.

The preparation of thin and ultrathin inorganic layers by treatment with an ion beam or plasma also works on other heteroorganic materials. Suitable components are readily covalently integrated into organic layers, and form compounds stable with respect to oxygen and humidity at room temperature. In an oxygen plasma, highly stable oxides are observed.

Tab. 2 In situ generation of ultrathin SiO_x-layers from Si-containing polymers

Matrix element	*Oxide*	*Property*
C	CO	gaseous
	CO_2	gaseous
H	H_2O	volatile
O	–	
Si	SiO	non-volatile
	SiO_2	stable solid

3.3.5 Immobilization of Nanoparticles

In addition to the ultrathin layers, nanoparticles are interesting both as materials and for nanostructure technology. Nanocomposite materials and nanoparticles are already established in materials sciences in several areas. Thus, the special properties relating to the small size, such as size-dependent light adsorption or the huge specific surface area, are utilized, when a large number of particles will be involved.

In addition, the individual particles are of interest in nanostructure technology. Therefore, methods are being developed to manipulate small ensembles and even individual nanoparticles. One approach is based on the principle of reducing the dimensions. Nanoparticles are deposited in mono- or submono-layers on planar solid-state surfaces, where they bind on the surface in a similar manner to molecules. However, larger particles form non-specific bonds when in contact with surfaces, based on their relatively large contact area, and these numerous non-specific forces overcome the few specific bonds; smaller particles exhibit a greater influence of individual chemical bonds compared with ensembles of weak and non-specific bonds. This enables the controlled deposition of nanoparticles onto microstructured areas, resulting in microchips with spots covered with nanoparticle monolayers. A prerequisite for this specific binding is the existence of complementary-reactive groups on the surface of both the nanoparticle and substrate.

The reactions of nanoparticles with each other is also possible with appropriate functionalization of the surface. This extends the surface chemistry with nanoparticles in solution to synthetic chemistry. The specific hybridization of single-stranded DNA can be used for the connection of gold nanoparticles exhibiting thiol-functionalized oligonucleotides with complementary sequences [24][25]. These structures can be deposited as synthetic nanoparticle modules. Analogous operations are possible with organic nanoparticles, e. g., polymeric nanoparticles.

3.3.6 In Situ Formation of Inorganic Nanoparticles

Nanoparticle formation is observed in the initial phase of various layer deposition processes. The growth of galvanic, electrodeless-electrochemical, sputtered or reactive deposited layers usually starts with seeds statistically distributed over the surface. If the deposition process is stopped before coalescence, that is before the merging of islands or seeds into a continuous layer, nanoparticle layers are observed. The coverage depends strongly on the ratio of seed formation to growth rate. This ratio is controlled by deposition rate and surface mobility of the deposited particles. The density of the nanoparticles increases with decreasing surface mobility, so it rises with the interaction energy between the particle and surface, and is inversely proportional to surface temperature.

A local formation of thin layers with immobilized nanoparticles, mono- and sub-mono-layers of nanoparticles is possible through deposition and transformation processes. The in situ fabrication of nanoparticles in thin layers uses the segregation of compounds in a molecular-disperse system; one example is the recrystallization of $AgNO_3$ nanoparticles in poly(vinyl alcohol) [26].

3.4 Structure Generation and Fabrication of Lithographic Masks

3.4.1 Adhesive Mask Technique

The transfer of a small structure onto a substrate or a functional layer is carried out with a wide variety of different sources, ranging from 1:1 initial images to digital data sets. The various techniques applied in this context are combined in the term microlithography, with photolithography being a widely utilized method [1][3][27][28]. The larger proportion of microlithographic techniques apply a resist layer to create adhesion masks on the substrate or on the functional layer.

The adhesion mask technique is a key aspect of planar technology. In this way the structure of every functional layer is patterned into a mask layer (the resist) prior to pattern transfer in the functional layer. A resistance of the mask layer to the applied etch media is required. This requirement often narrows the choice of mask material.

Resists based on organic polymers are preferred as adhesion masks. These resists show a local change in solubility in a certain solvent (developer) due to local exposure to electromagnetic radiation. The change to higher or lower solubility differentiates the resists into positive and negative ones, respectively (Fig. 35). Steep edges in the resist are a prerequisite for the fabrication of small structures and high aspect ratios. Therefore, the resist should exhibit a high gradation, so that a small difference in dose results in a larger change in solubility (Fig. 36).

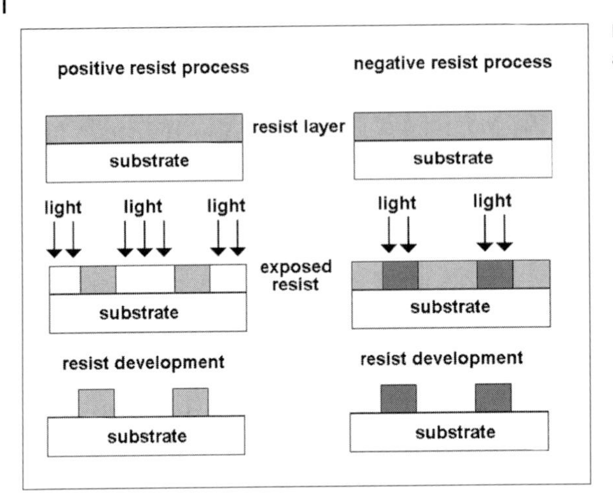

Fig. 35 Principles of positive and negative resist processes

The requirement of high gradation is of particular importance in nanotechnology due to the proximity effects that are observed. These effects are usually associated with external parameters, such as the wavelength of electromagnetic radiation (a proximity effect due to diffraction) or the energy of fast particles (a proximity effect due to ionization and X-ray emission). These effects result in a distribution of the locally deposited energy of a structural element instead of a sharp transition along the image boundaries. Hence the transformation product, which is generated locally by the exposure, also exhibits no sharp distributions (Fig. 37). This product dispersion results in a lateral gradient of the solution rate, which in turn increases the deviation from the upright flanks. This blurring effect decreases with a steeper gradation curve of the resist material.

There is a variety of demands on resist structures. The resist must be stable against liquid process media with respect to dissolution and rate of chemical reaction. As well as the resist, the interface between the resist and the substrate also has to show a high stability. This requirement is not trivial, because the interfaces are usually quite different from the resist regarding their composition. If the process solution reacts with chemical groups in finest cracks or dissolves components of the interface, a shift of structures or even an undercutting (and therefore a loss of the mask) is observed.

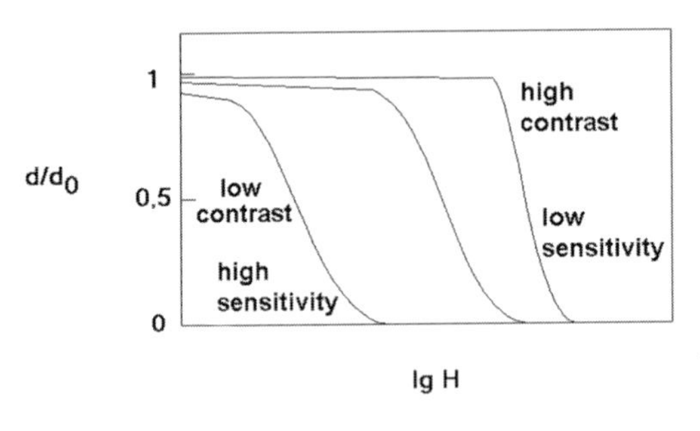

Fig. 36 Schematic gradation curve for positive photoresist: Remaining resist thickness d related to the initial thickness d_0 dependent on the exposure dose H

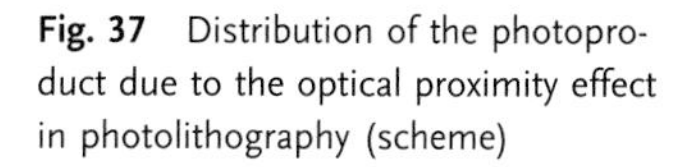

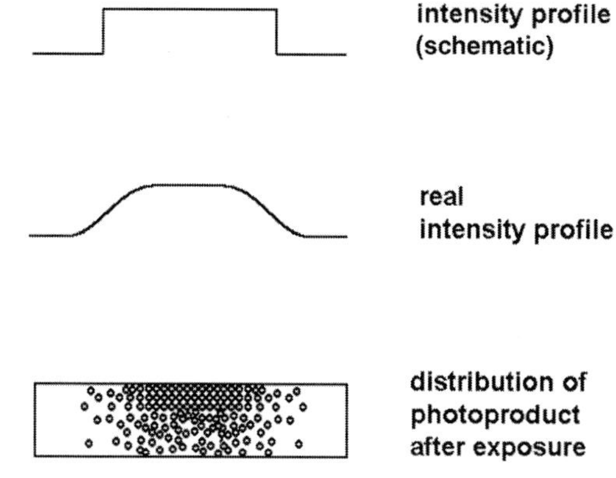

Fig. 37 Distribution of the photoproduct due to the optical proximity effect in photolithography (scheme)

Adhesive masks should therefore establish a stable bond with the substrate. This is achieved by a certain density of strong covalent bonds or by a higher density of coordinative, dipole–dipole, hydrogen bridge or Van der Waals interactions. As long as established bonds are not opened by the process solution, weaker polyvalent and unspecific interactions between resist molecules and surface groups of the substrate can be as efficient as individual strong bonds. Van der Waals interactions between polymers and hydrophobic surfaces are very capable of avoiding the intrusion of aqueous etch baths between the substrate and resist.

The stability of the interface is of lesser importance for pattern transfers in the gas phase, especially for the case of the interaction of directed particles with the surface. On the other hand, this situation requires a higher resistance of the layer to the impact of particles on the surface. This requirement is hard to accommodate, because the kinetic energy of the particles exceeds the binding energy of the resist often by some orders of magnitude.

The choice of resist has to fulfill three conditions. The first point is the stability of the material with respect to the process that includes the pattern transfer. For all lithographic processes that utilize some type of radiation, the resist has to be sensitive to this radiation, which means that the exposed part undergoes a local chemical change, but the remaining area is unaffected.

Secondly, this local chemical change has to result in the development of a relief in the resist. Therefore, the change has to significantly affect the removal rate in an appropriate development medium. In the simplest case, exposed areas are removed completely (positive resist) in the development process compared with the preserved remaining parts, or the surface areas without exposure are removed by the development and the exposed parts remain (negative resist).

Thirdly, in subsequent processes the remaining parts have to exhibit a protective property with regard to the underlying material, so that the functional layers can be etched by dry or wet techniques, or so that the processes for lifting of the layers deposited on the adhesion mask are possible.

Also the integration of solvent by insoluble structural elements of the adhesion mask affects the fabrication of small structures. Swelling results in failed geometries, especially in the case of negative resists. Negative resists are usually based on cross-linking of linear or slightly branched polymers by an irradiation step. The resulting network in the exposed areas significantly decreases the solution rate in a solvent, but the affinity of the resist to the solvent is preserved to a large extent. The solvent also diffuses into the exposed regions, increases their volume and so deforms the preserved structures. Negative resists are therefore in most cases not suitable for the realization of high aspect ratios and very small structures.

The molecular length is significant in the case of the nanotechnological application of long-chained polymers. A molecular mass of 10^6 results in a length of several micrometers for an extended molecule, and a length in the upper nanometer range is still observed for a molecular mass of 10^5. Although the molecules are usually in a coiled state, the ends are exposed in the development process and generate a gel-like zone in the solvent. This zone also induces a smoothing of the structures and an increase in the radius of curvature of the edges.

The majority of the effects discussed induce a flattening of the resist edges and so a decrease in resolution especially for thicker resist layers. Therefore, thinner resist layers are preferred for nanostructures. Besides the protective function required in particular in plasma processes, there is another negative aspect for an extremely thin resist layer. For the typical case of multilayer arrangements, resists must often planarize previously structured surfaces or at least cover the edges of the structures in a subsequent step. This function requires an adequate resist thickness.

The so-called reversal processes counteract the effect of flattened resist edges. These processes invert the contrast values of the resist. For positive resist processes in particular, the common orientation (perpendicular to the substrate) of decreasing locally deposited energy and solvent attack can be transformed into opposite processes which compensate each other. Positive resist-based reversal processes apply a fixation step after the masked exposure, so that the exposed regions are stabilized. Then the whole substrate is irradiated, so that the previously unexposed regions can be dissolved (Fig. 38). In addition to the simple conversion of the contrast values of masks, this procedure is ideal for the realization of steep or even overhanging flanks, as is often required for dry etch processes or an additive generation of structures.

The choice of resist thickness is influenced by the level required for sufficient protection, the topological relief of the substrate and the expected size of particles from the environment. Although media and air are filtered, this cleaning procedure has a cutoff and the number of smaller particles is reduced but not negligible. So there is a given probability of smaller particles residing on the substrate. The negative effect of this event increases with the density of nanostructures and the miniaturization of the structures. Although the probability of a defect in a given nanostructure is minimal, for higher coverage of substrates with nanostructures the probability of defect structures increases. Defects not covered by the resist are usually failures, so a decreased resist thickness results in increased malfunctions due to particles.

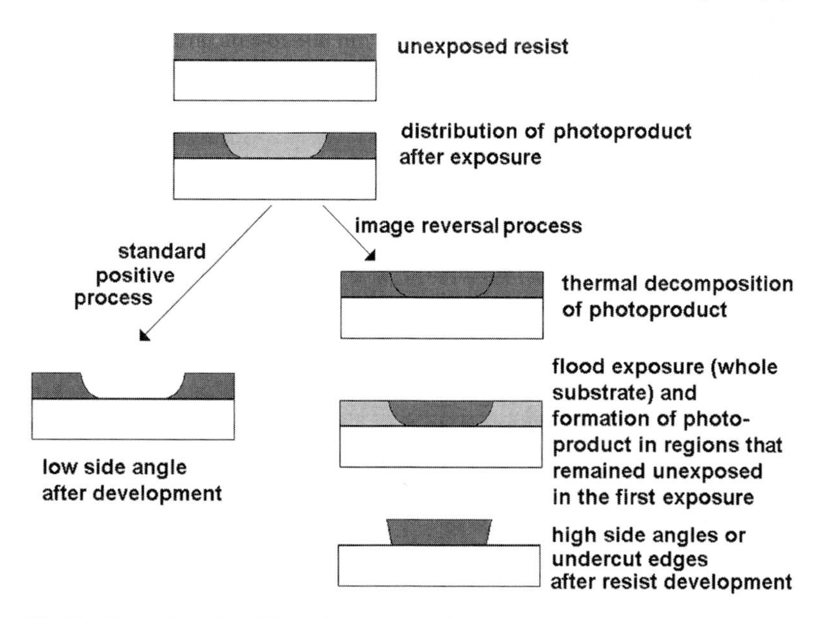

Fig. 38 Comparison of positive and image reversal process

3.4.2
Role of Resist in Photolithography

Photolithography is the classical method for the fabrication of reliefs on substrates by pattern-transfer processes, and will therefore serve as an example for a discussion of the generation and function of adhesion masks. Pattern transfer occurs by light. Light transfers the pattern from the original mask, e. g., a shadow mask, onto the substrate which incorporates a light sensitive layer. For a photolithographic process, this layer is a photoresist, which means it must adsorb at the particular applied wavelength and undergo a photochemical reaction. A transfer without optics occurs either in direct contact (contact lithography) or in close proximity (proximity lithography) of the mask and substrate. The use of an optical system allows the scaled or distorted projection of the pattern onto a resist layer. With respect to the optical image transfer, photolithography corresponds to the principles known from photographic techniques. The difference lies in the result, which is a relief and not a change in color or brightness. Because photolithographic transfer does not involve single structures but whole images or at least groups of structures, photolithography represents a process for group transfer (cf. Section 3.4.4).

Ideal materials for photolithography are resists based on relatively short-chain phenol or cresol resins ("Novolak"). The phenolic OH groups lead to stable networks with dyes such as naphthoquinone diazide, which release nitrogen under irradiation prior to a reaction with water to give carbonic acids. This reaction increases the solubility of the resist in weakly basic solutions by several orders of magnitude. Moreover, the phenolic OH groups can be involved in interactions leading to hydrogen bridges to protic groups on the substrate surfaces, so that substrates with surface OH groups (as are often found in inorganic materials) show a good resist binding. Another point is the potential for interactions with organic surfaces due to the aliphatic substituents on the phenol cores. The chemical protection of Novolak extends down to very low pH values, allowing the use of processes with strong acids. The resist is only unstable in the higher pH range. Novolak has advantages in the protection against plasma and energetic particles compared with resists based on aliphatic polymers. This stability can be explained by the relatively low hydrogen content and the aromatic groups.

3.4.3
Serial Pattern Transfer

The geometrical data of micro- and nanotechnical devices are usually generated in a digital format. These data are arranged according to functional layers, similar to the principles of planar technology. They represent a set of tables corresponding to two-dimensional images. The third dimension is only included indirectly: when considering guidelines for the thickness of the respective layer, in the information on process-related deviations from the original dimensions and the respective corrections in the layouts of the individual layers. These data have to be converted into images. In principle, this process could be carried out by imaging procedures, such as techniques based on arrays of lamps, apertures or mirrors. These techniques are rarely applied, due to the low lateral resolution and the limited amount of data. This situation could change in the future with the latest developments in optoelectronics and micromechanics, leading to inexpensive integrated luminescence diodes, micro apertures and micro mirror arrays.

A higher density of data and higher resolution can be accomplished through serial processes for structure fabrication. Therefore, the images are broken down into individual simple standard elements. These elements are subsequently transferred onto a substrate, e. g., into a resist layer. Typically, microtechnology applies serial structure generation to fabricate an intermediary mask that is applied to fabricate substrates in a serial manner. Serial structure generation can also be used for a direct structure generation of individual or a number of substrates, in particular to allow for changes in the mask design between subsequent processes, as is typically the case in research and development. This is one reason for the importance of serial structure generation in nanotechnology. Another reason is the higher resolution usually provided by serial processes compared with parallel techniques, hence the fabrication of nanostructures often requires a serial technique.

Serial processes have an inherent time requirement. Very short times are needed for one element, and these times add up for large areas. If a square of 1 μm requires, for example, 10 μs (10^{-5} seconds), then an area of 70 cm^2 with 10 % coverage will take as much as 7000 s (about 2 hours). Through reducing the square size down to 0.1 μm the overall time grows to 700 000 s (more than one week), which is usually not feasible.

Nanotechnology requires structure sizes below 0.1 μm, which is in principle within reach of techniques working with focused beams. However, because of the extreme time requirements, a serial- or mass-production of whole substrates is not possible. A drastic reduction of the positioning and exposure time per element would be needed.

Often the required small structures cover only parts of the substrate. Nanotechnical structures are often embedded into microtechnical arrangements with lower requirements regarding lateral resolution. The application of a variable beam diameter is much more efficient compared with a focused beam.

Microtechnology is already using shaped beam geometries. In its simplest realization, two apertures generate rectangular areas which are illuminated by the beam (Fig. 39). Shifting the apertures varies the dimensions and the size ratio of the shaped beam over a wide range, so that complex structures can be put together in a rational manner (Fig. 40). This principle is utilized by the optical pattern transfer in a so-called pattern generator, with a lateral resolution of only 1–2 μm. Smaller dimensions can be realized in a rational way with shaped electron beam procedures. Here, the pattern increment of the structures is usually between 50–200 nm.

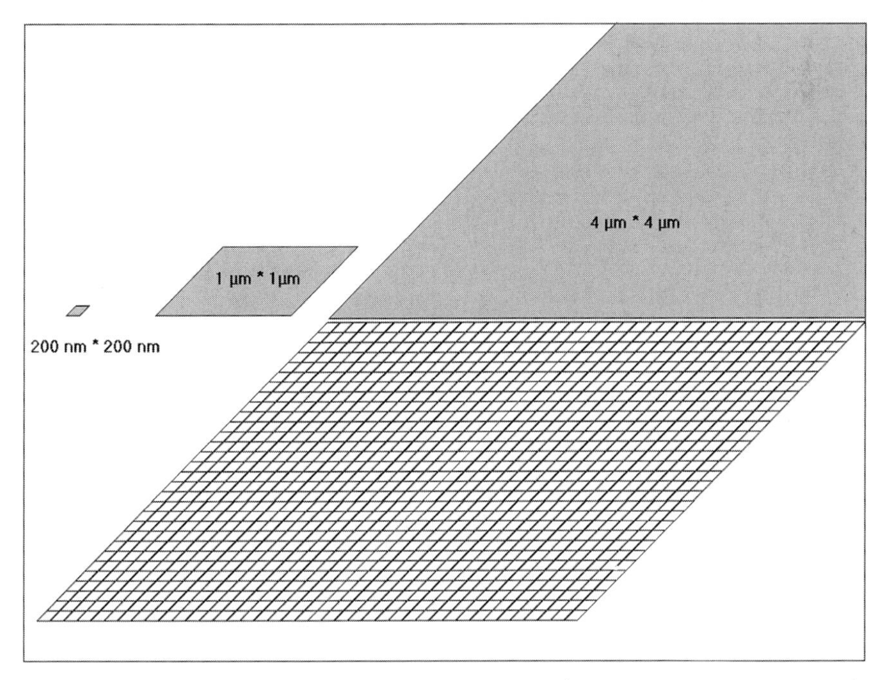

Fig. 39 Efficiency of shaped beam techniques. Scaled drawing of the areas simultaneously exposed by a stamp, from a 0.2 μm increment over a 1 μm^2 shaped beam to a squared 16 μm^2 shaped beam

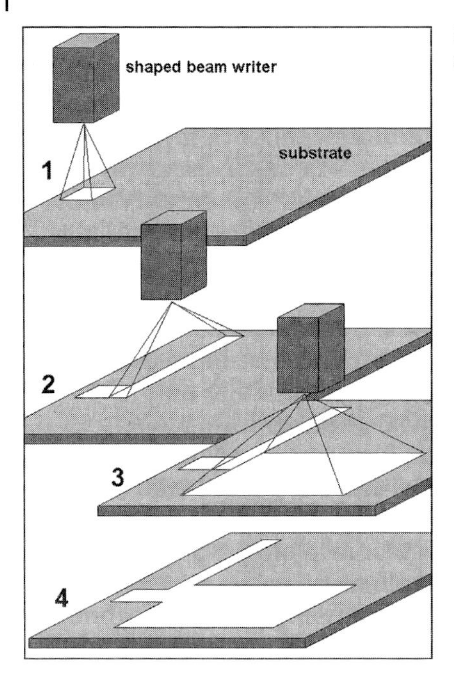

Fig. 40 Assembly of complex areas with a variable shaped beam in a serial pattern generation

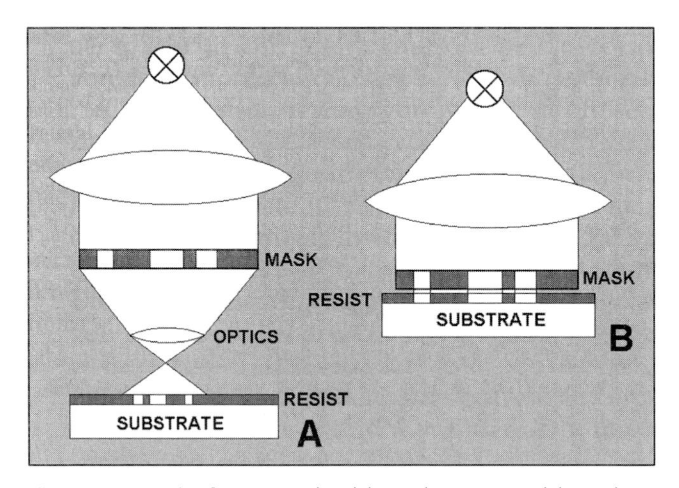

Fig. 41 A, Principle of projecting photolithography (projection lithography). B, Principle of contact and proximity lithography

3.4.4 Group Transfer Processes

Parallel transfer processes, also denoted as group transfer, are much more efficient than shaped or focused beam processes. When all structures in one plane are represented in an image, parts of or even the complete image containing a multitude of structures are transferred at once. Photolithography is a typical example of such a process (cf. Section 3.4.2).

Parallel image transfer can be achieved using two principles: the shadow technique and the optical projection (Fig. 41). The shadow techniques require the mask and the image-receiving layer to be in close proximity, the remaining distance determines the diffraction limit of the lateral resolution. Light-optical proximity techniques do not address submicrometer or even nanometer structures. However, in the case of direct mask–substrate contact (contact lithography), the optical near-field that is utilized provides sub-micrometer resolution. The optical near-field is not limited by diffraction. However, being optically-near is only realized by distances of less than the wavelength between the aperture and the location of photochemical reaction. Often the substrate relief and the resist thickness are limiting, because the relief is in the order of the wavelength, and the resist exceeds this value. For substrates with no or very low relief the resist thickness can be minimized down to about 10 nm, hence enabling contact lithography with optically near-field methods.

The transfer scale of structures for shadow techniques is 1. Optical transfer techniques allow almost any scale, photolithography usually applies to transfer ratios between 1:1 and 10:1. The mask and substrate are at a certain distance. The lateral resolution b is determined by the numerical aperture NA and a factor k (with a theoretical value of 0.5 and in practice values of some 10% higher):

$$b = k \cdot \lambda / \mathrm{NA} \tag{3.4}$$

The maximum numerical aperture is not only limited by the efforts for the optical system, but also by the required depth of focus, which decreases with an increase in the aperture. The optimization of both parameters becomes critical when approaching the specific wavelength. This is why photolithography thrives on the application of shorter wavelengths in microelectronic fabrication. Today minimal structure widths of 130 to 80 nm can be reached by wavelengths that approach vacuum-UV ($\lambda < 200$ nm). So although photolithography covers the lower submicrometer range, it is not clear if this principle will also be valid for structures in the lower nanometer range.

The extreme-UV range (EUV lithography) is under discussion as an alternative mass production technology at these dimensions. Group transfer processes are also possible with very short electromagnetic radiation, such as X-ray, and with particle beams. Hence dimensions in the medium and lower nanometer range can be addressed (cf. Section 4.3). However, owing to the requirements of highly sophisticated equipment and an extraordinary mask technology, they are only applied in special applications. One such application is the X-ray deep lithography for structures with high aspect ratio, e. g., in the LIGA process (lithography with ionizing radiation and galvanic forming). LIGA utilizes highly parallel synchrotron radiation (energetic X-rays) to overcome focus and diffraction limitations. The radiation writes structures with high aspect ratio in thick resist layers. In a subsequent step, these structures are transferred by galvanic deposition into metal structures [29][30][31].

3.4.5 Maskless Structure Generation

Microstructures can be realized without adhesive masks, when a pattern transfer from an external source is directly coupled with a local ablation or the removal of a functional material. Therefore, high energies have to be applied locally on the substrate surface, which is not really possible for larger areas. The primary limitation is not the power of the external source, but rather problems with both energy dissipation and the chemical temperature plus the stability of the shape of the substrate. These disadvantages usually exclude group transfer processes from maskless structure generation.

A simple implementation for maskless techniques is a local removal process induced by an intensive beam of either laser, electrons, ions or nanoparticles. Because a local removal of material requires high power densities, focused beams are preferred. A suitable process is the direct-writing lithographic technique in a serial approach. The removal usually occurs as a result of a combination of mechanical impulse transfer and local thermal effects. Specific chemical processes play only a minor role. As the diameter of the probe and the lateral radius of its expansion determine the resolution of direct writing procedures, use of such procedures with particle beams can be utilized for the fabrication of nanostructures. For light-optical procedures, e. g., by the use of lasers, application is limited to the micrometer and sub-micrometer range.

3.4.6 Soft Lithography

Processes for the mask generation by local transfer of minute amounts of material (acting as the resist) by stamps of soft material are denoted as "soft lithography". This term includes all procedures without local exposure prior to a subsequent development step. Such techniques are discussed as potential alternatives to conventional photolithographic methods.

"Soft lithography" transfers microstructures by direct mechanical contact, usually of elastomeric materials. These materials adapt well to the micro- and nanoroughness of the substrate, and have the ability to swell, which can be used as reservoir for the transfer of small volumes of liquids. Micro contact printing (μCP) is a technique that has been introduced which utilizes elastomeric stamps made of poly dimethyl siloxane (PDMS) [32]. PDMS is applied as a liquid onto a master structure prior to hardening as a result of chain extensions and cross-linking. The master structure is usually fabricated by conventional lithographic processes (Fig. 42). Natural struc-

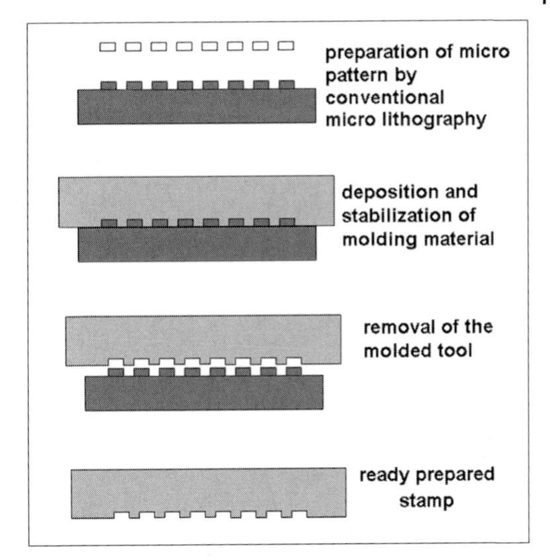

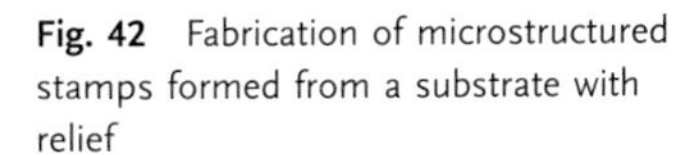
Fig. 42 Fabrication of microstructured stamps formed from a substrate with relief

tures can also be replicated by the stamping procedure. Cross-linked PDMS is particularly well suited for the intake of low-molecular mass liquids. A well-known example for the application of PDMS stamps is the generation of laterally microstructured monolayers of alkyl thiols on heavy metal surfaces. One introduced substrate is gold, which yields dense layers (Fig. 43). Structure sizes down to 35 nm are attainable through μCP [33]. In contrast to classical microlithography, μCP is also compatible with curved surfaces by the application of complementary curved stamps [34].

Soft lithography exhibits several advantages. It is simple and inexpensive, and is easy and can be applied quickly. The transfer of three-dimensional structures is easier compared with conventional lithography, and it is compatible with a wide range of substrate materials. By the inclusion of curved and more complex substrates, the generation of complex three-dimensional microstructures is feasible, e. g., chained tapes and rings by stamping on a cylinder surface [35]. Soft lithography is hampered by an increased failure rate, by possible distortion of structures due to elastic properties of the stamps and by the missing technology for precise alignment in order to address multiplane structures [32][36].

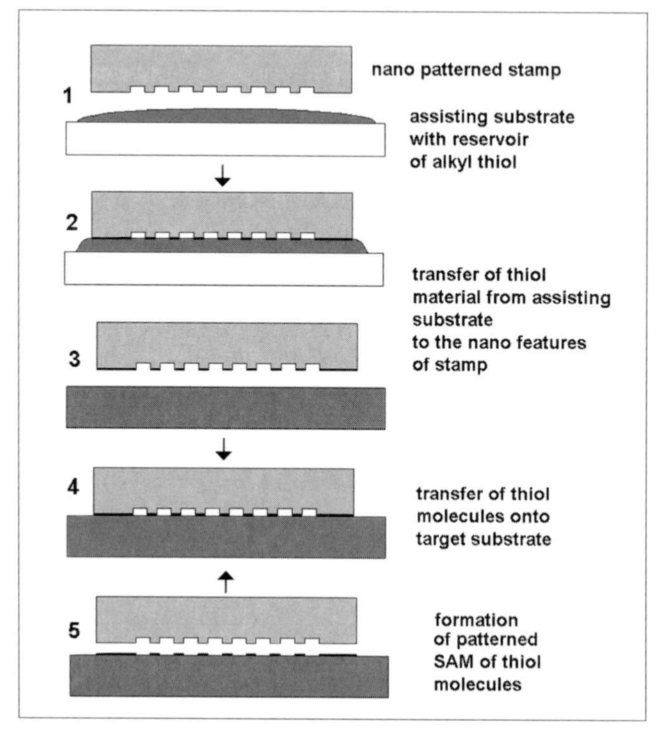

Fig. 43 Fabrication of lateral microstructured molecular monolayer by transfer of SAM-forming molecules from a reservoir onto a substrate using a microstructured elastomeric stamp [32]

3.5 Etching Processes

3.5.1 Etching Rate and Selectivity

Microtechnology is dominated by subtractive processes of structure generation. This explains the importance of etching techniques. Etching steps remove locally – usually in the apertures of a resist mask – the functional material. Etching techniques are divided in wet and dry methods depending on whether the material removed enters the liquid or gas phase. All etching processes require that the material removed from the surface of a solid-state substrate is transformed into a mobile state. The liquid phase requires soluble species; the transfer into the gas phase needs readily evaporable compounds or the effect of strong impulses to remove atoms or clusters from the substrate [37]. Thus most etching processes include a chemical transformation of the material. The etching rate is determined by the provision of reaction partners by the mobile phase, the chemical reaction at the solid state surface, or the removal of reaction products, depending on the rate of the individual steps. The etch rate r_{etch} is defined as ratio of removed thickness d_{etch} per unit of time t_{etch}:

$$r_{etch} = d_{etch}/t_{etch} \tag{3.5}$$

The mass of material removed m_{etch} includes the etched area A and the density ρ:

$$m_{etch}=r_{etch} \cdot t_{etch} \cdot A \cdot \rho \tag{3.6}$$

Microdevices usually contain several materials, so lithographic processes should exhibit a selective etching behavior. The etch rate of the material to be removed should be significantly higher than the rates of the other materials in this stack of layers.

This selectivity requirement is dependent on the layer thicknesses. To selectively etch a very thin (e. g., $d_1 = 10$ nm) on a thicker ($d_2 = 1$ μm) layer with a selectivity of 10, the loss in thickness of the second layer is only 0.1%, a negligible amount in standard processes. In the opposite situation of the selective etching of a thicker layer in the presence of a thin layer, a selectivity of at least 200 is needed to preserve at least 50% of a 10 nm layer during the process of etching a 1 μm layer.

Wet-etching processes achieve very high selectivities, where often the material to be preserved is fully inert. In dry-etching processes, especially when energetic particles are utilized, a certain change or removal of the material may be observed as a result of the etching particles. Controlling the changing selectivities is a problem. This can be achieved by appropriate selection of the etching procedures, and an arrangement of the functional layer according to the technology or a problem-oriented resist-technology.

3.5.2 Isotropic and Anisotropic Etching Processes

A differentiation of the etching procedures with respect to their spatial effect is of particular importance for the generation of small structures. The spatial distribution of the etching results determines (in addition to the mask geometry) to a great extent the three-dimensional shape resulting from etching-technical treatment. The spatial efficiency of the etchant depends on the local rate-limiting factor. If a chemical reaction on the surface is involved, the local distribution of the limiting material property is projected. This is the basis for etching processes applied for the generation of material textures, e. g., as used in microscopy. The local geometries created for amorphous, partially amorphous or polycrystalline materials often only exhibit short characteristic lengths, the orientation of the grains and crystallites and sometimes also their shape are distributed irregularly and are therefore not applicable for a controlled shaping process. On the other hand, anisotropy can be used for shaping in the case of monocrystalline materials. More than three orders of magnitude difference in etching rate between different crystal orientations can be controlled by selection of the etch bath

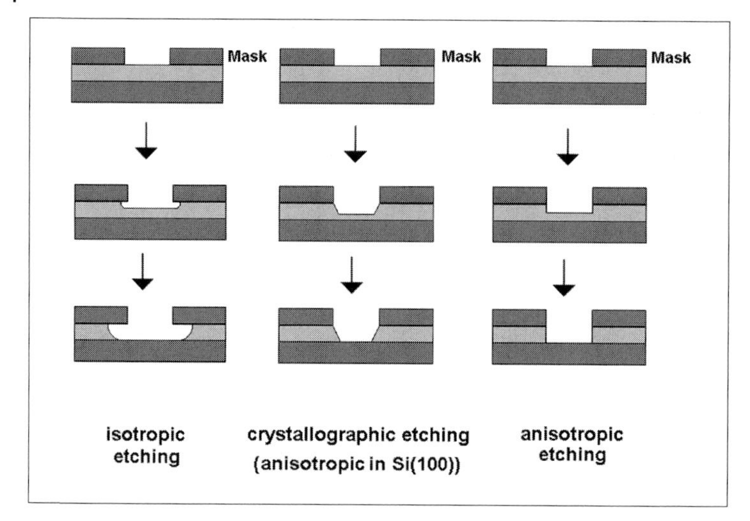

Fig. 44 Formation of etch geometries for isotropic and anisotropic etching

composition. One example is the highly reproducible etching of Si(111) facets of monocrystalline Si.

When the transport of particles to and from the surface is rate-limiting (rather than the surface reaction), the distribution of the direction of this transportation determines whether isotropic or anisotropic etching dominates. In the case of frequent strikes between particles involved in the etching, which means it is controlled by diffusive transportation, their trajectories are isotropic. Therefore, an isotropic etching occurs, yielding hemispheric profiles under the mask edges. For an isotropic concentration distribution, the attack of the etchant occurs perpendicular and parallel to the substrate plane at the same rate (Fig. 44). Isotropic etching is observed for many wet etching processes, but also for etching procedures with reactive gases or plasmas, as long as the pressure is not too low and the kinetic energies of the particles are predominantly in the thermal range.

3.5.3 Lithographic Resolution in Etching Processes

Etching processes transfer atoms or molecules from the solid into a mobile phase. Therefore, chemical species are generated, which are moved away from the solid substrate by transport or diffusion. This connection between chemical transformation and the shaping process is an interesting feature of nanotechnology. Etching processes achieve high resolution due to the chemical transfer of individual atoms or molecules. The individual reaction steps involve single molecules, ions or atoms. Therefore, materials are removed by their smallest unit. For macromolecular materials, etching processes are applied that disintegrate the molecular backbone, resulting in groups of atoms being removed. In general, etching processes allow a resolution down to atomic dimensions. Moreover, etching processes can be chemically selective. Many etching techniques remove materials of different composition with different rates. This is of particular importance for thin mask layers that protect thicker functional layers. Thin mask layers are essential for many nanolithographic processes with high resolution, because the lateral resolution is determined by the thickness of the mask layer. Hence a selectivity of the etching process is essential for lithographic nanotechnology.

3.5.4 Wet Etching Processes

Wet etching processes include a transfer of the removed solid state material into the liquid phase. Therefore, a chemical species is generated that is soluble (and therefore mobile) in the liquid phase [37][38]. Metals and semiconductors are transformed into cations or soluble coordinated compounds of their cations, through the involvement of local electrochemical processes. The standard open-circuit etching process compensates the local currents by reduction of oxidizing agents at the metal or semiconductor surface (Fig. 45).

For isotropic wet etching processes, the local removal rate is equal for all spatial directions. It is determined by diffusion, and therefore the minimal structure size correlates strongly with etching depth. Ideal isotropic transport means equal rates of material removal in both the lateral and vertical direction. Underetching of the mask edges occurs to the same extent as the layer material is removed in vertical direction. Therefore, the maximum aspect ratio for structures realized through an opening in a mask is 0.5 (related to the dimensions of the structure). Individual upright structures overcome this limitation in the case of precisely controlled underetching of the mask edges on both sides. This process is very sensitive to changes in the process, and is therefore barely reproducible. Moreover, such processes are only applicable for isolated structures. For periodical arrangements, the distance between two structures should be more than twice the etch depth.

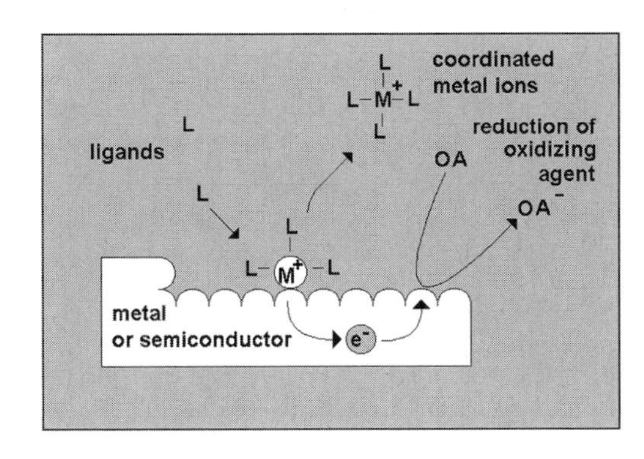

Fig. 45 Anodic and cathodic process during the electrodeless etching of metals and semiconductors (M, metal; L, ligand; OA, oxidizing agent)

This disadvantage of isotropic etching processes becomes critical in the case of nanometer structures in relatively thick layers. Thicknesses of 100 nm, for example, only allow structures down to 200 nm width. The limitation is less critical for very thin layers, because the lateral etch width (the underetching along the mask edges) is minimal. The extreme selectivity of wet etching processes therefore allows the application of monomolecular layers as a mask. Patterns generated in such masks can be transferred with minimal variations into a functional layer, as long as this layer is sufficiently thin. An alkylthiol mask with a thickness of typically 2 nm, for instance, allows wet etching of thin gold layers resulting in etch depths of significantly greater thickness compared with the mask thickness. An Au thickness of 10 nm and a desired structure width of 40 nm require an alkylthiol mask opening of 20 nm in order to compensate for isotropic underetching on both sides.

Anisotropic wet etching procedures need an isotropy of the etch rate in the material to be removed. This effect occurs in wet chemical procedures, that are controlled by chemical surface processes instead of diffusion. Such a procedure is utilized in the crystallographic etching of monocrystalline semiconductors, especially in silicon micromechanics (Fig. 46).

The basic chemical etch principle, which differentiates between the various crystallographic planes, can be transferred into nanometer dimensions. A prerequisite is a nanostructured mask with sufficient resistance towards the anisotropic etch baths.

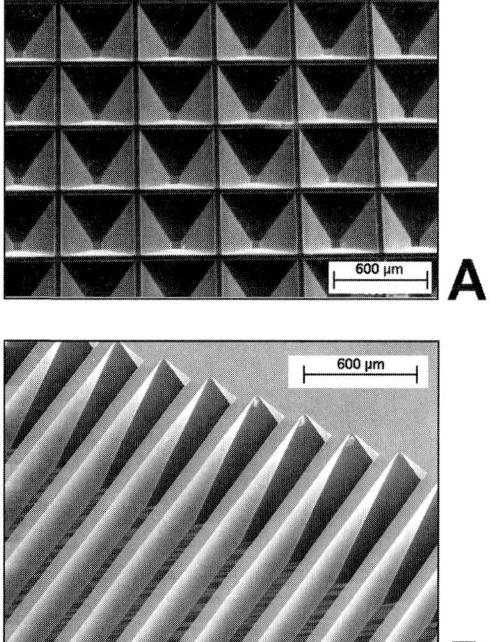

Fig. 46 Anisotropic etched pits in single-crystalline silicon substrates. A, Pyramidal pits in Si(100). B, Parallel channels with upright side walls in Si(110)

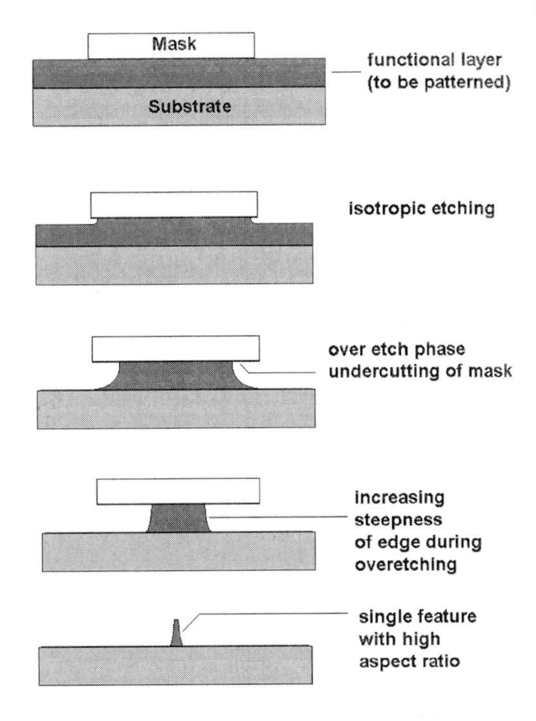

Fig. 47 Preparation of embossed discrete features with steep flanks by time-controlled isotropic overetching

In nanotechnology, wet etching is utilized to structure ultrathin layers in a highly selective manner. Isotropic wet etching is not applicable for nano-trenches at thicknesses of several tens of nm, due to underetching on both edges of the dimension of the etch depth, so that the final structure has dimensions of a twofold of the etch depth. Elevated structures, such as thin lines, are only achieved by isotropic wet etching when the distance to neighboring structures is larger than the horizontal etch component. So, controlled underetching yields steep edges (Fig. 47). A prerequisite is control of the process time, because underetching is determined by the etch time, but not by selectivity. Anisotropic wet chemical etching, e. g., on monocrystalline substrates, allows the realization of nanostructures with greater aspect ratios.

Wet etching is preferred for the transfer of pre-differentiated structures into a relief. Local implantation, electron or ion beam damages change the wet-chemical etch rate significantly. Based on this effect, focused particle beams induce a locally (lower nanometer range) confined implantation or damage prior to removal of the modified regions by a highly selective etchant (Fig. 48). So the basically isotropic etching combined with the previously generated anisotropy yields an anisotropic process for nanofabrication.

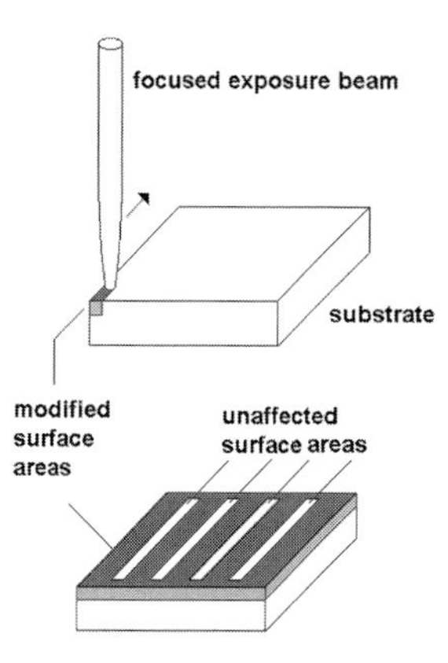

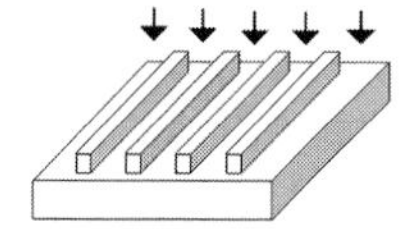

Fig. 48 Direct-writing fabrication of nanolithographic structures by local modification of the material by focused ion beam and subsequent selective etching

3.5.5
Dry Etching Processes

Etched nanostructures of higher complexity with aspect ratios larger than 0.5 are realized by dry etching. This process transfers the removed material from the surface of a solid-state substrate into a gaseous chemical species, e. g., by highly reactive vapors or by plasmas [39]. Therefore, the transport of the etching species to and of the etch product from the surface is diffusion-limited. Such processes are (comparable to the isotropic wet etching) thus limited by a maximum aspect ratio of 0.5.

Dry etching procedures with a preferred orientation regarding the removal of material are of particular interest in nanotechnology. This effect is realized, when the etch rate is determined by the density of particles with higher energy reaching the surface in a preferred orientation.

Non-reactive dry etching methods, such as physical sputtering (PE, physical etching) and ion beam etching (IBE) utilizing inert ions, are able to structure any material. Ions with a kinetic energy many times that of the binding energy are accelerated onto the surface. The local transfer of the mechanical impulse knocks individual atoms or clusters out of the surface and they are transferred into the gas phase prior to transport by convection (Fig. 49).

Standard sputter etching generates and applies energetic ions in the same reactor chamber. Therefore, an inert gas under reduced pressure (typically argon at 1–10 mtorr) is exposed to a high frequency source, so that a plasma is generated. Because of the much higher mobility of electrons compared with cations, spatial charging zones are generated, originating from the plasma and reaching the electrodes and the walls (sheet zone). Here the cations are accelerated towards the electrodes (and

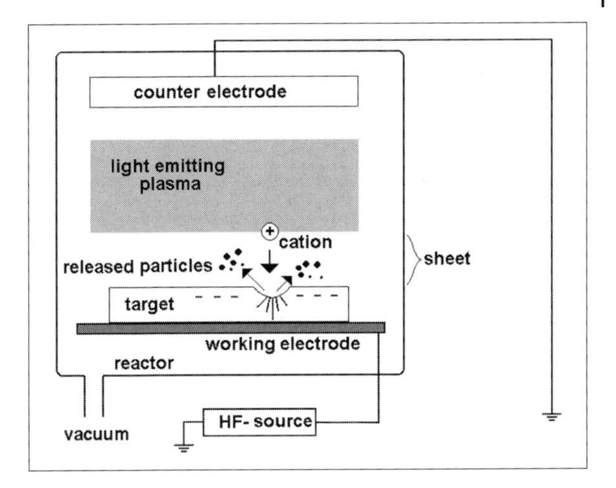

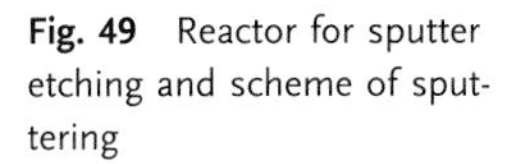
Fig. 49 Reactor for sputter etching and scheme of sputtering

thus the substrate to be etched). This so-called self bias effect can be enhanced by an external DC field. Typical potentials are some hundred volts, so that monovalent cations are accelerated to energies of several hundred electron volts.

As the lines of force of the electrical fields arrive approximately perpendicular to the surface of the substrate, the sputtering effect is induced in the same direction. Therefore, sputter etching is ideal for anisotropic etching. Upright mask structures can be transferred into functional layers with high precision and usually minimal undercutting. Tilted mask planes are problematic, being enhanced by faceting or leading to the reflection of particles into narrow cleft structures and thereby to an underetching with bent edge profiles. Because the mask material in sputter etching usually shows non-negligible etch rates compared with the etch rates of the functional layer, tilted mask edges result in a projection of their angle into the edge angle of the functional layer. However, sputter etching is well suited for lithographic masks with steep edges, to facilitate small structures with high aspect ratios and are therefore often used for the fabrication in the medium nanometer range.

Sputter etching as well as reactive ion etching are usually conducted in reactors with two parallel electrodes. This arrangement ensures a relatively homogeneous distribution of the plasma. The ion energy and the density of effective sputtering particles depend in particular on the pressure and the introduced power.

The conditions for etching by energetic particles can be controlled more effectively when the ion source is separated from the ion generation. This principle is realized in ion beam etching (IBE). An ion source creates and accelerates the ions, and the reactor is separated by at least one electrode. Creation, acceleration and the movement of the energetic particles onto the target are controlled separately. The spatial separation allows for a precise control of the ion energies and a high density of ions being achieved without thermal or plasma damage to the substrate. Moreover, the energetic ions can be neutralized in order to etch with neutral energetic particles. Ion etching allows any orientation of the target substrate with respect to the direction of the particles, so that structures with defined edge angles are possible.

For lower and medium aspect ratios, the achievable resolution of sputter and ion etching is mainly determined by the dimensions and the etch stability of the mask. The latter is the key problem in non-reactive dry etching. As a result of the variability of the processes, practically all mask materials exhibit a significant removal. This limits in particular the application of thin mask layers, resulting in limitations to the lateral resolution in the direct lithography of the mask layer.

Reactive dry etching processes are applied to achieve high etch rates with high anisotropy but also sufficient selectivity. They are realized either by replacement of the noble gases by reactive gases (reactive ion etching – RIE) or by the use of reactive gases for the ion generation (reactive ion beam etching - RIBE). Another possibility is the insertion of highly energetic inert particles (such as noble gas ions) into the reaction zone with the simultaneous delivery of reactive thermalized particles (chemical assisted ion beam etching – CAIBE).

3.5.6
High-resolution Dry Etching Techniques

Dry etching processes are ideal for the realization of ultra-small structures, due to their anisotropic effect on homogeneous and isotropic materials. Thus upright or other chosen flank angles can be realized. They are required for higher integrated small structures with aspect ratios below 0.5. While wet etch techniques dominate in micro mechanics and for structures in the medium and upper micrometer range, dry etching becomes interesting with decreasing structure sizes. Therefore, dry etching processes play a key role in microelectronics today. They are also the most important group of techniques in nanotechnology. The directed acceleration of energetic particles towards the substrate surface is required for an anisotropic etch process with a high precision of pattern transfer, even for small dimensions. Sputter and ion etching along with the respective techniques with reactive gases, RIE and RIBE, are the most important techniques for the etching of nanometer structures.

To etch with a high rate and simultaneously achieve a high selectivity of the dry etch process with respect to mask and other materials, a reactive etch process above a certain gas pressure is required. Thus the non-specific sputter effect is minimized compared with the reactive removal. Under these conditions, the isotropic component in the process is hardly avoidable. It leads to an underetching below the mask edges and prevents upright flanks being obtained. To overcome this problem, a reactive etch gas with a component that induces a stable chemical passivation of the flanks is applied. This passivation by the deposition of a thin and etch-resistant film prevents the lateral removal of material. However, the etch parameters have to ensure that this passivation layer is limited to the flanks. The flank passivation can be controlled by the composition of the etch gas. An increase in the percentage of trifluorotrichloroethane in an etch gas mixture of SF_6 and $C_2Cl_3F_3$ from 20% to 27% and then to 32% leads to a transformation of isotropic to partial anisotropic and to anisotropic etching of silicon. An increase in etching time was needed for the formation of an efficient flank protection [40]. If necessary, a periodic change in etch parameters is required to induce alternating etch and protection steps.

Depending on the material and the required etch depth, ion etching allows the transfer of very small structures from the mask into functional layers. Hence lines with less than 10 nm width were transferred from a PMMA mask into a GaAs substrate in a reactive ion etching with an $SiCl_4/Ar$ mixture. Using an AuPd mask for a silicon substrate, individual structures with diameters as low as 7 nm were obtained with a mixture of $SiCl_4$ and CF_4 [41].

In addition to mask-based etching techniques, direct etching methods with focused probes are also applied. Ion beams in particular can be well focused and controlled by electronic optics. Therefore, focused ion beam (FIB) processes are especially well suited for to obtaining small structures by probe-induced etching. Instead of ions, molecular cluster and noble gas atoms can also be applied in nanostructure fabrication [42]. Clusters of gas mixtures are ideal, when they or their products are easily transferable into the gas phase and when no secondary mask effects occur. The gas cluster are created by condensation of the respective atomic or molecular disperse gases through expansion, prior to an ionization of the cluster and the beam generation using an accelerating electrode with an aperture and additional electrodes if required.

The clusters usually include some hundreds up to some thousands of atoms or individual molecules. Apart from clusters of inert gases such as Ar or mainly inert molecules such as N_2, N_2O or CO_2, reactive gas clusters such as SF_6 or O_2 can also be utilized [43] .

Damage to the mask or damage of the layers or interfaces below the removed material are a problem of etching techniques using energetic particles. This makes the application of monomolecular layers or single supermolecules as the etching mask difficult. Subsequent layers are hit by the energetic particles when the functional layer is removed. Because of local variation in the layer thickness and etching rate, there is no absolute control of the etching process, and the subsequent layer is at least partially exposed to the etching process. In the case of sensitive surfaces, RIE and RIBE processes result in significant damage or even removal. However, chemically rather inert structures, such as inorganic tunneling barriers, are also affected due to their lack of thickness. Additionally, the impact of energetic particles results in failures of the solid substrate, which can change the electronic properties, such as the local conductivity.

Tab. 3 Examples of lithographic masks in nanotechnology

Method of primary structure generation	*Resist material*	*Etch resistance*	*Defect tendency*	*Thickness homogeneity*	*Edge steepness*	*Resolution*
EBL	resist	good	low	good	moderate	high
	SAM	partly good	moderate	very good	moderate	high
	LB film	poor	high	very good	moderate	high
FIBL	resist	good	low	good	good	high
	SAM	partly good	moderate	very good	good	high
	LB film	poor	high	very good	good	high
XRL	resist	good	low	good	moderate	moderate
STM	SAM	partly good	moderate	very good	good	very high
cond. AFM	LB film	poor	high	very good	good	very high
AFM	SAM	partly good	moderate	very good	good	moderate
	LB film	poor	high	very good	good	moderate
SNOM	photoresist	very good	low	good	poor	low

3.5.7
Choice of Mask for Nanolithographic Etching Processes

For all etching techniques that utilize lithographic masks, the quality of the mask is a key factor to the excellence of the structure obtained. In addition to the lithographic resolution, which means the minimal structure width in periodic patterns, the etch resistance, the defect density or tendency towards defects, the homogeneity in layer thickness, the roughness and the steepness of the flanks determine the quality of pattern transfer into the functional layer.

3.6
Packaging

Microtechnical structures are part of the technical world with interfaces in the centimeter to decimeter scale. One main reason for these dimensions is man, who can master the manipulation of objects that are 1 cm or larger, but who has difficulties with smaller structures. The main part of our technical environment is adapted to people – e. g., cars and household equipment, buildings, but also office and computer equipment, such as the keyboard buttons that are adapted to the size of the fingertip. Therefore microtechnical devices, in addition to the internal interface with dimensions of the individual microcomponent, but also demonstrate an external interface with respect to the exchange of energy and the flow of signals, materials and heat between the classical technical dimension and the micro dimension. Because of the standardized production of many microtechnical components required for mass production, most microsystems are modular. The arrangement of these components into systems occurs at the millimeter to decimeter scale, necessitating interfaces in these dimensions. The integration of microcomponents, their supports and their elements into an external technical environment is the purpose of packaging. This term describes two aspects: the geometrical and the functional connection. Nanotechnical systems are usually integrated into a microtechnical environment, which is in turn coupled to the external technical world. While the microtechnical integration of nanostructures is realized by microtechnical methods (chip technology, planar techniques), the external interfacing into the macroworld is achieved using packaging [44][45].

Assembling techniques are needed to achieve the transportation, the alignment and the connection of parts and devices. A variety of robots and machines exist for the transport and the connection of chip elements with each other as well as for the integration of chip elements onto substrates. To achieve a high throughput of elements, standardized methods are utilized, e. g., roll systems. For individual elements, a programmable pick-and-place instrument is best suited.

The precision of tables and mechanical robots can be adjusted for small elements for nearly every precision necessary. The classical mechanical techniques allow for manipulation in the sub-micrometer range. When translation is coupled with precise distance measurements (e. g., laser interferometry), a precision in the lower nanometer region can be achieved. With the help of piezo actuators, sub-atomic precision in the picometer range is possible. Such techniques are especially suited for standardized elements, for example, chip elements with thicknesses of between 0.2 and 1.0 mm and lateral dimensions between 2 mm and 5 cm, and many robots, and holding and transport systems exist for these elements. There are also several techniques available for the separation of larger substrates into chip-sized pieces. Examples of these are sawing techniques with a precision in the lower micrometer range and with sawing gaps of 50–100 μm, but also breaking techniques for microlithographic single crystal- and glass-substrates with resolutions below 100 μm.

A separation by sawing is always connected with the potential of contamination through the rinsing liquid and the creation of particles, which are not acceptable for some functional elements, such as sensitive micromechanical structures or microfluidic channels. Therefore, breaking techniques are applied instead of sawing, facilitated through lithographically created trench structures representing predetermined breaking lines. Because the breakage precision is especially high for monocrystalline substrates such as silicon wafers, lithographically based breaking techniques lead to, in principle, significantly higher precision than sawing. The general application of breaking is hampered by the large mechanical stress to the substrate as a whole.

Regarding its precision, microtechnical substrate separation techniques are comparable to high-precision mechanical methods, but they are much less precise than the lithographic fabrication of microstructures on a chip. The concept of interfaces between chips requires, therefore, much higher tolerances than the lithographic microstructures. Here a general technical principle becomes clear: the absolute dimensions of objects should not be too far away from their tolerances. The potentially more precise dimension has to take into consideration the tolerance of the less precise one. Therefore, contact pads or microfluidic connections are lithographically structured

– but with dimensions representing the tolerances of the separation and assembling techniques. What is valid in microtechniques for the interface between the microworld (lithographic structures) and the macroworld (chip holder, chip assemblies) can also be found in an analogous manner for the interface between micro- and nanostructures. To embed nanostructure elements into a microstructure environment, tolerance dimensions have to correspond to the requirements of the surrounding microworld.

Microstructures are usually fabricated on whole substrates. On the other hand, systems consist of individual chips instead of whole substrates. There are sometimes up to hundreds, sometimes even thousands, of similar chips on every substrate.

For certain applications, substrates are connected to other substrates before separation into individual chips. This approach is often observed in micromechanical sensor applications or in microfluidics, which require a stack of several substrates. Typical examples are acceleration sensors or micropumps. Channel structures and reaction chambers for microfluidic and micro reactor applications are always assemblies of two or more substrates. Therefore, a connection of the substrates before separation is much easier than afterwards. However, such stacks of substrates also have to be separated into chips.

The mechanical connection of devices and assemblies is the basic interface between systems. It provides a spatial fixation, and assists all other interfaces. Besides a passive connection, a mechanical coupling could also fulfill functional tasks, such as the transfer of mechanical forces by mechanical actuators. There are two classes of methods for the connection of chips and other platforms for microtechnical devices and assemblies: (a) techniques which rely on the use of connective materials such as solder and adhesives, and (b) connection without such materials. The first class has corresponding techniques in classical connection systems. In the collecting of microtechnical assemblies, adhesion techniques are much more prominent than mechanical connections such as screws and clamps. Adhesives are usually kept as thin as possible – typically some ten or even only just a few micrometers. The amount of glue needed for each device is often in the microliter range, requiring a special microdispenser.

When two monocrystalline solids with atomically smooth surfaces without any surface contaminations are brought into direct contact, they fuse into a single solid. The connection of very smooth silicon wafers with chemically clean surfaces is an excellent proof of this principle [46]. The fewer surface failures are observed, the lower the temperature requirements for such a bonding process.

However, highly processed surfaces are not always ideal for such direct connections. As a result of small failures or particles, direct joining of similar high melting material can often also require relatively high temperatures, although only a few atom layers have to be moved. A prerequisite for such a process is that the temperature and the applied pressure are sufficient to induce a surface movement of the atoms so that the two substrates fuse.

An overlay of thermally activated movement of atoms by a second component that contributes to the impact of the two bodies in each other provides ideal conditions. The anodic bonding of glass with silicon utilizes the migration of alkaline ions of the glass in an electrical field to assist the intermingling of the two materials. Typical voltages are between 0.3 and 1 kV, and temperatures are usually about 400 °C. Owing to the increased process temperature, an adjustment for the different thermal expansion coefficients of glass and silicon is required to avoid bending or even destruction of the assembly.

The precision requirements are not critical for merely mechanical connections between the chip elements themselves or between the support and devices. The requirements of the electrical connections usually determine the mechanical site. A typical range is between one and several micrometers.

A self-organization principle can be used for the assembly and three-dimensional arrangement in the micrometer range. Hence tensed thin film structures, which are locally separated from the substrate, can arrange microstructured optical elements in the upright position on the wafer into compact optical arrangements with several assemblies and complex optical paths [47]. Pre-tensed bilayer structures can even create closed three-dimensional bodies out of the planes, when the dimensions and the tension of the planes are carefully planned [48].

Optical connections call for significantly higher requirements with respect to the precision of positioning compared with the mechanical and electrical connections. Glass fibers as well as light guides need coupling precisions of better than 1 μm, sometimes even better than 0.1 μm. Such requirements are usually not within the range of mechanical robots or assembly lines. So assisting structures are needed to help with the positioning of the chip, or of fibers with respect to a chip or other structures. These structures are obtained lithographically to ensure a high precision of the structures as well as a high accuracy of the relative positioning of the miniaturized elements in relation to the positioning structures.

A variety of different techniques have been developed for the fabrication of specific functional connections. Small series of electrical contacts are preferably achieved through wire bonding. This process of welding a wire with an electrically conductive thin layer is, for example, based on ultrasonic activation. Owing to their key role in signal and energy transfer, electrical connections are important in microelectronic and micro system technology. For processes with small production numbers and large tolerances with respect to the positioning of devices, wire bonding is still applied. It requires bondable contact pads on both parts.

For a large number of contacts between individual chips or between chips and supports, the flip-chip technology is applied. Here metal material is arranged in a series of contact spots. For a mirror-symmetrical arrangements of the two parts, hundreds and up to thousands of electrical contacts and a mechanical fixture are realized after positioning and a brief melting of the contact material.

The importance of strategies for wiring of individual elements based on microstructured connections rises with increasing integration, but also with the larger complexity and diversity of electronic devices. Microtechnical wiring is therefore also important as contact periphery for future generations of nanoelectronic devices [49].

3.7
Biogenic and Bioanalogue Molecules in Technical Microstructures

Today, in addition to synthetic molecules, molecules produced by living organisms also play a part in technical microstructures. Such molecules are denoted in the following as "biogenic molecules", which means biomolecules in the narrowest sense. These are proteins and nucleic acids, and also small molecules such as steroids and alkaloids. From a technical point of view, all synthetic molecules that are identical with the real biogenic molecule in structure and composition are also included, e. g., nucleic acids produced by isothermal amplification outside of living cells or synthetic di- and oligopeptides.

"Bioanalogous molecules", biomolecules in a wider sense, should include molecules that are derived with key components, properties, functions and structural parameters from biogenic molecules or models based on them, but created synthetically without an identical model in nature. These bioanalogous molecules should also include synthetically modified biogenic molecules. On the other hand, there are also biomodified substances. These substances are modified biologically, e. g., by microorganisms or enzymes, and should not be regarded as biomolecules.

The arrangement of small molecules in arrays of laterally microstructured monolayers created a novel field in miniaturized combination chemistry. Examples are chip arrays of peptides fabricated through light-controlled synthesis, using the photocleavage of protection groups utilizing photolithographic masks [50][51].

Biomolecules and small bioactive molecules are screened for pharmaceutical relevance in large substance libraries. Therefore, microstructured supports are an interesting alternative to arranging and processing the libraries with minimal amounts of substances. A technical solution is the nanotiterplate with integrated sieves on a silicon base [52][53][54]. In the meantime, a variety of microreactors has been developed, which allow for the processing of chemical substances and biomolecules in small (μl–pl) volumes.

Since the development of DNA chips, nucleic acids have been introduced into microtechniques [55][56][57][58]. Synthetic DNA chips consist of a support with a two-dimensional microstructured array of oligonucleotide monolayers. The oligonucleotides bind (hybridize) complementary DNA from the solution, and markers (such as fluorescence dyes) visualize the binding location (and thus the sequence). In addition to synthesis, also stamping or spotting of DNA is utilized for the preparation of microarrays. Such arrays are usually less integrated due to the larger spot size [59].

Recent years have witnessed the development of a variety of chip reactors for nucleic acids analysis. These reactors are either applied between sample preparation and chip incubation or as an alternative to them. Examples are chip thermocyclers [60][61][62][63], electrophoresis chips [64][65][66], and integrated DNA analysis systems [67][68].

As well as carbohydrates of higher molecular weight, carbohydrate derivatives (such as cellulose) and synthetic polymers [such as polyacrylamide, silicon rubber and polyvinyl alcohol], protein layers are utilized in various microsensors for immobilization. Typical examples are alginate, collagen, gelatin and agar [69]. However, proteins are not only used as the immobilization layer, but they are mainly applied as the functional layer (e. g., in enzyme sensors). The transduction into electrical signals requires the presence of redox active species. Therefore, redox-active enzymes are preferred in amperometric and potentiometric biosensors. Such enzymes are also important for miniaturized detectors based on bioluminescence.

4
Preparation of Nanostructures

4.1
Principles of Fabrication

4.1.1
Subtractive and Additive Creation of Nanostructures

The predominant lithographic microtechnique is subtractive fabrication, which removes a deposited material from certain locations (cf. Section 3.1). Subtractive principles can be transferred to nanotechnology. An analogous layer deposition can be used, but with reduced thicknesses. However, methods with a high lateral resolution and a high precision with respect to positioning are required for fabrication. These requirements exclude standard optical projection methods as applied in photolithography. A drastic reduction of exposure wavelength would be necessary. Beam probe techniques are particularly well suited for the direct fabrication or transfer at the nanoscale (cf. Section 4.3).

The requirements for the process of removal of the functional material depend on its thickness. For typical microtechnical layers of 0.1–1 μm thickness, isotropic etching processes have to be excluded, because they do not allow aspect ratios larger than 0.5 and so the lateral dimensions will not reach the nanometer range. However, this does not apply for the subtractive patterning of a molecular monolayer or just functional groups on a plane solid surface. As a result of the dimensions in or below the lower nanometer range, nanometer structures are possible even with small aspect ratios. Wet chemical and plasma chemical methods of subtractive patterning, but also isotropic gas phase processes are suitable for such fabrication steps. In this manner, lithographic masks are applied in an analogous way to that known from microtechnical lithography. The ultrathin functional layer is therefore covered by a resist layer that is structured by a lithographic process. This pattern is transferred into the ultrathin layer, prior to removal of the mask.

The subtractive patterning of ultrathin layers is comparable to microtechnical etching, because it is also based on a series of individual reactions that remove the layer in the apertures of the mask and transfer the material into the gas phase. A significant difference exists in the amount of material and the required reaction rate. Because process times are in the second to minute range, the removal of ultrathin layers re-

Nanotechnology. M. Köhler and W. Fritzsche

ISBN: 978-3-527-31871-1

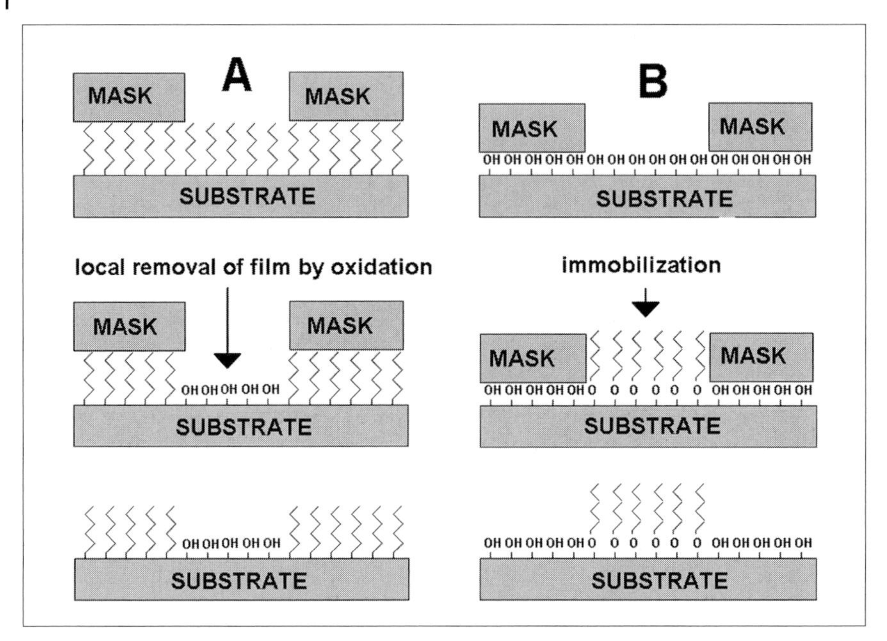

Fig. 50 Fabrication of lateral microstructured molecular monolayers by a lithographic process. A, Local oxidation of alkylated surfaces in the windows of a mask. B, Local alkylation of an OH-group rich surface in the windows of a mask

quires only a low rate of removal. The transport of etch-active particles to and the removal of reaction products from the surface is usually not critical, in contrast to microtechnical etching processes related to thicker films.

The removal of ultrathin layers sometimes occurs in one single step. This is the case for cleavage of a small group of atoms or the local removal of molecules from a molecular monolayer (Fig. 50A). This process is not comparable to microtechnical etching, and is better described as surface chemistry with a mask-guided process. Such nanolocal surface manipulation methods do indeed represent the backbone of nanotechnology.

Mask-guided processes of surface chemistry are not limited to the removal of molecules or atoms. The binding of new atoms or group of atoms to the surface is also possible (Fig. 50B). Because the new material is added, such a synthetic process is an additive pattern transfer. In the case of ultrathin layers, and especially for monomolecular layers, there is no clear dividing line between subtractive and additive processes.

Such analogous additive processes from microtechnology can be applied in nanotechnology for structures consisting of several layers of atoms or molecules. As regards the subtractive methods, the steepness of the mask edges has to be sufficient in the case of thick masks, to facilitate lateral structures in the nanometer range. An example is galvanic fabrication, which is also applicable on the nanometer scale.

4.1.2
Nanostructure Generation by Lift-off Processes

The so-called lifting is a specific process (Fig. 51). In a similar manner to the additive processes, a mask is generated. The functional layer is deposited on top of the mask layer, so that the functional material is only in direct contact with the support in the mask apertures. In contrast to galvanic methods, where the functional material only develops in the aperture, the functional material covers the whole substrate. However, after removal of the mask layer, all parts of the functional layer on top of the mask material will be removed ("lifted"). From the standpoint of deposition, lifting is more a subtractive than an additive process, due to the whole substrate surface being covered by the functional material.

A great advantage of lifting is the independence of the process from the chemical solubility of the material that represents its etching behavior. The applied etching bath is independent of the functional material, because it only has to remove the masking material. Therefore, lifting is a universal process. It has the potential to fabricate very small structures with high precision. The quality of the mask edge and the deposition of functional material inside the aperture of the mask material determine the edge quality of the final structure in the functional layer. A change in the dimensions of the structure and bent or round edges, as is often observed in the case of isotropic or partially isotropic etching processes, are minimized. In particular, the fact of preserving the dimensions makes lifting an ideal process for submicron and nanostructures.

There are some requirements for successful lifting. The mask layer covered by the functional layer has to be removed in a certain time frame. So a sufficient number and

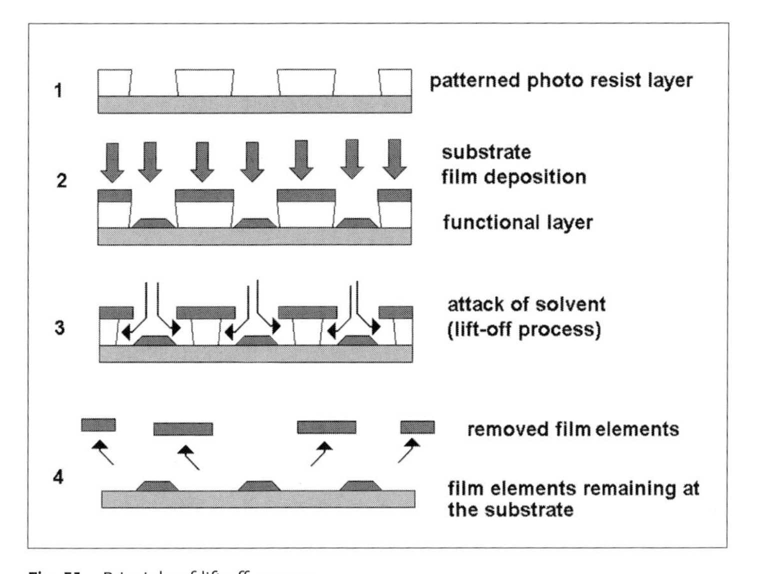

Fig. 51 Principle of lift-off process

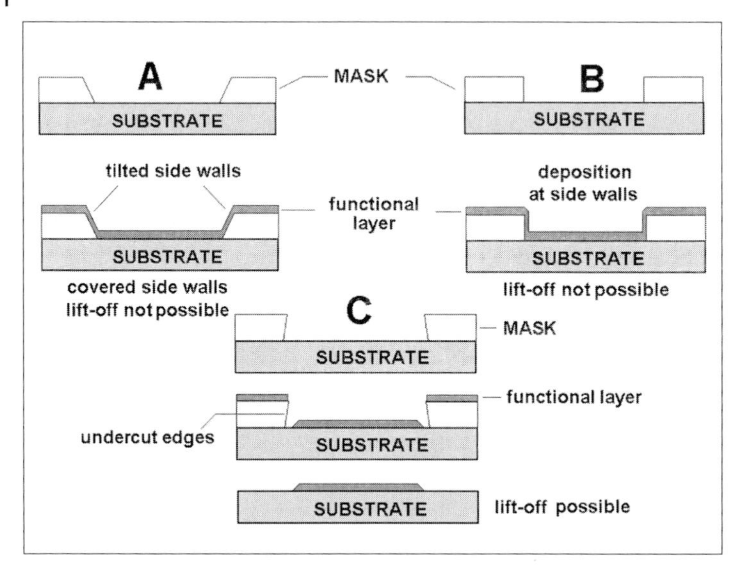

Fig. 52 Limits of lift-off process

density of holes in the functional layer are needed to ensure the access of the solvent onto the mask layer material. This requirement is met for small lateral dimensions of the structures, for densely packed structures and when the functional material does not cover the edges of the mask layer (Fig. 52, A and B).

Lifting can be also applied in the case of nanoporous functional materials, so that the solvent and the resulting product can penetrate this layer. In this process intermediate structures of layer elements can often be observed, which are free of the substrate but still connected laterally. Owing to the mechanical instability of this formation, the application of ultrasonic agitation leads to a breaking of these formations and removal of the resulting smaller particles from the surface.

Continuous nanopores are often observed in ultrathin inorganic layers, especially after evaporation and sputtering. They are less abundant for epitactic deposition and for reactive gas phase deposition than in CVD processes. The density of pores decreases sharply with thicker layers.

Lifting processes are not applicable for compact layers with thicknesses greater than the thickness of the mask layer. They are also problematic for layers with minimum pore density for low mask flanks, so that continuous areas of functional material with high density are formed above the flanks of the structures, hence access of the solvent is hindered.

Lifting processes are assisted by underetched mask flanks (Fig. 52 C). Also the direction of impact of the deposited atoms or molecules influences subsequent lifting. Through isotropic distribution, mask edges are usually covered, thereby slowing down the access of the solvent.

Anisotropic deposition in fact yields shadow effects, especially for underetched edges, where the solvent will have easy access into the cleft between the mask and functional material during the lifting process.

For thin layers, both the tendency to form nanopores and the low ratio of functional to mask layer thickness makes lifting particularly interesting. Hence it has a greater importance for nanotechnology compared with the microtechniques. It is often applied for specific problems and especially in research and development. For mass production, two problems exist. One point is no or incomplete solubility of the functional material, resulting in particles remaining in the lifting bath. So the bath is saturated with material from the functional layer, which can lead to defects especially in highly integrated structures. Another point is the time necessary for the process, which, even for thin layers and ultrasonic agitation, is still in the range of some minutes, resulting in a low productivity.

The lifting represents a process that includes aspects of both an additive (deposition of functional material inside the windows of a mask layer) and a subtractive (removal of undesired material as particles) process.

The question arises, as to whether – in analogy with the molecular disperse methods – a reverse process of this nanodisperse procedure is possible to achieve an additive process. Such an approach would require the transport of nanoparticles from a suspension and their integration into layers with the formation of bonds. For a sufficient number of small particles, such processes are possible, when interparticle bonds occur only at the surface. Electrical charges of the particles which are removed that are in contact with the substrate stabilize the suspension and also promote fixation onto the substrate, thereby supporting such a deposition process.

4.1.3
Principles of Nanotechnical Shape-definition and Construction

The integration of single particles is much more prominent in nanotechnology than in microtechnology, even greater than the problem with continuous layers. This prominence is because of both functional and purely technical reasons. The functional aspect relies on the same scale of functional structures and larger molecules, so that these molecules can be considered as individual elementary units themselves.

The technical aspect for the preferred use of individual particles in nanotechnology is explained by the ease of manipulation in this size range. In conventional mechanics and construction, the elementary units are large enough to be manipulated by hand or with standard tools. In microtechnology, this handling is significantly hampered, structures with micrometer dimensions are no longer accessible for manual operations, and even machines have difficulties. Therefore, microtechnology developed planar technique as a technical platform, enabling the efficient fabrication of microstructures but also a functional and material connection (cf. Section 3.1). Planar techniques normally fabricate and manipulate ensembles of structures, instead of handling the individual assemblies. The required changes in the local material arrangement are realized by establishment or cleavage of chemical bonds. For each structural element and assembly, a large number of such changes occur. Planar techniques define local areas or volumes for these bond changes, without exact definition of the position and geometry of single atoms, molecules or bonds. On the contrary, it does not seem likely that microparticles are transported and specifically bound by chemical manipulations in microtechniques.

In nanotechnology, the structural dimensions describe not only functional units, but also natural units that are accessible for individual manipulations. In particular these manipulations concern the material transfer and the specificity of chemical bonds, which are significantly more favorable for nanoparticles and molecules than for micrometer objects.

It is questionable whether the term technical manipulation of nanoparticles can be transferred from classical mechanics to nanotechnology by simple scaling. Such an extrapolation ignores the fact that the analogous approach in precision mechanics is of no importance, and the reason for this failure increases with the smaller structure sizes. It is not additional mechanical translations in a nanorobot, but rather new and dimension-adapted principles of nanomanipulation that are effective in nanotechniques. These nanomanipulations are assisted by the existence of natural structures with the appropriate dimensions on the one hand, and the individual differentiation by chemical interactions on the other.

Besides the forced transportation of particles, other basic construction principles in the nanoworld are spontaneous material transport by diffusion, thermally activated intramolecular movement and the hierarchy of chemical bond strength. The trick of nanotechnology is the limitation and control of the three translational and the three rotational degrees of freedom in the spontaneous movement of particles in such a way that the desired functional geometry is achieved. Management of the interactions of the units, such as nanoparticles, molecules and atoms, has to be achieved. The fabrication of some atomic or molecular functional groups that can be still considered as solid-state structures is possible in a way that is analogous to microtechnical approaches (cf. Sections 3, 4.3 and 4.4). In addition to this rather simple case, one has to consider that the geometry and topology of units are not identical. The topological context of nanotechnical construction is a required, but not sufficient of a prerequisite for shape definition in the nanoworld. The generation of topological connections limits the degrees of freedom in movement of the units. Construction in the nanoworld means specifically therefore the generation and enforcement of connections between units. The degree of connectivity determines the geometrical definition of a structure. This approach uses a series of limitations of degrees of motional freedom by transferring topologies of low connectivity into topologies of higher connectivity. Above a certain state of topological connectivity, the geometry of a nanoobject is clearly defined. This point of view applies both for the scaling of solid state techniques and the chemical–synthetic generation of nanostructures. In the first case, the motion of the particles on surfaces depends on the kinetic energy in the gas phase and the ratio of the surface temperature of the solid to the binding forces, the generation and the nanoporousity of clusters, and therefore the particle distribution or layer morphology reflects the role of the connectivity between particles and the limitation of the degrees of freedom in a controlled process regime. In the second case, the strength and density of different chemical bonds is successively introduced into the molecular architecture.

Limitations to the motional degrees of freedom as an approach to nanogeometries will at present, and in the foreseeable future, not solely be realized by lithographic or chemical processes. Only by the interplay of formation and cleavage of selective bonds ("molecular self-organization") under the constraints of geometries given by lithographic structures will a nanostructure technology be achieved. It opens new possibilities for shape-generation in the nanoworld, thereby fulfilling the expectations of nanotechnology.

The formation of chemical bonds positions the molecular units, or also nanoparticles, relative to each other. The unambiguous nature of the geometrical assignment depends on number, durability and the spatial position of the formed bonds as well as the degrees of motional freedom within the molecular units. A bond formation always generates a new topological association. This process is equivalent to processes in chemistry. The formation of new bond topologies at conventional chemical processes is independent of the number of molecules involved.

The arrangement of a functional – which implies technical – system always involves a point when only the internal ordered components are integrated into an external determined geometry. In nanotechnology, as often in technical processes, the assembling of units occurs through surface processes. The integration of a mobile unit in an area preserving this mobility already represents a first limitation to the degrees of motional freedom. The number of possible directions for translations is thereby reduced from three to two. This general technical principle is the first step in a reduction of the dimensions in order to manipulate and integrate objects following geometrically determined concepts.

Another restriction to translation requires structures on the surface that interact in specific ways with the object. The simplest case is the definition of lines that are used to position the object. Through this positioning, the second degree of translational freedom is also suppressed. To facilitate this approach, lines have to be defined. Because the lithographic processes with the highest resolution, such as electron or ion beam lithography, are limited to about 10 nm resolution, it is difficult to fabricate line structures for the positioning of small and medium molecules directly. This is only possible for small individual structures by means of scanning probe techniques (cf. Section 4.4).

A confinement in lines could also be realized by a two-fold binding of an object onto two adjacent areas (Fig. 53). The generation of a line as a separation between two adjacent surface areas is a simple method of confining the binding of mobile objects in an external geometry. This approach also applies with small dimensions of the objects and therefore is suitable for micro- and nanotechnology. Without specific requirements for the miniaturization of the binding areas themselves, only the mutual boundary of the areas defines the position for the binding of the objects. Hence no particular requirements regarding lithographic resolution or projection of small structures are needed for such a nanotechnical operation. Only the precision in positioning of the boundary line defines the precision of the integration of the mobile objects into the solid substrate.

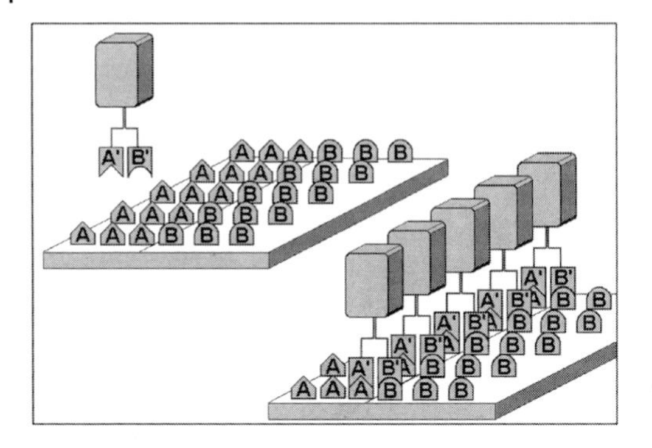

Fig. 53 Binding of bi-functionalized molecules at the borderline between two different complementary modified surface regions

The realization of positions by bonds for technical use has to meet two requirements. The binding strength in the areas has to be low, so that no binding occurs. The binding strength along the boundary line has to be high, resulting in durable bonds. So at least for the duration of the objects being contained in the medium (step 1) or of a subsequent rinsing step of the solid substrate (step 2), the ratio of binding strength in each area compared with the overall area has to be so small that the objects exhibit no binding on the homogeneous areas, but that there is sufficient interaction with the line. The binding constants K_{sf1} and K_{sf2} are determined by the ratio of the rate constants for adsorption and desorption k_{1ad}/k_{1des} and k_{2ads}/k_{2des}, respectively. For the binding constants of step 1 the equilibrium state applies:

$$K_{sf1} = k_{1ad}/k_{1des} << 1/N_{b1} \text{ (area 1)} \tag{4.1}$$

$$K_{sf2} = k_{2ad}/k_{2des} << 1/N_{b2} \text{ (area 2)} \tag{4.2}$$

with N_{bi} describing the number of binding places in the areas. On the other hand, the binding probability on the boundary line K_L should be high:

$$K_L = k_{Lad}/k_{Ldes} >> K_{sf1} \tag{4.3}$$

and

$$K_L = k_{Lad}/k_{Ldes} >> K_{sf2} \tag{4.4}$$

Because the adsorption constants k_1 for areas and line are comparable, the given relationships require significant differences in the desorption rates.

For a selective removal of species adsorbed onto areas whilst preserving the bonds to the line, a kinetic strategy can also be applied without equilibrium conditions. In this case, the rate constant for desorption has to be sufficiently low.

The third degree of translational motion can be constrained by the shortening of the binding line to a point. With respect to the line, such a point does not require a direct lithography with these dimensions. A point could be defined as the crossing point of two lines, so that the dimensions of the point are given by the width of the edges, which

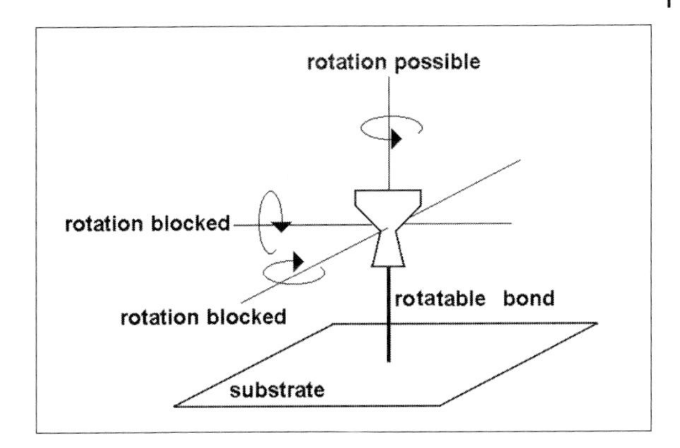

Fig. 54 Restriction of rotation for an immobilized particle by individual bonds

are sufficiently small in the case of monomolecular layers. The precision in the positioning of the point is determined by the edges. Such a crossing point can only be used when compatible binding molecules are provided, which need at least three specific binding groups. The binding process and, if necessary, desorption have to ensure that not only two but three bonds are established to immobilized binding partners, so that only the crossing but not the edges or even the areas are occupied by bound objects.

The binding of mobile molecules on solid substrates introduces constraints between the translational and also the rotational degrees of motional freedom. Even a single bond suppresses the two rotations parallel to the substrate plane (Fig. 54), so that only the rotation perpendicular to the substrate normally remains. If two or more bonds are formed, this rotation is also prevented. Hence only the internal degrees of motional freedom remain, such as conformational changes and isomerization in so far as they do not interfere with the surface-attached binding regions of the molecule.

A variety of chemical approaches can be utilized for the generation of nanostructures. Molecular units can either be pre-synthesized in solution prior to binding onto solid surfaces, or they can be added successively to structures growing on the surface. Two parameters determine the choice of the procedure: solubility and rigidity of the molecular units. In solution pre-synthesized units have to remain in solution, so also larger assemblies should be solvatized. This property is a required condition to avoid unintended aggregation, the precipitation at interfaces or the condensation into small solids or liquid heterophases. The potential for aggregation increases with the compactness and rigidity of the molecules.

Highly mobile linear polymers exhibit an extremely high solubility in suitable solvents. On the other hand, larger molecules with the potential to develop a large number of different conformations tend strongly towards nonspecific binding on surfaces, through the simultaneous formation of multiple weak bonds to the substrate surface. The tendency to adsorption increases drastically with the variety of chemical functional groups and subunits in such macromolecules. Hence the multifunctionality of such molecules conflicts with the desired, highly specific coupling onto surfaces. Therefore, a compromise has to be found for such molecular constructions between variability, solubility and rigidity of the molecular units.

4.2 Nanomechanical Structure Generation

4.2.1 Scaling Down of Mechanical Processing Techniques

The scaling down of classical mechanical processes into the nanoworld is not simply a reduction in size. While in precision mechanics large numbers of molecules and atoms are processed, nanomechanical methods include only limited numbers. So the characteristics of such methods approach chemical processes, and the surface and interface effects play a dominant role.

Moreover, the dynamics of the processes are changed. With decreasing dimensions, heat is transferred much faster compared with the macro- and microdimensions. This is a great advantage for many methods, so that, for example, thermal energy can be carried away faster and thermal damage is avoided.

Another aspect is the sharpness of the edges. In the nanometer range, no sharp edge is ideal. In the ultimate case, the radius of curvature is the radius of an atom. Macroscopically sharp edges often appear from an ultramicroscopic point of view as mountain ranges of material, and even the sharpness revealed by light microscopy can contain hidden areas with sizes of several atom diameters. Therefore, edge quality should be described by incorporating the radius of curvature and the distribution of the radii if necessary. The sharpness of edges depends on mechanical and chemical stress and especially on the mobility of particles in the material itself. Because of attempts of the atoms and molecules situated on the surface to increase the number of interactions with other particles in order to achieve minimal energy, all materials show a tendency towards rounding edges. Surface energies and mechanical relaxation processes influence the geometries of nanostructures [1]. At solid–liquid interfaces, the interaction of particles in the liquid phase counteracts this tendency. On the surfaces of gaseous phases, only the limitation of the particle mobility on the solid substrate limits the rounding of the edges. Because the particle mobility depends greatly on temperature, this parameter is decisive for the radius of curvature formed. The rounding by spontaneous transport of material is initially relatively fast; the speed decreases with increasing radius. However, after a longer duration or thermal stress a further increase in radius can be observed. The geometrical stability of nanoscale devices and assemblies is determined by the surface mobility of the particles of the applied materials and the external activation, such as heat, electrical fields and radiation.

The application of hard stamps for nanoimprinting opens the possibility of generating very small structures. Thus, diamond and SiC stamps generated by nanolithography can be used for mechanical imprinting in metals such as nickel. K. A. Lister et al. [2] demonstrated the formation of sub-10-nm structures by such imprinting processes with a diamond stamp, and of kline widths of 14–30 nm using an SiC stamp.

The mechanical generation of nanostructures by an external tool requires a high precision for the relative positioning of the tool and structure. Positioning systems are needed with sufficient precision in all translational and rotational degrees of motional freedom. Using pre-adjustment, four degrees can be a constraint, so that a movement in two directions has to be controlled with nanometer precision. However, a freely movable arrangement is preferred. Tables are available that meet the nanometer precision requirements for all six axes. Extreme precision is achieved by piezo actuators, but the overall movement is limited. Therefore, combinations of classical mechanical with piezo actuators are applied. So a compact six axis table was developed for micro- and nanomechanical processes, with a precision of 1.25 nm (x,y) and 8.5 nm (z) as well as 0.5 and 0.97 μrad for the rotational axes [3].

4.2.2 Local Mechanical Cutting Processes

Cutting processes are typical for classical mechanical processes. With adequately small tools, these principles can be scaled down to the micrometer range. To cut real metals requires a tool with an edge sharpness much less than the thickness of the shaving to be removed, so that shear forces can overcome the local cohesive forces of the material. Therefore, cutting is limited by the edge sharpness of the tool. In principle, tools with sufficient edge sharpness and length are possible to generate structures in the medium nanometer range. Theoretical considerations indicate that shavings as small as 1 nm are possible [4]. However, in contrast to the micrometer range, no tools have yet been developed with such a high precision.

In general, cutting methods work in a serial manner, so that they are not suitable for highly integrated nanotechnical assemblies. Hence these methods will be limited to specific individual or to simple regular structures.

4.2.3 Surface Transport Methods

The local removal of material by a probe is possible with inhomogeneous materials, especially in the case of a thin layer with weak interactions to the support. A moderate pressure of the tool induces a shifting or removal of parts of the layer. For manipulations in the nanometer range, scanning probe techniques are preferred due to the existence of nanometer tools (the scanning tips) and a positioning system with sub-nanometer precision (cf. Section 4.4).

For additive surface transport processes, the writing probe can be used as reservoir for the material to be deposited. The so-called "dip-pen nanolithography" holds surface-active molecules on the tip of a scanning force microscope. Contact of the tip with a substrate which has a high affinity for the adsorbed molecules results in a transfer of these molecules via a water meniscus to the surface, and a nanostructure is formed according to the x–y movement of the tip relative to the substrate (Fig. 55). A prerequisite is sufficient mobility of the molecules on the tool surface and a high affinity for the substrate surface. The latter is given for compounds forming self-assembled monolayers (SAM) on surfaces. Examples are alkylthiols such as octadecylthiol on

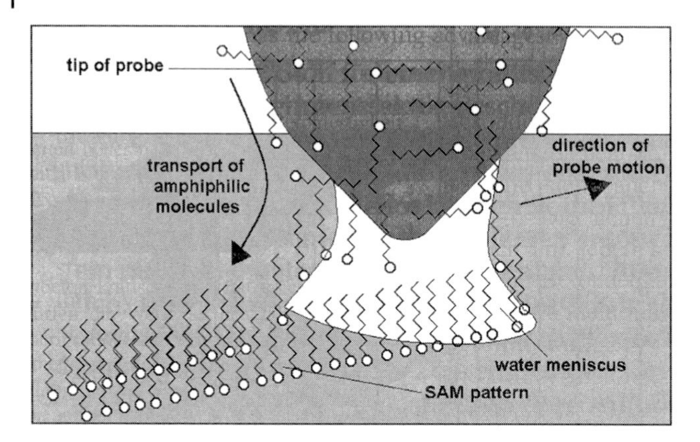

Fig. 55 Transfer of amphiphilic molecules from a scanning probe tip onto a substrate surface resulting in nanostructures of a molecular monolayer (dip pen lithography) (R.D. Piner et al. 1999)

a gold substrate, yielding a monolayer of nanometer thickness. Spots of about 15 nm diameter and structures with widths of 50–70 nm were realized with 16-mercaptohexanoic acid or octadecanethiol on gold [5][6].

The principle of transport on surfaces can be scaled down to the atomic range. Instead of mechanical probes, fields can be applied for force transfer. Atoms with a tendency towards high electrical polarization, such as neutral alkaline metal atoms, are moved by electrical fields on monocrystalline surfaces. In this way the manipulation of Cs atoms on GaAs surfaces has been demonstrated [7]. Electrically conductive micro- and nanoparticles show an even better surface mobility in the electrical field. The speed of electromigration increases with substrate temperature and particle size [8].

4.2.4
Reshaping Processes

In contrast to cutting methods, reshaping procedures are much better suited to manipulations in the micro- and nanometer range. With decreasing structure sizes, the material transport distances with reshaping processes become shorter. Although reshaping requires a precision for the tool geometry better than the precision in the resulting structure, there are no special requirements for the tool regarding hardness and edge sharpness. Moreover, no ultrathin layers are needed, so that thicker or bulk material can also be processed. Thus a polymer layer prepared by spin-off can be nanostructured using imprint processes (Fig. 56).

Hot embossing of plastics is established in macro- and microtechnology. Therefore, the material to be shaped as a raw work piece and the shaping tool are pressed together at a temperature above the softening temperature of the material. The plastic adjusts to the shape of the shaping tool, yielding a three-dimensional replica of the tool. The shaping is limited to shapes without underetched edges.

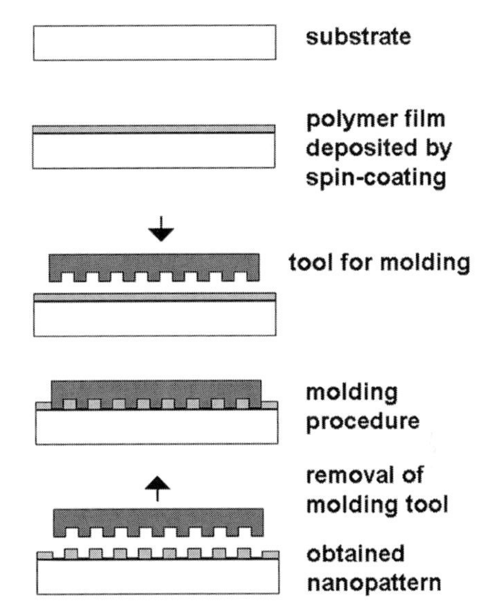

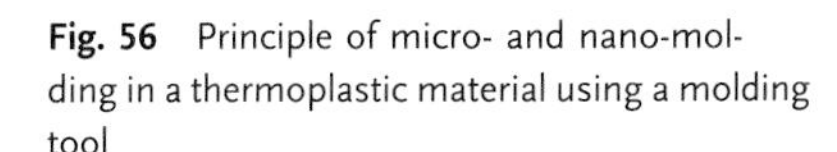
Fig. 56 Principle of micro- and nano-molding in a thermoplastic material using a molding tool

Hot embossing is of special interest for a cost-efficient mass production of work pieces. The process itself is usually very fast (typically below one second). The tool can be recycled many times. The temperature depends on the softening behavior of the material and is typically in the range of 150–300 °C.

Very small structures can also be hot embossed, so that this technique is well suited to the nanometer range. Holes with diameters below 25 nm were structured into a PMMA layer (T_g 105 °C) of 55 nm thickness. PMMA is particularly appropriate, due to its good separation from silicon wafer surfaces and its low shrinkage. The polymer structures were at 200 °C with a nanolithographically fabricated tool made from a structured SiO_2 layer on a silicon wafer [9]. Shaping tools for nanoscale hot embossing can be prepared, for example, by electron beam lithography and dry etching (Fig. 57).

Polymer nanostructures created by hot embossing can be applied as masks for the lithographic manipulation of the subsequent functional layer. This method belongs to the family of soft lithography. This “nano imprint lithography” (NIL) uses resist techniques similar to those known from organic photo- or beam-resist procedures (cf. Section 4.3). The only principal difference is that hot embossing results in residues in the aperture of the mask material, which have to be removed before further processing. For pure organic mask materials, e. g., silicon-free resists, a short oxygen plasma treatment is sufficient. Shaping techniques can be combined with multilayer resist technologies. In analogy with optical multilayer techniques (cf. Section 4.3.4), two- or three-layer systems are possible. Thus the lowest layer is a planarizing layer to level the topographic relief of the substrate [10].

Ultrasmall structures in different materials are possible through a combination of hot embossing with lift-off. So structures with a width of 6–10 nm have been realized in a metal layer [11].

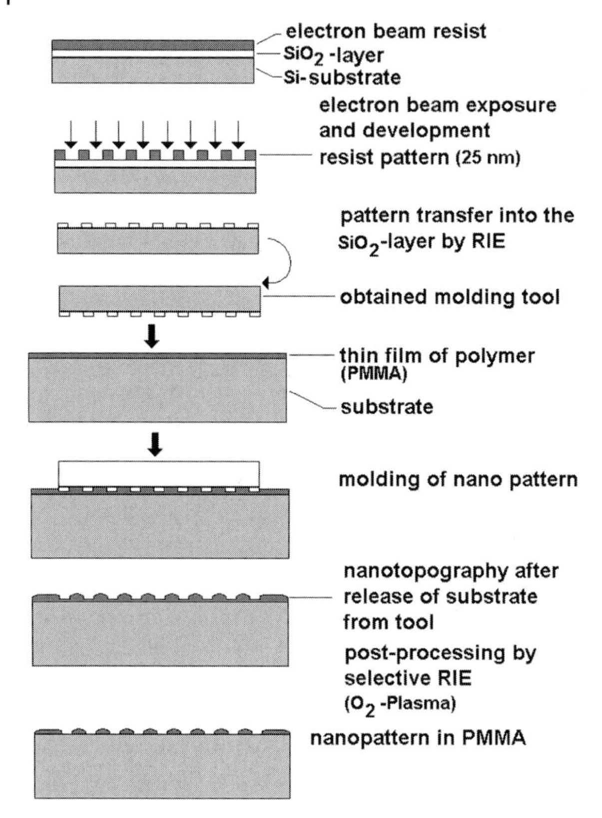

Fig. 57 Fabrication of continuous nanostructures by molding using an electron-beam fabricated tool and subsequent processing by time-controlled reactive ion etching [93]

Molding of nanostructures is not limited to increased temperatures. Also a local softening of the substrate material by the solvent allows embossing procedures to be used [12]. The required solvent can be provided by the stamp (Fig. 58). Another approach is the use of a liquid as starting material that fills the nanocavity of the stamp and is subsequently hardened. The adhesion of the layer on the substrate is enhanced when a primer layer is activated on the substrate (Fig. 59).

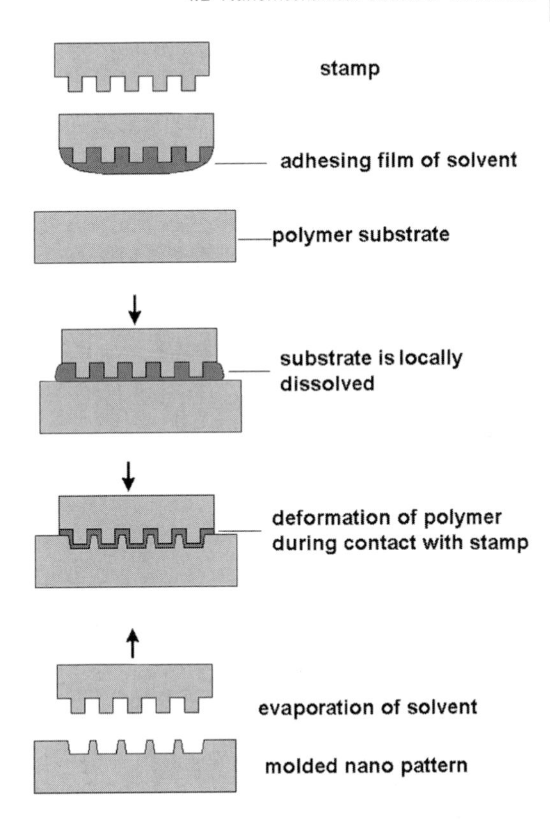

Fig. 58 Principle of solvent-assisted nanomolding

4.2.5
Soft Lithography for Nanopatterning and Nanoimprinting

Nanomechanical techniques include all processes that incorporate a local transfer of material from a tool onto a substrate, when either the tool or the substrate is pre-structured. Soft lithography is well suited to nanostructure fabrication (cf. Section 3.4.6). Nanostructured stamps can transfer material from the stamp through a mechanical contact onto the substrate; in this way the stamp shows the mirror image of the desired structure. An important prerequisite for stamping techniques in the nanometer range is that the transfer of material takes place in ultrathin layers, molecular monolayer are particularly suitable.

Advantageous for the realization of a high precision is the application of a nanoporous stamp material to ensure the supply of material for transfer during the contact with the substrate surface. An ideal stamp material is poly dimethyl siloxane (PDMS), a cross-linked polymer that incorporates different liquids through swelling. Moreover, its elasticity results in a good mechanical contact with the substrate, thereby minimizing wedge errors and problems related to wavy substrates. PDMS stamps show excellent transfer behavior for organic liquids. The layer formation can be enhanced in the case of molecules with the potential for self-assembly and subsequent

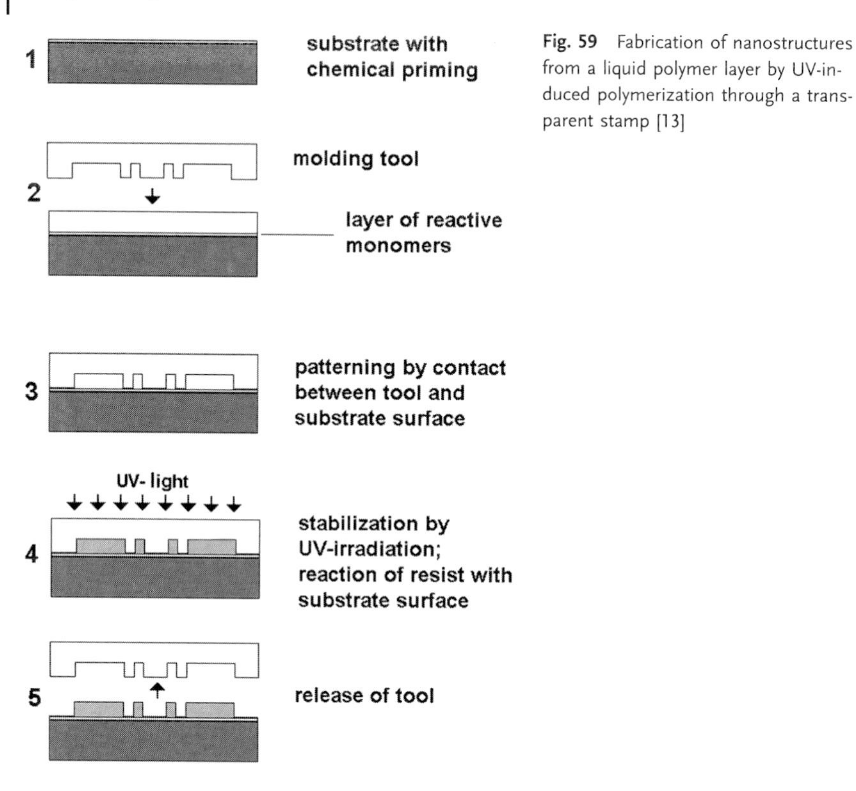

Fig. 59 Fabrication of nanostructures from a liquid polymer layer by UV-induced polymerization through a transparent stamp [13]

Tab. 4 Nanolithographical structure generation by mechanical techniques [14]

Method	*Resolution achieved*	*Reference*
Mechanical nanomachining	100 nm	[15]
Nanoimprinting	25 nm	[16]
Nanoimprinting and lift-off	6 nm	[16]
Nanoscale replication	50 nm	[17]
Nano contact printing	35 nm	[18]
Replica shaping	30 nm	[19]
Solvent-assisted nanoshaping	60 nm	[12]

monolayer formation on suitable substrates, e. g., in the case of alkyl thiols on gold substrates [18].

For the formation of closed nanostructured layers through stamping procedures, the molecules of the liquid phase have to form stronger bonds to the substrate than to the stamp or the solvent. This does not call for strong chemical bonds, as long as the ratio of lower solvation of the molecules in the stamp material on the one hand and a stronger adsorption onto the substrate on the other hand is given. Nanostamp techniques always rely on a combination of preferred molecular interactions with lateral structure generation.

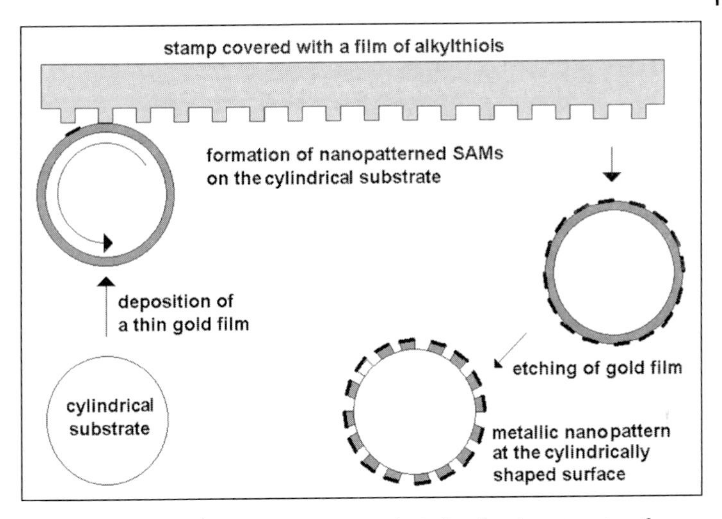

Fig. 60 Fabrication of nanostructures on a spherical surface by pattern transfer of an SAM from a planar stamp onto a cylinder

A nanolocal transfer of material is also possible when the substrate is structured instead of the stamp. Then a plane printing tool such as a non-structured roller can be used (Fig. 60). For local selective transfer, no relief on the substrate is needed if there are surface regions with different chemical activity. So a molecular monolayer or nanoparticles can be transferred with plane tools onto micro- and nanospots that differ in their binding properties from the remaining substrate. Transfer techniques utilizing planar stamps are able to modify and structure even cylindrical surfaces by unrolling and printing transfer of the nanostructured self-assembled monolayer [20].

Because soft-lithographic procedures are not resolution-limited as in photolithography, they are discussed as a replacement technology for the production of integrated circuits, which approach critical dimensions below 100 nm. They are also applicable to bent surfaces in optical applications [19].

Imprint lithography (IL), or nanoimprint lithography (NIL), is of particular importance for the generation of micro- and nanostructures of devices on flexible substrates. Polymeric materials such as transparent polyethylene-terephthalate (PET) are well suited as substrates for micro- and nanoimprinting. The application of a polybenzyl methacrylate film allows the nanofabrication of metal and oxide films on PET substrates. For this, the UV-curable monomer benzyl methacrylate is applied as a thin film. After mechanical imprinting into a thin film of the polymerized but non-cross-linked polymer, a Ti/Au film is deposited by evaporation and patterned by a lift-off process of the patterned polybenzyl methacrylate film using acetone as lifting solvent. In this way, metal structures with line widths down to 100 nm may be obtained [21].

Imprinting lithography can also be applied for the generation of more complex three-dimensional micro- and nanostructures if more than one mold is used. H.-Y.

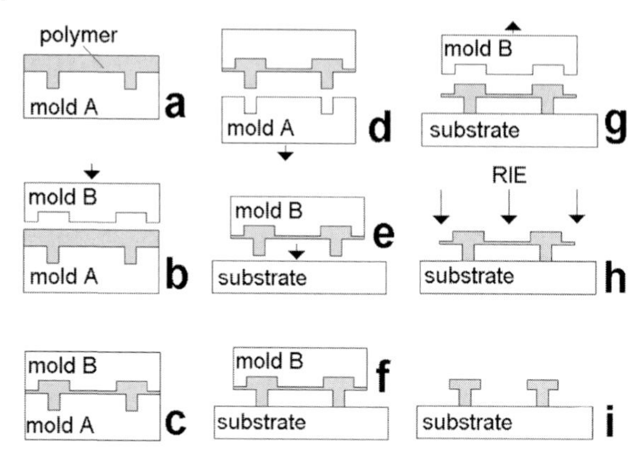

Fig. 61 Combinatorial mold imprinting lithography: a) the polymer material is deposited on the surface of the first mold, b) a second mold is applied from the top (c), d) the first mold (bottom) is removed and the polymer is transferred (e, f) to the final substrate, g) the second mold (top) is removed, h) thin residues of polymer material between the lines are removed by reactive ion etching, i) finally, nanostructures with Z- or T-shaped cross-sections are obtained [22]

Low et al. [22] reported a combinatorial mold NIL technique with two complementary molds for the generation of nanopatterns with Z-shaped or T-shaped cross-sections. For this, the polymer film was imprinted from both the upper and lower sides. A residual polymer film in the contact plane between the two molds was ultimately removed by RIE (Fig. 61). Circular cross-sections can be made by reflow of nanopatterned polymer films. After release from the surface, polymer nanofibers are obtained [23]. Special forms of nanoimprinting over topographic substrates and the combination of multi-step NIL with metal deposition permit the generation of complex, three-dimensional nanostructures [24][25]. The application of nanoimprinting as a multilayer technology also allows the fabrication of suspended nanostructures, as are required for nanomechanical elements for NEMS. For this, both the functional layer as well as the sacrificial layer for suspending of free-standing nanostructures are patterned by imprinting [26].

Capillary force lithography (CFL) is a special type of imprinting lithography. In this technique, patterns are transferred from a master (template) to a polymer film by the strong sucking effect of capillary forces at small structures on the master when it comes into contact with a polymer-containing liquid film. The capillary pressure is size-dependent and can increase above atmospheric pressure when the structures are in the submicron or nanometer range. Structures of less than 100 nm line width have been generated by CFL in poly(urethane-acrylate) (PUA). For this technique, either the master or the substrate must be flexible [27].

Nanoimprint lithography can be used like standard lithography for the patterning of different materials. Thus, zirconium nitride has been successfully patterned by imprinting-supported local oxidation and subsequent RIE [28]. The universality of

NIL has been demonstrated by W. Zhang and S. Y. Chou [29] in relation to the fabrication of a MOSFET with critical structures of 60 nm, which was patterned exclusively by NIL.

NIL is also suited for application to larger substrates. J. Tallal et al. [30] demonstrated the generation of a submicron pattern by imprinting a 4 inch (10 cm) wafer using a three-layer system consisting of a chemically amplified resist, a thin titanium layer, and a PMMA film. The generation of gaps down to 30 nm was thereby demonstrated.

The application of DNA for the generation of nanopatterns in polyvinyl alcohol by NIL has been demonstrated by T. Othake et al. [31]. For this, the DNA was adsorbed at a poly-L-lysine-treated glass surface and fixed by UV exposure before the application of polyvinyl alcohol.

4.3 Nanolithography

4.3.1 Structure Transfer by Electromagnetic Radiation

There are three strategies for shifting the resolution of optical lithography into the nanometer range: 1. reducing the wavelength, 2. increasing the numerical aperture of the optical projection system, and 3. using resolution-enhancing effects such as phase-shifting masks, etc. The application of these measures has determined the development of optical lithography during the last two decades. This has allowed a reduction of critical feature sizes from dimensions equating to the wavelength of visible light (436 nm, for example) down to the sub-100-nm range. Whereas about 30 years ago it was assumed that optical resolution would have a fundamental limit at around 1 μm, there is now the hope that the principle of optical lithography can be further extended into the deeper nanometer range.

The fabrication of stable structures requires changes in bonds between atoms of a solid or a resist layer. Therefore, the locally introduced energy has to overcome the activation barrier of the respective chemical reactions. If this activation relies just on temperature, only the total energy or the energy per area and time (power density) are of interest, but not the energy of the individual particles. Locally introduced thermal energy is easily dissipated on the micro- and nanoscale, so the required power density increases with the decreasing diameter of a structure of interest. High power densities are realized by lasers that are used routinely for structure fabrication in microtechnology. However, there are several problems related to the application of lasers in the visible or UV range for nanostructure technology:

a) The required power densities become very high.
b) The focus size is limited by the wavelength of light.
c) The heat transfer into the surrounding area broadens the structure.
d) The material specificity of bond activation is low.

Therefore, processes with particles each providing sufficient energy for bond activation are preferred on the nanoscale rather than methods based on the integral effect of many particles.

Typical bond energies of materials that are stable at room temperature are in the order of magnitude of 100 kJ mol^{-1} (about 1 eV per bond). For bond cleavage by individual activated particles or photons, these particles have to exhibit at least the same energy. To achieve chemically selective effects, the introduction of resonant energy at the bond is advisable. This phenomenon is utilized in photochemical processes, but their resolution in optical projection is restricted by the diffraction limit. Fabrication techniques also apply particles of higher energies, so that one particle is able to cleave a group of bonds.

The energy required for bond cleavage can be provided as photons. To achieve activation energies of about 1 eV, photons of wavelength 1.2 μm and below are needed. This energetic reason limits the range of the electromagnetic spectrum that is usable in micro- and nanostructure technology to lower wavelengths. For particularly stable bonds, the required energies reach the range of 1 MJ mol^{-1}. Hence light from the far UV into the near vacuum UV is applied.

Besides the energetic factor, the wave character of light limits the application of electromagnetic waves for structure generation in nanotechnology. When photons are applied as a free beam or by a projecting optical system, diffraction limits the local resolution.

To fabricate nanostructures with typical dimensions below 100 nm using photons, a wavelength below 200 nm is required (for $k = 0.5$ and $NA = 1$). Nanotechnology with optical projection requires the use of vacuum UV or even shorter radiation, such as X-ray or extreme UV (EUV). Soft X-ray radiation is especially suited for structure fabrication. A wavelength of about 1 nm is sufficient to achieve structures in the medium nanometer range for standard apertures without a diffraction limit. The photon energy is about the 100- to 1000-fold that of typical bond energies. Thus the activation volume of bonds is in the same range as the wavelength.

The situation changes with the transition to highly energetic radiation, such as hard X-ray or γ-radiation. For energies in the MeV range, the wavelengths are in the pm range, therefore of subatomic dimensions. Millions of bonds are activated by a single particle, yielding activation volumes with diameters of 10 nm and more. Secondary emission of energetic electrons and hard X-ray radiation lead to the growth of such activated regions of up to 1 μm^3. Thus, not all bonds in this volume are activated, but changes to the bonds distributed in this volume occur. Therefore, an upper energetic limit for photons regarding applications in nanotechnology should be defined. This value can be found in the medium keV range. It is only an estimation, due to the complexity of the interactions of energetic particles with the target material and the large differences dependent on the composition of the target. Under certain conditions, the application of highly energetic photons can be useful for nanostructure fabrication (cf. Section 4.3.6).

One solution to the diffraction limit of resolution is the structure fabrication in the optical near field. Near-field processes rely on very small distances between the volume element of the target and the light source or light-forming element such as the edges of masks, tools or structures. These critical dimensions should be below half of the applied wavelength. Optical near-field probes reach distances of 1 nm (cf. Section 4.4).

A special method of direct-writing structure generation utilizes focused atom beams in standing electromagnetic waves. The standing wave acts as a periodic lens for the beam, so that the beam is projected with a period of half the wavelength onto a substrate with a radiation-sensitive layer. Line gratings with a line width of 28 nm and a period of 213 nm were fabricated using a light wavelength of 425 nm [32].

Standard pattern generators in microtechnology apply two apertures arranged at an angle to each other. The form of the structural elements is controlled by changing the aperture size and – if possible – the angle. Visible light of shorter wavelengths or UV is used to project the aperture onto the light sensitive layer; the resulting diffraction limit hampers any application in nanolithography.

A simple method for the enhancement of pattern density is the transformation of structure edges into separate structures. This principle can be realized by special deposition, molding, and/or etching technologies after exposure. Alternatively, it may also be applied during the exposure itself. For this, the local change in refractive indices is used. The locally deposited dose of light (local concentration of photoproduct) in a resist is lowered if light is scattered from a certain resist volume into the neighborhood. This is exactly what happens at rough edges of transparent lithographic patterns. The effect can be used in the exposure of a thin photoresist layer through a transparent but patterned near-field mask that is brought into direct contact with

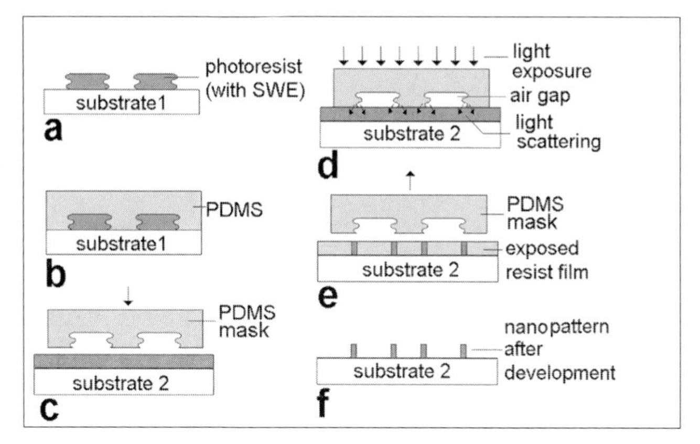

Fig. 62 Optical generation of nanostructures by transforming one line in the mask into two lines in the resist through local reduction of deposited light energy by scattering (after Kostovski et al. [33]): a) fabrication of a template bearing submicron structures (mold) by monochromatic photolithography under conditions leading to the formation of rough edges due to the standing wave effect (SWE), b) structure transfer into a PDMS mask, c) application of the PDMS mask with lithographic topography onto the resist film to be exposed, d) exposure to light incident in the normal direction but with scattering at the rough PDMS structure edges, e) distribution of photoproduct after exposure through the PDMS mask, f) nanopattern in the resist obtained after development (positive process)

the resist film. The scattering of light in the region of the structure edges in the mask leads to a reduction in the light exposure under both edges of a single line in the mask. As a result, two lines of resist are obtained after development of the positive photoresist instead of the one line in the mask (Fig. 62). The principle has been demonstrated using a pre-structured PDMS mask [33].

Two-photon absorption techniques can help to overcome the diffraction limit in optical lithography. For this, very high light intensity is required in the resist or in a functional film otherwise modified by radiation. Y. Lin et al. [34] showed that sub 100-nm patterns could be produced by a pulsed femtosecond laser with an emission wavelength of 800 nm. They were able to apply this technique for the patterning of $GeSb_2Te_4$ films by dipping the exposed substrates in NaOH solution after local light-induced recrystallization of the semiconductor film.

To transfer the principle of the pattern generator for use in nanotechnology, radiation with shorter wavelengths (such as extreme UV or X-ray) would have to be applied.

4.3.2 DUV- and Vacuum-UV Lithography

Serial processes are time-consuming and therefore not applicable for a cost-efficient and highly productive fabrication of devices with small structures. Therefore, large complexes of structures are transferred together in the size of a chip or larger areas. Such group transfer processes require a high optical precision and highly homogeneous illumination. Again, the wavelength limits the resolution.

So a process of continuous stepwise improvement of equipment and technologies ranging from visible through the near, medium and far UV to the vacuum UV is leading from a micro- to a true nanotechnology. Such a development was unbelievable several years ago. It seems possible, that with the extension of optical projection of small structures into the extreme UV range that further developments will addresses the medium nanometer range (20–50 nm) and so the optical lithography will reach a seamless transition into EUV and X-ray-lithography.

The application of vacuum UV light for optical lithography depends on the availability of resist materials with sufficient transparency in this spectral range and with sufficient sensitivity, in other words a response behavior that results in readily developable exposed areas. Many different polymers are in use or under investigation for this purpose. Particularly, partially fluorinated polymers meet the requirements, beside such classical DUV resist materials as polyhydroxystyrene (PHOST) and PMMA [35][36].

The sensitivity as well as the optical resolution can be improved by the application of UV resists with chemical amplification (CA). T. Itani et al. [37] demonstrated the fabrication of sub-100-nm features using an ether-based fluoropolymer as an acid-sensitive material in a CA resist system. Sub-50-nm half pitches were fashioned by using polyhydroxystyrene resists with high solubility sensitivity towards acids, which could be formed during the UV exposure [38].

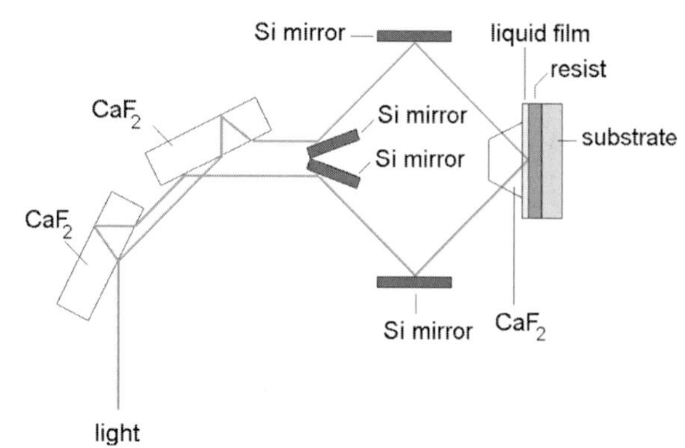

Fig. 63 Optical arrangement for the preparation of regular patterns with sub-half-wave-length resolution using vacuum UV interference lithography with an immersion liquid [41]

The quality of pattern transfer by optical projection lithography can be improved by using polarized light [39]. In their study, the use of light of the optimal polarization direction led to a 28 % enhancement in precision and an 11 % improvement in pattern uniformity.

The use of high numerical apertures for achieving maximal microscopic resolution has been coupled with the application of liquid media with enhanced refractive index between the final lens of the amplification system and the object to be studied. This strategy of immersion can also be used for the enhancement of optical resolution in lithography [40]. The increase in resolution is dependent on the magnitude of the refractive index of the immersion liquid. Line space patterns with a feature size of 30 nm have been generated by interference immersion lithography with excimer laser exposure at 157 nm [41]. Fig. 63 shows the optical arrangement for optical interference lithography using an immersion liquid. Calcium fluoride elements and silicon mirrors were used to generate a high-resolution interference pattern in a DUV photoresist film. Perfluorinated polyethers (PFPEs) were found to be best suited for the role of inert immersion liquid. A resolution of 38 nm was achieved by means of immersion lithography using water as the immersion liquid at an exposure wavelength of 193 nm [42]. The lithographic applicability of zone-plate arrays in connection with optical immersion has been demonstrated at the deep sub-micron level. Zone-plate-array lithography (ZPAL) may also be further developed to a promising technique for optical nanostructuration [43]. The fabrication of sub-100-nm structures is possible with light in the DUV wavelength range (244 nm) if optical interference is used for pattern generation [44]. Interference lithography using vacuum UV light allows the fabrication of line space patterns with structures at the 50 nm level. T. M. Bloomstein et al. [45] demonstrated the fabrication of 50 nm lines in spin-on glass by means of interference lithography involving exposure to a fluorine excimer laser. The nanostructures of the spin-on glass could be transformed into InP structures by organometallic vapor-phase epitaxy (OMVPE).

4.3.3 EUV and X-ray Lithography

The wavelength of electromagnetic radiation can be decreased until it limits no longer the resolution. Extreme UV radiation (EUV) includes the medium nanometer range of wave lengths. In recent years, values around 10 nm have also been reached, the region of the transition to soft X-rays. The lower nanometer and sub-nanometer range is addressed by X-ray lithography (XRL) [46][47]. Using X-rays, sharp local distributions of photoproducts are realized for structure dimensions in the medium and lower nanometer range, which are not accessible even for multilayer mask processes with UV radiation (Fig. 64). Such radiation transfers structures in the medium and lower nanometer range, so that this technology is applied for the fabrication of quantum dot contacts [48].

The progress in EUV lithography is mainly connected with the development of catoptric systems instead of refraction optical arrangements. The interaction of EUV radiation with all materials required an optical solution without the use of refractive elements. Therefore, reflection optical systems are favored. The recent semiconductor roadmap suggests the introduction of EUVL for the fabrication of LSI circuits at the 32 nm level by 2009. So, the development of EUVL systems for IC production is now at a high level [49]. The introduction of demagnification reflective systems allows the concept of demagnifying projection lithography, which has been very successful in optical lithography during the last decades, to be transferred to EUV lithography at the mid-nanometer level. A possible arrangement for the demagnifying transfer of

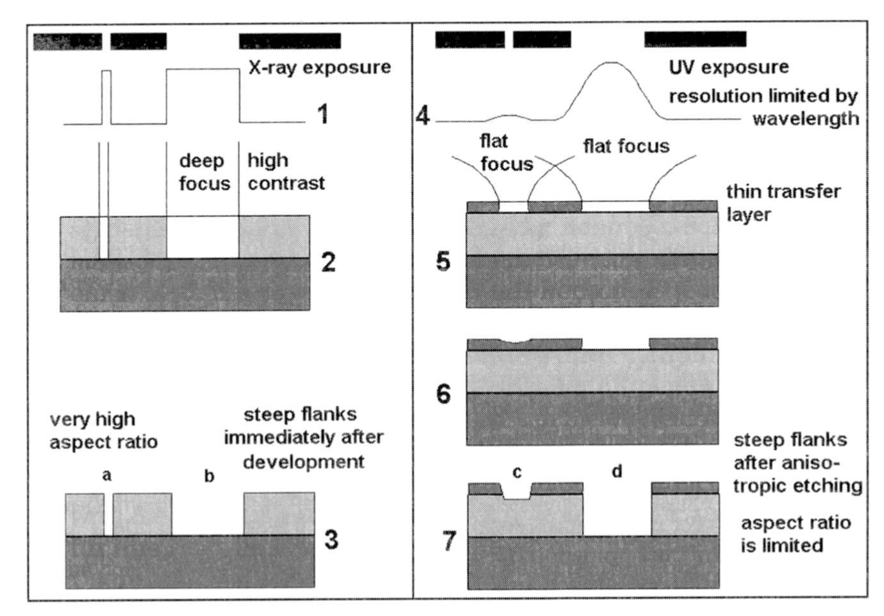

Fig. 64 Comparison of the principle possibilities for nanostructure generation by X-ray lithography (left) and diffraction-limited UV-lithography in multilayer resist technology (right)

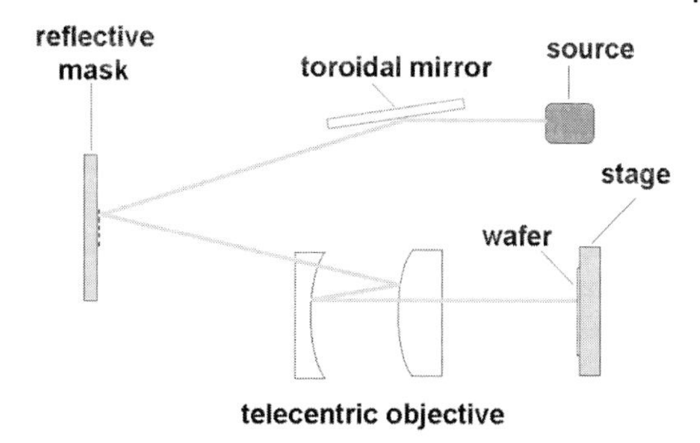

Fig. 65 Principle of the arrangement of light source, mask, optical system, and substrate in optical lithography using extreme deep-UV light (EUVL) with a reflective demagnification system (after H. Kinoshita [49])

mask patterns to a wafer by EUV lithography is shown in Fig. 65. The mask pattern can be scanned by a EUV beam in much the same way as in the reflection lithographic systems that were developed for XRL about 20 years ago (Fig. 66). The magnitude of the reduction step becomes clear upon considering the exposure areas at the typical critical dimension levels for DUV, vacuum UV, and EUV lithography (Fig. 67). The 32 nm level can be addressed by lithography using a sodium laser as radiation source. The development of optical lithography from short visible wavelengths through the near-UV, mid-UV, DUV, and EUV ranges is well reflected by the dependence of a figure of merit that includes the typical depth of focus, DOF, and the mini-

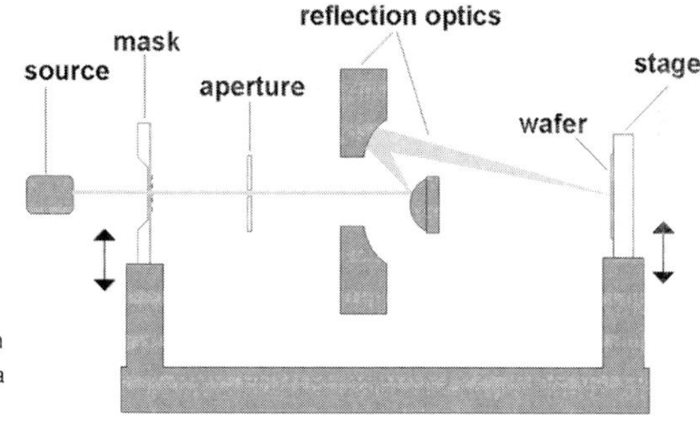

Fig. 66 Principle of an arrangement for EUV exposure using a scanning system and a reflective optical system for pattern transfer (after H. Kinoshita [49])

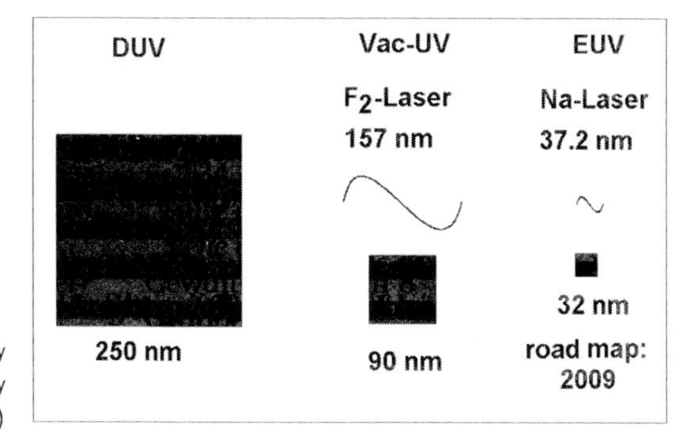

Fig. 67 Comparison between typical exposure area sizes in DUV lithography (250 nm level), vacuum UV lithography (90 nm level), and EUVL (32 nm level)

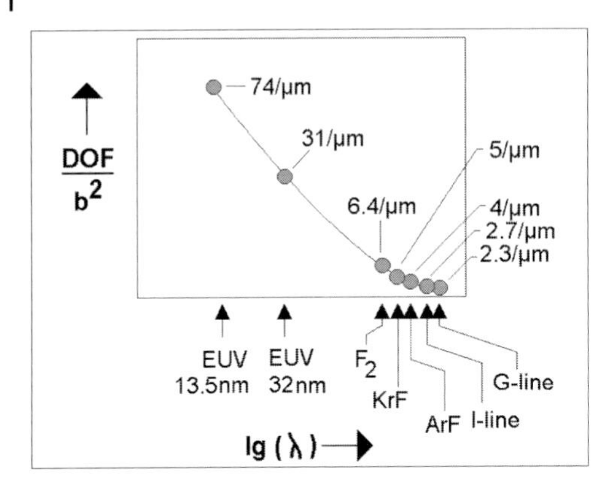

Fig. 68 Dependence of the lithographical figure of merit DOF/b^2 for optical projection lithography on the exposure wavelength (DOF = depth of focus, b = critical feature size)

mum feature size, b, on the exposure wavelength (Fig. 68). The ratio of DOF to b^2 gives a characteristic value for the description of the generation of small-area structures with high aspect ratios. The increment in the value of this parameter was small in the development from G-line (436 nm) projection lithography to I-line technology (365 nm). A more significant improvement was achieved by the introduction of excimer laser lithography in DUV and vacuum UV lithography. However, the progression to EUVL represents an enormous improvement. The extension of the principle of optical lithography into the EUV range opens up the prospect of using optical exposure in the generation of medium nanostructures [50]. In the future, the further development of EUV lithography and the stepwise reduction of the wavelength of the applied light could close the methodological gap between optical lithography and XRL.

The transition from UV/VIS to EUV and XRL transfers the technological problems from the limitations due to diffraction to those for materials. Considering the optical aspects, materials with extremely large differences in optical transmission exist –the absorption for a nanometer thick metal layer is the same as for a kilometer long glass fiber – absorption coefficients in the soft X-ray region exhibit only small differences at a maximum two orders of magnitude. The fabrication of transmissive and mechanically stable lithographic masks is therefore difficult and costly. Also the realization of high-resolution optics for lithographic projection systems is difficult. These are the reasons that technological breakthroughs in this field are still missing.

The development of XRL for application in nanopatterning was started a long time ago in order to overcome the diffraction limit of optical lithography with visible or near-UV light. Parallel X-rays are particularly well suited for the 1:1 pattern transfer in shadow mask procedures [51][52]. Therefore, gaps of 10 µm or 5 µm are applied for copying masks at the 100 nm or 25 nm level of design rules in the realm of proximity X-ray lithography (PXL). It can be assumed that the resolution of this pattern transfer method may be further extended below the 25 nm level [53]. Meanwhile, mask technologies as well as an X-ray stepper are available for PXL. The device is specified to have a resolution of 45 nm and should allow resolutions below 35 nm [54].

X-ray radiation is extremely parallel, especially for synchrotron sources. An exceptionally parallel beam and large depth of focus allows the penetration of even thick layers of resist. Therefore, X-ray radiation is suitable for the direct writing of nanometer structures with high aspect ratios into X-ray sensitive material. Using a special plasma source with wavelengths of 1.2 and 1.7 nm, columns of 70 nm diameter and an aspect ratio of 4 have been structured directly into the radiation resist SAL 601 [55]. Aspect ratios of up to 13 were achieved for structure widths of 20 nm in PMMA (layer thickness of 250 nm) or a PMMA/MMA layer (layer thickness of 265 nm) using Au and W absorption layers [56][57]. Optimization of dosage allows for even smaller mask structures [58]. High aspect ratios were also reported for the EUV range. For a wavelength of 13.4 nm, line-space patterns for 80 nm structure widths have been generated in a resist layer of 600 nm [59].

4.3.4 Multilayer Resist Techniques with Optical Pattern Transfer

Multilayer resist processes are applied when a single resist layer cannot meet the requirements for the resist [60]–[64]. One problem is the sharpness of the contours of projections into the resist layers of certain thickness, when the thickness cannot be minimized due to the relief on the substrate or protection required against etching processes. Very high numerical apertures are applied for the projective optics to enable high resolution, so that the depth of focus is significantly reduced. Even in the sub-micrometer range the height of the structures often exceeds the depth of focus, so that a high-resolution projection throughout the resist is hampered. Thus either a planarizing step is integrated into the technology or the resist has to fulfill this planarizing function for the lithographic process.

Multilayer resists provide a separation of functions: the actual collection of the optical image is achieved by a thin layer that is on top of a thicker layer, which both levels out the topological relief and protects the structural elements. Multilayer resist technology requires a precise and anisotropic pattern transfer from the primary structure of the thin upper layer into this planarizing sublayer, which is usually realized by dry etching. Although silicon-containing or other doted organic resist layers were developed to make them more resistant to an oxygen plasma, the precision and anisotropy of such two-layer resist systems are not sufficient for high quality lithography.

Therefore, triple-layer systems with an additional thin layer as the transfer mask are applied in most cases (Fig. 69). The thin intermediate layer ("transfer layer") is precisely structured under medium selectivity from the projected pattern in the upper layer. A material is chosen that allows for a highly selective etching of the planarizing layer compared with the transfer layer. Thereby the transfer layer is preserved in the subsequent structure transfer process. Hence precise resist structures with high aspect ratios are realized, which are also applicable for demanding etch processes down to the nanometer range.

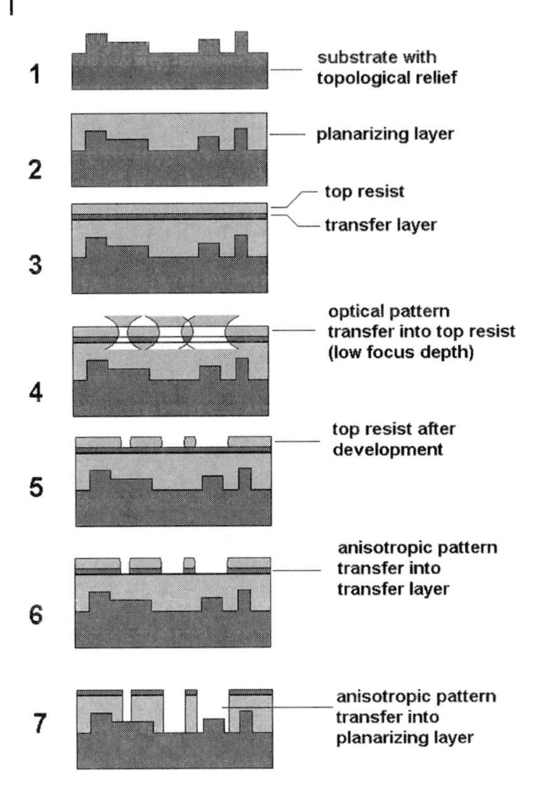

Fig. 69 High-resolution photolithography by multilayer resist technology. Planarization of the topological relief ("bottom layer"), projection into a small layer volume with low focus depth in a plane, thin light-sensitive ("top resist") layer and pattern transfer using an etch-selective layer ("transfer layer")

4.3.5
Near-field Optical Micropatterning Techniques

Both pattern generation and projecting group transfer processes are resolution-limited due to diffraction, as a result of the application of an optical system between the mask/aperture and the collecting resist layer. This limitation does not apply for pattern transfer in the optical near field. In the case of direct contact of mask/aperture and the resist, electromagnetic radiation is applicable for lithography in the optical near field.

In reality, the quality of the mask, the thickness of the resist, and other problems make it difficult to achieve optical resolution below half wavelength in the near field by near-field lithography (NFL).

In principle, even the contact exposure techniques applied early on belong to the group of near-field optical processes. Here a mask is pressed against the resist to provide for a short light path between the edges of the mask structure and the image-collecting volume element of the resist. Therefore, both mask and substrate must be extremely even, particles capable of acting as spacers have to be excluded and the resist layer thickness should be below the wavelength to minimize diffraction effects inside the resist layer. Because these three requirements are not usually met for most contact exposure techniques as the proximity methods, the resolution of a near-

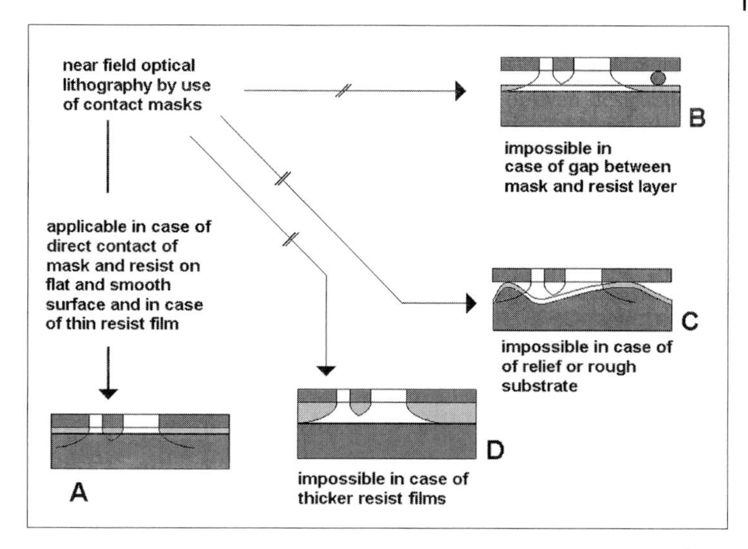

Fig. 70 Restrictions of optical near-field lithography in contact exposure processes

field optical projection is not achieved (Fig. 70 B–D). However, ultrathin resist layers and a direct contact between the resist and mask enable near-field optical group transfer processes to occur. They allow the realization of nanostructures, even by use of light with a wavelength much greater than the structures obtained (Fig. 70 A).

T. Ito et al. [65] demonstrated the fabrication of a resist pattern in an MLR process using Cr and Si as light-absorber materials down to a half pitch of only 32 nm under exposure from an I-line light source (365 nm).

The excitation of surface plasmons can be used for the generation of lithographic features with dimensions below the exposure wavelength [66]. A precondition for such surface plasmon-assisted lithography (SPL) is direct contact between the photosensitive film and the thin, light-receiving metal layer. Array structures with feature sizes of about 60 nm (120 nm period) have been obtained by the illumination of a silver-coated photoresist through a quartz mask at an exposure wavelength of 365 nm (I-line). The generation of structures with periods down to about 50 nm seems to be possible through G-line exposure (436 nm), if the SPL technique is combined with interference in the framework of the so-called SPRINT process [67].

4.3.6
Energetic Particles in Nanolithographic Structure Transfer

In addition to mass-free photons, energetic particles which have a mass can also be applied to the generation of nanostructures. They have the advantage of small characteristic wavelengths λ_b (de-Broglie wavelength) of the particle wave function even for relatively low energies. The wavelength decreases with increasing mass m_b and increasing velocity v_b of the particles:

$$\lambda_b = h/(m_b \cdot v_b) \tag{4.5}$$

$$\text{with } v_b = \sqrt{(2 \cdot E/m_b)} \tag{4.6}$$

Even for slow or low mass particles, such as low-energy electrons, the wavelength of the particles does not limit the lateral resolution (cf. Table 5).

The possible lateral resolution is determined by the energy of the individual particles, influenced by the following factors:

- penetration depth of the particle
- diffraction of the particle
- induced secondary processes
- reactivity of target material

The mask processes dominating the microtechnology also play a role in nanotechnology. Additionally, processes with direct structure generation are important for nanostructures. As well as division according to the applied projection (mask transfer, writing mask, direct-writing), structure-generating processes using energetic particles can be divided according to the type of particles.

The dimensions and scale of probes in the sub-micrometer and nanometer range utilized for structure generation are compiled in Fig. 71. Particle beams can be focused down to 1 nm, thereby exceeding focused UV beams. Hence they can be applied for

Tab. 5 Parameters for particle beam lithography particles

Particle in beam	*Kinetic energy*	*Velocity*	*Wavelength*
Electron e^-	1 eV	ca. 600 km s^{-1}	1.2 nm
	10 eV	ca. 1800 km s^{-1}	0.4 nm
	1 keV	ca. 18 000 km s^{-1}	40 pm
Proton H^+	1 eV	ca. 14 km s^{-1}	28 pm
	10 eV	ca. 44 km s^{-1}	9 pm
	1 keV	ca. 440 km s^{-1}	0.9 pm
	1 MeV	ca. 14 000 km s^{-1}	28 fm
Argon ion Ar^+	1 eV	ca. 2 km s^{-1}	5 fm
	1 keV	ca. 30 km s^{-1}	0.16 fm
	1 MeV	ca. 2000 km s^{-1}	0.005 fm

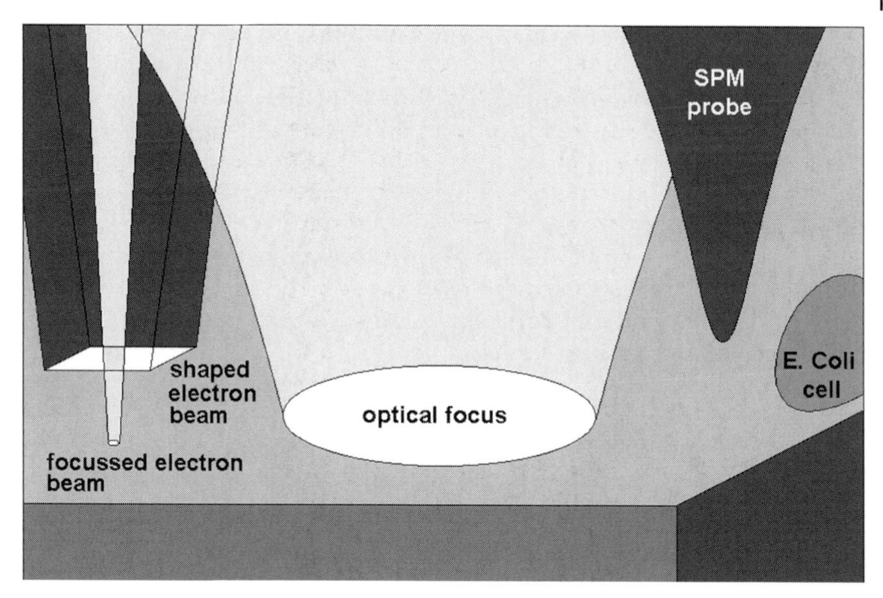

Fig. 71 Size comparison of lithographic probes and natural nanoobjects

the precise reproduction of complex structures. However, the resolution of a highly focused particle beam is limited by the mechanisms of the interactions with the resist material.

4.3.7 Electron Beam Lithography

Electron beam techniques apply fast electrons for the generation of structures [46][68]. Because of their electrical charge, electrons can be easily accelerated and deflected. The two-directional deflection of the beam, as is known from cathode tube display technology, is also applied in lithography to project structures. The energies of the electrons are farway above the binding energies, so that an individual electron will induce a cascade of reactions after hitting the target. The beam voltage is determined by the geometrical requirements of the beam, and not by the chemistry of the target material. Because a cloud of negatively charged particles creates the beam, a tendency towards broadening due to the interparticle electrostatic repulsion can be observed. High acceleration voltages and focusing only in the proximity of the substrate surface minimize this effect. High beam voltages suppress the beam divergence due to the short time interval the electrons are at high density.

An electron beam is always subject to inference by fields from the surrounding environment. Accelerated electrons are sensitive both to electric and magnetic (Lorentz effect) fields, resulting in complex influences from the environment. The higher the beam voltage, the lower the influence of external fields. High voltages result in a relatively tight beam. Typical values for the beam voltage are about 20 kV, and sometimes also 30–40 kV.

The beam current, that is the number of electrons per time interval, influences the quality of the beam. High beam compactness and good focusing are supported by lower beam currents, when the local electron density and therefore the interelectron repulsions are lower. Low currents lead to longer exposure times and therefore conflict with the productivity of the process. Highly sensitive resist materials (facilitating the high efficiency of the solubility-changing reaction per penetrated electron) provide the means for sufficient writing speed also for low currents. Higher voltages also increase productivity, because then the energy per electron is higher and therefore also the number of induced secondary reactions.

The resolution of electron beam lithography (EBL) does not really depend on the beam diameter. The mechanisms of the interactions of the energetic electrons with the atoms and molecules in the electron-sensitive layer are important. Owing to their large energy, the beam electrons do not usually interact with the outer electrons (which are responsible for chemical reactions), but knock electrons out of the inner shells. Because the primary (beam) electrons lose only part of their energy in this manner, one primary particle is able to initiate several such ionization processes. The removed electrons have a sufficiently high energy to ionize the material in their surroundings, so that secondary processes can be observed. The direction of these processes may differ from the direction of the primary electron beam. The secondary processes lead to a broadening of the area affected by the beam. The range of this effect depends strongly on the atomic number of the target material. Elements with fewer electrons show smaller interaction cross-sections with respect to the primary electrons, and lower energies are required for ionization. This leads to larger penetration depths and a lower diversion of the beam effect by creation of secondary electrons, compared with heavier atoms.

More important for the dimensions of the interaction volumes is a second effect: after the ionization of target atoms, an electronic relaxation occurs by transfer of electrons from the outer shells to the free places of the inner shells and an energy emission as radiation. Owing to the large differences in energy, X-ray radiation is generated. This radiation is readily absorbed by the adjacent material. However, the expansion distance is much larger than the diameter of the electron beam. Under the influence of direct electronic excitation and the secondary effect of the X-rays, an excitation volume element develops in the target material around the impact area of the electron beam, which extends to a larger penetration depth and is therefore pear-shaped (Fig. 72). Because X-rays induce ionization and therefore chemical reactions, the resist solubility in this volume changes. The lithographic resolution of electron beam lithography is mainly determined by the dimensions of this excitation pear, which is based on the target material and beam parameters (especially the acceleration voltage). The deposition of beam energy around the actual beam area causes the so-called proximity effect of electron beam lithography. This region can reach dimensions of 0.5–1 μm for low-atomic number elements, as in the case of organic resists.

The proximity effect results in concentration gradients of the exposure product starting from the center of the region. Instead of a defined area separating regions of high

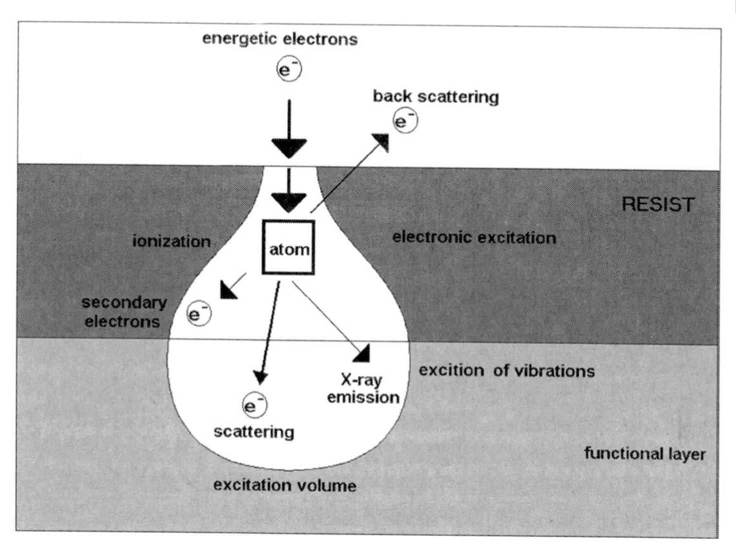

Fig. 72 Proximity effect by formation of an excitation volume in electron beam lithography

to regions with low solubility, a transition region with dimensions for a multitude of the beam diameters is observed. Thus the development leads to low edges in the resist structure. Without a correction of the dose, small structures receive insufficient doses of electrons, resulting in incomplete development of the resist and often the complete loss of the structure. So the proximity effect is usually the resolution-limiting factor in electron beam lithography.

Because the excitation pear is much larger than the beam diameter and the precision of its positioning, a partial compensation is possible. Therefore, higher doses are deposited at the edges of structures or at small structures, compared with lower doses inside extended areas.

The most important application for electron beam lithography is the generation of masks for photolithography with critical structure dimensions in the sub-micrometer range, extending below 200 nm. However, owing to the size reduction applied in optical projection lithography, the mask structures are usually much larger than the critical dimensions of this lithographic technique. Mask generation usually utilizes shaped beams. Thus the beam is not fully focused, but an area is covered homogenously. According to the lithographic requirements, this area element can be varied, e. g., in a range of widths between 0.2 and 6 μm, yielding a ratio of the resulting areas of 1:900, which clearly shows the potential for the productivity of the electron beam lithography with a shaped beam. These stamps are used to assemble structure fields with dimensions in the millimeter range of exposure. On a third level of hierarchy, the substrate stage is moved, so that the individual structure fields can be used to cover areas as large as 300 mm diameter with sub-micrometer structures. A prerequisite is that precision for the mechanical movement is better than the electronic precision. This condition is usually realized by an extra system to determine the exact position of the stage using laser interferometry, and measurements for corrections to the electron beam control.

For structures in the medium and lower nanometer range, a focused beam is required instead of the shaped beam. Moreover, the proximity effect and its consequences have to be minimized. The local electron dose has to be adjusted, so that the proximity effect barely contributes to the exposure, and the local changes in resist material are mainly due to primary electrons and the energetic secondary electrons. Therefore, only the neck of the excitation pear should be used. The nominal local doses are higher than for area exposure. The resist composition and development procedure has to ensure that the solubility of the volume elements exposed to a dose below a certain threshold is low. To realize small structures, resists with a steep gradation curves, which means high contrasts, are required (Figs. 73, 74).

The conditions for working in the neck of the pear are well accommodated in the case of thin freestanding membranes, when the membrane thickness is small compared with the typical diameter of the excitation pear. Thus no backscattering and X-ray exposure of material below the resist is possible. A significant reduction in the proximity effect is possible through the combination of a small resist thickness with a support of heavy elements (Fig. 75). Because the slowing down of the electrons and also the transmission of the X-ray radiation depends strongly on the atomic numbers, layers of for example gold or platinum reduce the proximity effect.

Shaped and focused beam techniques are ideal for converting sub-micrometer and nanometer structures from digital data sets into 2D-structures. For monolayer-resist processes, aspect ratios of larger than 1 are also possible (Fig. 76). Both techniques require significant time to fill larger areas with small features. Even when a 2D-element of 0.04 μm^2 requires only 1 μs, a whole wafer (100 cm^2) would result in 2.5×10^5 seconds (nearly 3 days). To overcome this problem, parallel exposure procedures were developed. They apply masks with the desired structure as opening or on a thin membrane, and achieve in principle sub-micrometer down to nanometer structure sizes.

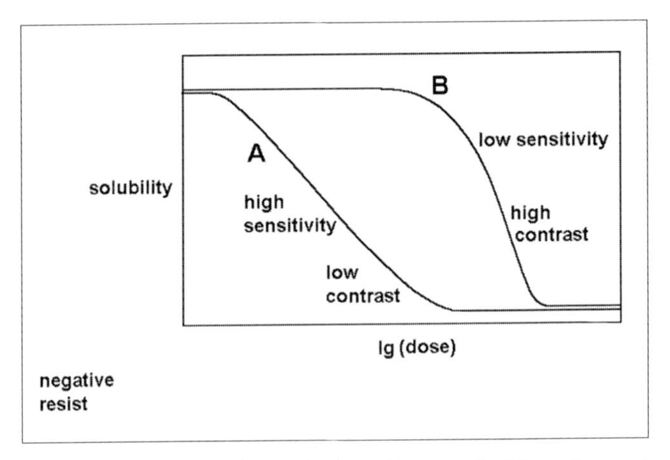

Fig. 73 Gradation curve of a sensitive (A) and less sensitive (B) negative resist

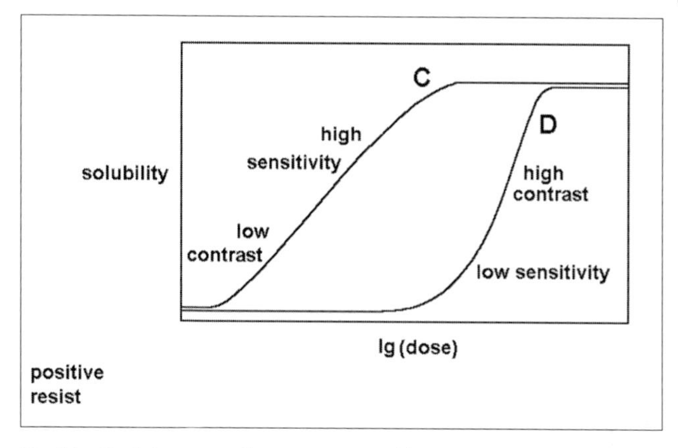

Fig. 74 Gradation curve of a sensitive (C) and less sensitive (D) positive resist

However, the requirements of the mask material are very high, and the fabrication of freestanding membranes is not yet a routine procedure. Therefore, such techniques have no practical relevance.

While the semiconductor and microsystems industry applies electron beam processes for mask generation rather than for direct structure generation, the electron point beam method is a key technology in nanolithography to fabricate small structures on the substrate directly [69][70][71]. Even by 1981 structures of only 20 nm width had been realized in a 30 nm thick PMMA layer. The beam diameter was below 1 nm

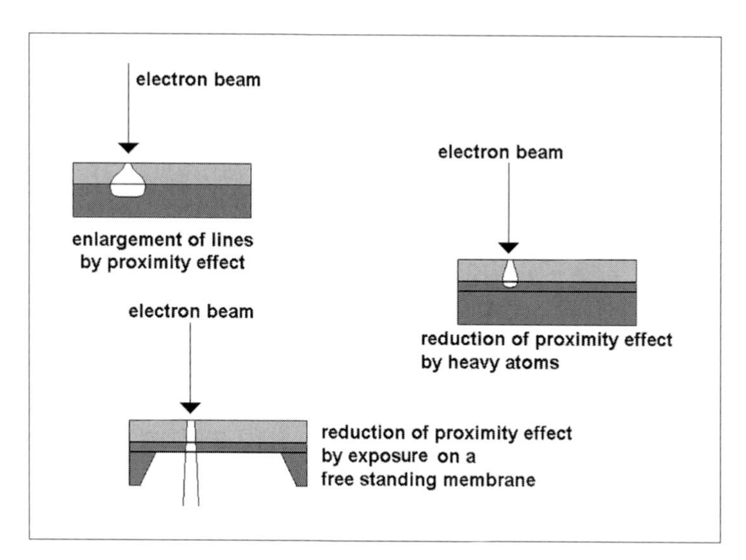

Fig. 75 Minimization of the proximity effect by heavy elements in the target material or by freestanding membranes

Fig. 76 Nanometer resist structures (negative resist ma-N 2400, Microresist Technology: 80 nm lines, aspect ratio 2.25) fabricated by electron beam lithography (courtesy of H. Elsner)

for an energy of 50 keV [72]. Development of negative resists with high resolution, which are chemically related to Novolak-based positive photoresists, enabled the fabrication of electron beam lithographically prepared nanostructures with high aspect ratios. Acid-induced cross-linking of oligo cresols with tris(hydroxyethyl) benzene and a 50 kV exposure yielded 80 nm structures with aspect ratios of nearly 4 [73], in the resist ZEP520 for a thickness of 50 nm in structures of only 10 nm width [74]. Under optimized conditions, periodic structures can be prepared in thin resist structures with repetitions down to 20 nm, e. g., by point beam exposure of line gratings and two-dimensional arrays in PMMA prior to development in isopropyl alcohol (Fig. 77). These structures are transferable in metal films (e. g., Au, AuPd, Pd) by lift-off techniques [75][76]. A highly focused electron beam (diameter below 1 nm) enables the smallest structures with extreme aspect ratios to be created, such as structures of less than 30 nm width in PMMA of 400 nm thickness, resulting in an aspect ratio of up to 15 [77]. Similar structural periods and line widths of up to 10 nm were realized in calixarene-based resists [78][79]. Also with negative resist such as XP90166, the fabrication of structures with an aspect ratio of 10 and structure widths below 40 nm has been shown [80]. The highly sensitive and chemically enhanced electron beam resist (NEB 22a) was able to demonstrate aspect ratios up to 7 for a structure width of 75, and better than 3 for a resolution of 36 nm, respectively [81]. Addition of fullerenes to EBL resists increased the thermal stability [82].

Beside organic polymers, metal fluorides such as AlF_3, FeF_2, CrF_2, LaF_3, BaF_2 and SrF_2 are potential inorganic electron beam resists. From this series, aluminum fluoride exhibits a particularly good resolution in combination with the resist stability in plasma and an acceptable development method. After exposure, the layers are developed in water. Structural periods of down to 50 nm were realized in 50 nm thick layers, so that a maximum aspect ratio of 2 was achieved [83].

To avoid large aspect ratios in the lithographic mask, one can – in addition to very thin spin-on layers – also apply monomolecular layers as an electron beam resist. On metal surfaces, especially on metals with easily polarizable electron shells (e. g., noble metals and heavy elements of the Main Groups III to V), the spontaneous assembly of

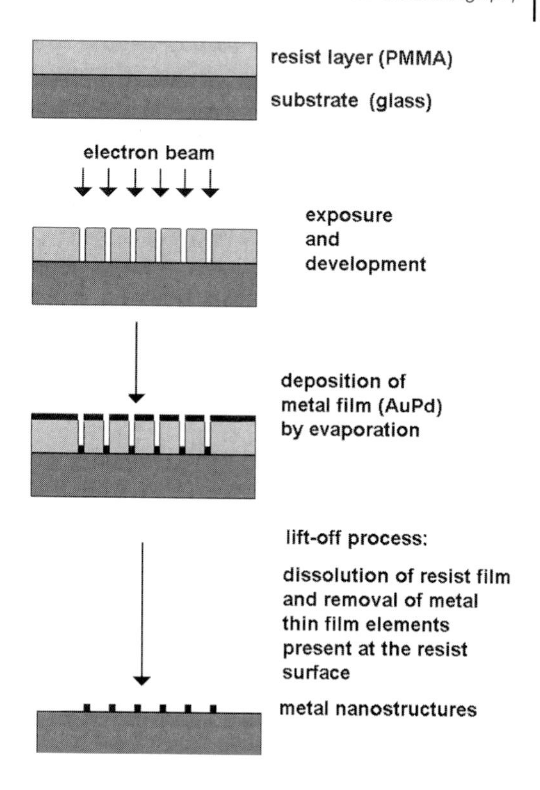

Fig. 77 Electron beam lithographic fabrication of metal nanostructures by a lift-off process (H.G. Craighead 1984)

closed monolayers of alkylthiols and dithiols can be used for such a mask technique. On oxidic surfaces, OH groups are created or existing OH groups are used to fabricate monolayers to act as a resist via a siloxane coupling, e. g., starting with trichlorosilanes or trialkoxysilanes. Also a perfluoroalkyl monolayer can be used as the electron beam sensitive mask layer [84]. Using an octadecylthiol layer on GaAs and an octadecylsiloxane layer on SiO_2, EBL structures of 25 nm width and 50 nm periodicity were realized and transferred by wet etching into the functional layer [85].

Regular structures such as line gratings with a constant periodicity can be realized without a mask or position-controlled beam: in analogy with the lithography with electromagnetic radiation, EBL can use the interference of electrons for structure generation. Applying this approach, gratings with line widths of 0.1 μm were structured into a 30 nm thick PMMA layer [86].

The generation of nanostructures by electron beams has also been successfully accomplished with low-energy systems. For example, a 50 nm line-space pattern and 66 nm dots were generated by means of a compact low-energy electron beam direct writing (LEEBDW) device [87].

The electron beam technique is particularly well suited for the direct deposition of material from reactive gases. The focussed particle beam permits the fabrication of deposited structures down to the sub-10-nm level. A serious problem, however, is the low process speed. Therefore, techniques are needed for the parallel application of multiple focussed electron beams. Such multiple beam sources (MBS) can be realized by using two-dimensional arrays of aperture lenses in devices for electron-beam-induced deposition (EBID). A system involving 100 parallel electron beams and employing an aperture lens array with 70 nm pitch in the electron optical system has been realized [88].

Despite all the progress in optical lithography, projection techniques for particle beam lithography still need to be developed in order to pave the way for future technologies combining very high lithographic resolution with high wafer throughput. Thus, electron projection lithography (EPL) tools are being developed for the fast transfer of lithographic nanostructures by EBL at the 65 nm level and below [89].

Because the EBL reaches its limits in the lower nanometer range, a special method was developed to fabricate tunneling barriers with critical dimensions of 1–2 nm. An EBL nanostructured shadow mask was positioned in close proximity to the substrate surface. A material deposition using a directed beam projected the mask apertures into an ultrathin metal layer. During this process, the width of the transferred metal lines was determined by the angle of incidence of the material beam. Lines of minimal width were realized by optimization of the angle. This technique can provide single electron tunneling transistors working at room temperature [90].

4.3.8 Ion Beam Lithography

In analogy with the use of electrons, ions can also be applied for lithography [91]. They too can be generated as well as accelerated by the electrode arrangements, filtered according to their energy and locally guided by electronic optics. Ion beams are utilized in four directions of nanolithography:

- to fabricate structures in ion-beam-sensitive resists
- for local implantation
- for structure transfer from structured resist layers in functional layer by ion beam etching
- to structure directly by local etching using a focused ion beam

Only the first point describes ion beam lithography in the narrower sense.

Comparable to parallel light, parallel ion beams can be applied for proximity lithography. In analogy with optical proximity lithography, this process is characterized by only a small distance between the mask and substrate; the distance is denoted as a proximity distance. A high resolution requires a constant distance between the mask and substrate, which means a highly parallel arrangement, and a low beam divergence. Because ion beams are virtually not diffraction limited, greater apertures can also allow for a high resolution. In addition, combinations of masks are applicable due to the highly parallel beam. Although this approach cannot decrease the dimensions of the periodic structures, small openings can be realized through the application of similar but slightly shifted masks (Fig. 78).

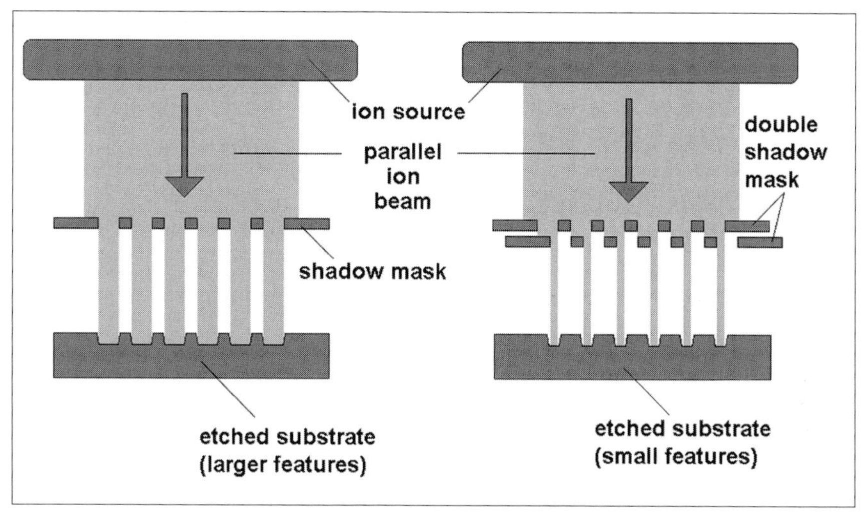

Fig. 78 Reduction of line width in ion beam lithographic etching by shadow masks using mask combinations

Owing to high resolution on the one hand and the possibility of an optical projection (possibly reduction) on the other, the development of projecting ion beam lithography is of interest. Ion beam lithographic projection exposure systems consist of a vacuum container with an ion source, an ion optical system, an alignment system and table for the masks and substrates. Such systems exhibit a much higher productivity for structures in the nanometer range than serial writing electron beam systems. An exposure rate of $12\,cm^2\,s^{-1}$ could be achieved with a MONTEC system with a resolution of 100 nm [92].

For the lithographic generation of structures in resists or the direct writing of small structures, lithography with a focused ion beam (focused ion beam lithography – FIBL) is of special interest [93]. Ion beams can be focused, and beam diameters in the lower nanometer range can be achieved. The interaction of energetic ions with the target material is more favorable than in the case of electron beams, because ion beams have only a reduced tendency to induce secondary electrons or X-rays. Therefore, the excitation pear is reduced, and the resulting proximity effect minimized. A beam of energetic ions can be focused and used for local implantation of foreign atoms without the need for masks. Therefore, the atoms to be implanted have to be used as the ion beam. With a focused ion beam (FIB), extremely small structures can also be realized, such as lines in GaAs of 2–3 nm width [94]. In addition, wave guides and luminescence devices of compound semiconductors can be produced by this approach, e. g., laser diodes on the base of GaInAsP/InP [95].

Focused and highly parallel ion beams are ideal for the fabrication of the smallest structures by local modification of the substrate material. So Co/Pt multilayers were radiated with N^+ ions through a mask, resulting in an array of magnetic domains [96].

Very high precision in the generation of electronic gaps has been achieved by the combination of EBL and FIB etching (FIB milling, FIBM). In this way, 10 nm openings in thin gold films were prepared, which are suited for the electronic characterization of small nanoparticles trapped in these gaps [97].

4.3.9 Atomic Beam Lithography

Instead of beams of energetic ions, atomic beams can be used to induce chemical changes in resists or to the structure directly. A disadvantage is the lack of a possibility for focusing and guiding by electrical or magnetic fields due to the missing charge.

However, atomic beams can be focused by intensive light. Standing waves of laser light act as an array of convex lenses and focus the atomic beam. So gratings with a period of half a wavelength can be realized. For an additive structure generation by neutral atomic beams, the atoms are directly projected onto the surface-projected intensity maxima of a UV radiation. So line gratings of aluminum with a width of 70 nm and a distance of 155 nm have been obtained [98]. Interference patterns that cross allow the generation of two-dimensional grating structures [99].

Apart from directly structured deposition, nanotechnical mask techniques are also accessible by atomic beams. A proximity radiation through a micro grating succeeded in the local removal of a molecular monolayer of helium atoms [99], and octadecylsiloxane layers on silicon were structured by neutral argon ions in a similar manner [100].

4.3.10 Molecular and Nanoparticle Beam Lithography

Instead of electrons and ions of individual atoms, molecular beams and beams of clusters or other nanoparticles can also be used for high-resolution lithography. Molecular clusters (MC) are nanoparticles consisting of a non-stoichiometric larger number of molecules, and are held together by bonds weaker than the internal molecular bonds. Owing to their kinetic energy, MC beams are able to remove layers. Therefore, shadow masks are applied to protect the parts of the functional layer to be preserved. A beam of clusters of about 1000 CO_2 molecules each was applied to etch layers of diamonds, silicon and glass. While the resolution of this mask technique is still in the micrometer range, it is an individual impact crater with dimensions of 10 nm that is of interest in nanolithography [101].

Clusters of volatile materials, such as gas clusters of CO_2 that are only stable at low temperature, are appropriate for etching processes, because they decompose even at low activation energies and their products desorb readily from the substrate surface, thereby removing any hindrance to the further action of the beam.

4.3.11 Direct Writing of Structures by a Particle Beam

Particle beams allow the generation of lithographic masks directly from the gas phase and without a resist layer. This has been demonstrated by the electron beam induced local deposition of a carbon-rich mask layer of polystyrene resulting in 30 nm structures and subsequent transfer into GaAs [102].

Particle beams are able to fabricate nanostructures without the generation of a mask. The non-reactive subtractive process accelerates particles of high energy focused onto the substrate and thus removes particles from the surface. Direct structure techniques with a focused ion beam (FIB) fabricate structures in the medium nanometer range. Because of the high precision regarding beam diameter and positioning, FIB is applied in local structure generation in a microtechnical production process, e. g., for the repair of lithographic masks.

In analogy with structure fabrication with neutral atomic beams, cluster beams also allow additive structure generation. Clusters arriving on a solid surface create layer elements with shapes determined by their relaxation behavior, their energy and the surface temperature [103].

Particularly suited to direct structure generation of metals is the decomposition of oxides and salts by a focused electron or ion beam, as in the case of a platinum oxide or palladium acetate layer. Such processes result either in metal structures as electrodes embedded in a less conductive metal oxide layer, or the unexposed areas of the layer are removed by a selective dissolving process, so that the metal structures remain.

Electron-induced Direct-writing Processes

Nanostructures can be written directly through chemical reactions under the influence of an electron beam. The material originates either from the substrate, from a layer on the substrate (subtractive direct writing) or from the gas phase (additive direct writing). In the first two cases, local areas of the substrate surface or the upper layer are modified under the electron beam. Thereby, solid-state material is directly transferred into the gas phase, e. g., by decomposition into gaseous products, or the material is converted in situ in such a way that it can be differentiated in a subsequent chemically selective etching step (Fig. 79).

In positive direct-writing processes, the exposed material is more easily dissolved or removed (e. g., in a selective plasma process) compared with the material without exposure. In negative direct writing, the exposed structure elements are conserved after the etching step, e. g., for the electron beam induced decomposition of noble metal oxide or salt layers such as palladium acetate (Fig. 80).

The additive direct writing is based on the introduction of the atoms of the generated structures in a state of easily decomposable gases into the reaction chamber. The so-called electron beam contamination techniques are widely used. Thereby, the beam decomposes organic molecules, present in the gas atmosphere and adsorbed on the surface. This effect creates surface deposition, which is usually not desired in electron microscopy. This technique readily generates lines below 100 nm widths (Fig. 81).

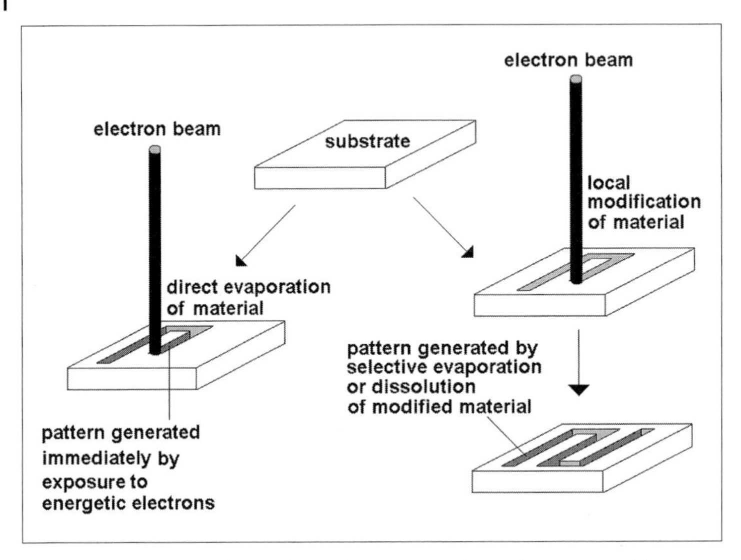

Fig. 79 Direct writing of nanolithographic structures by local material modifications using a focused electron beam and subsequent selective solubilization of modified regions

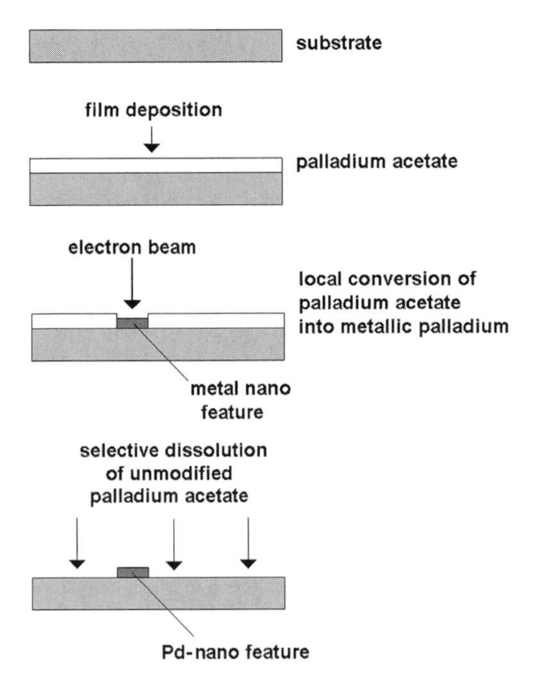

Fig. 80 Fabrication of metal nanostructures by electron beam induced decomposition of layers of noble metal salts or noble metal oxides

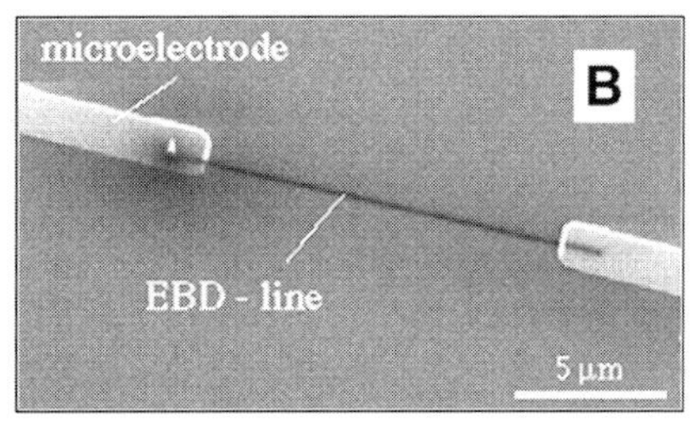

Fig. 81 Nanostructure between two microelectrodes fabricated by electron beam induced deposition (EBD) [104]

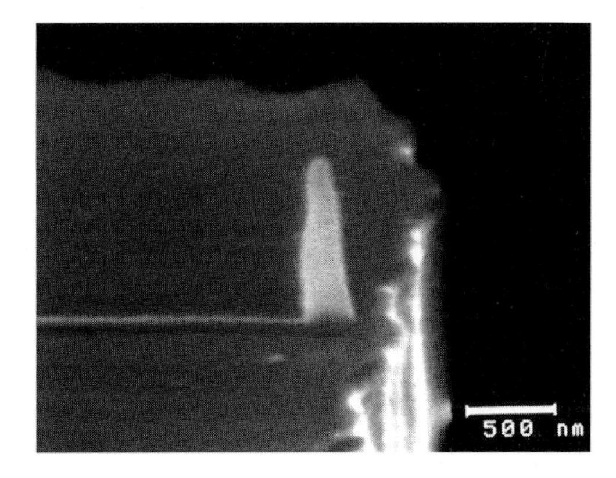

Fig. 82 Column structure as a scanning force microscopy tip, fabricated by the EBD technique

The development of improved pump oils and oil-free pumps reduced the original concentration of contaminants in electron microscopes significantly, so that today organic vapors are often introduced on purpose to generate additive structures under the electron beam. In addition to the gas phase, organic layers such as photoresist are also able to act as sources. An electron beam focused on a small area or even a point results in electron beam deposition (EBD) structures of high aspect ratio (Fig. 82). One application of such structures are AFM tips [105].

The fabrication of metal structures utilizes preferably metal halogenides, metal oxyhalogenides, metal carbonyles and metal alkyls [106][107]. Metal organic compounds and carbonyls often yield carbides instead of the metals [107]. Tungsten and chromium structures with sub-micrometer line widths were fabricated by local electron beam induced decomposition of WF_6, WCl_6 and $Cr(C_6H_6)_2$ (Fig. 83) [108]. This technique enables the creation of nanostructures with extreme aspect ratios, such as 100 nm high tungsten structures with a diameter of only 15 nm [109]. However, even single structures require process times in the minute range, so that this method is only applicable for individual preparations. Nanoscale metal structures with extremely high aspect ratios are of special interest to the fabrication of field emission sources [107].

Spin-on layers of poly(3-octylthiophene) yield conductive layers directly from polymers, because this material is itself conductive. Direct writing with a radiation of 30 kV

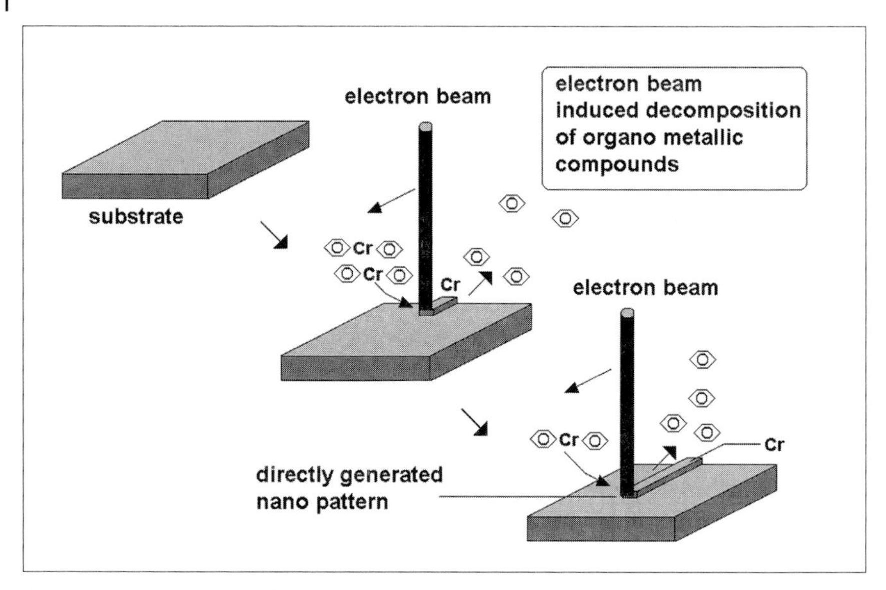

Fig. 83 Direct-writing of metal nanostructures using electron beam induced decomposition of metal–organic compounds

and subsequent development in chlorobenzene resulted in polymeric contact structures of 50 nm width and 30 nm height [110]. Also electron beams emitted by scanning probes are applicable for structure fabrication. A special technique uses low-energetic electrons that are extracted from scanning tunneling microscopy tips [111].

4.3.12
Nanostructure Generation by Accelerated Single Particles

In principle, beam techniques can be applied to individual particles. A significant removal of material is achieved by the impact of individual particles with very high kinetic energy in resist or other sensitive materials. One example of the effect of individual particles is the so-called ion track membranes: zones of damage that are generated by the penetration of individual particles through a polymer foil can be developed by selective dissolving into holes of several hundred nanometers. For standard particle sources, such as a nuclear reactor or radiating material, the positioned alignment of the work piece of nanotechnical interest is difficult. In principle, single ions can be activated to high energies in a particle accelerator and diverted with high accuracy by positioning systems. However, application in nanotechnology is hampered by the prohibitive costs.

In contrast to the subtractive structure generation by high-energy particles, the manipulation of charged particles by electrode optics also allows additive structure generation. The energy of such particles has to be just the amount required to set it down at the desired position. Beam techniques in structure generation usually apply lightweight particles, such as electrons and protons. Ions, such as Ar^+, are rather heavy particles for these applications.

Although the generation of chemical structures out of single atoms does not seem applicable, the generation of supermolecular aggregates using beams of larger molecules is an interesting approach. It could probably be realized by a combination of laser-assisted desorption and field-guided transport of molecules in the gas phase. Desorption of larger molecules by laser activation has been demonstrated. Using matrix-assisted laser desorption, even large molecules such as DNA and proteins with atomic numbers of some tens of thousands can be transferred into the gas phase without damage. Sensible biomolecules are transferred to the gas phase by matrix assisted laser desorption/ionization (MALDI) in high vacuum. This method is applied to separate the particles during the flight inside the electrical fields according to their charge-to-mass ratio. With this time-of-flight (TOF) spectrometry, species with molecular weights greater than 10^4 can be measured with high accuracy [112]. However, this method focuses on the time-of-flight, and not on the exact positioning. In electron and ion beam lithography, energy filters and positioning systems are applied, these handle small particles and position them with a precision in the lower nanometer range. So the development of additive working single particle deposition techniques for the generation of nanostructures seems feasible. If one succeeds in maintaining the distribution in both energy and direction of the particles sufficiently small, it should be possible (in analogy with ion beam lithography), after desorption, ionization and passing through an ion optical system, to achieve a positioning of individual, large molecules on solid surfaces (Fig. 84). This approach could open the way to a physically based supermolecular construction by use of individual molecules on surfaces in order to generate complex molecular structures on a chip.

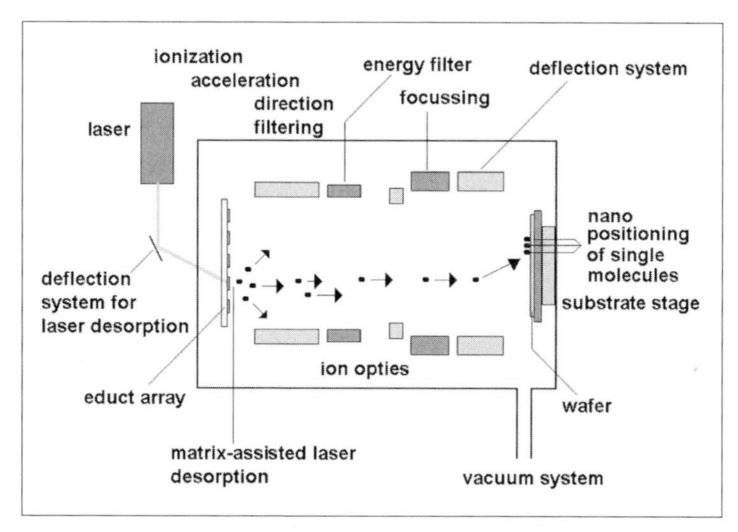

Fig. 84 Scheme of an instrument with direct-writing of larger molecules by laser-assisted desorption and ion optical deposition

4.3.13
Patterning by Local Chemical Conversion

The combination of lithographic patterning with self-organization is of particular interest for achieving higher degrees of integration. The idea involves creating an array of well-reproducible, self-organizing patterns from corresponding lithographically patterned materials. This strategy can be used, for example, for side-wall-directed patterning or for the in situ generation of nanotubes by catalytic processes.

Structures of nanometer dimensions can be generated by local conversion of thin films or by locally resolved conversion of thin surface areas of a bulk material. Two-dimensional structures may be obtained by the combination of two or more linear patterns. The conversion of metal or semiconductor films can be accomplished, for example, by local anodic oxidation. The process is chemically related to the direct writing of individual structures on metal with an electrically connected scanning tip anode under anodic polarization of the substrate. In contrast to this technique, a huge number of structures can be generated simultaneously if line arrays are used for local anodic oxidation [113].

The formation of more complex nanopatterns by local material conversion could possibly be further developed into techniques for the realization of three-dimensional nanoarchitectures at planar substrates. This could offer a way of overcoming the spatial restrictions of the strictly two-dimensional organization of planar technologies without loss of their technological advantages. The challenge for the use of local chemical conversion to generate three-dimensional nanopatterns becomes ever more important with the perspective approach of lithographic techniques at molecular dimensions.

4.3.14
Nanofabrication by Self-structuring Masks

Besides the application of focused beam probes to achieve local structures, mask material that spontaneously generate nanostructures can also be utilized in nanofabrication, using a homogeneous structure transfer method to transfer the pattern into a functional layer. Such a self-structuring mask is either applied as an adhesion or near-field mask directly onto the functional mask, or it is projected on it. Such masks are applicable for subtractive (etching mask) and additive (deposition mask) processes.

Regular gratings of hexagonal points and rings are easily prepared by densely packed nanospheres of similar diameters. Such monodisperse nanoparticles are available as metal (gold) or polymeric material. For structure generation, regular gaps between hexagonal densely packed spheres are used (hcp layers). Latex nanoparticles with diameters of 140 nm were spun-on as a monolayer and applied as a monolayer in a lift-off process after evaporation of the gold, yielding gold structures of about 100 nm in regular hexagonal arrangements. In addition to the direct projection of the gaps, hexagonal gratings created by ring-shaped structures were also observed. Such gold rings with a line width of 17 nm were generated by latex particles of 895 nm diameter. Combinations of ring and gap–dot structures are accessible through this approach [114].

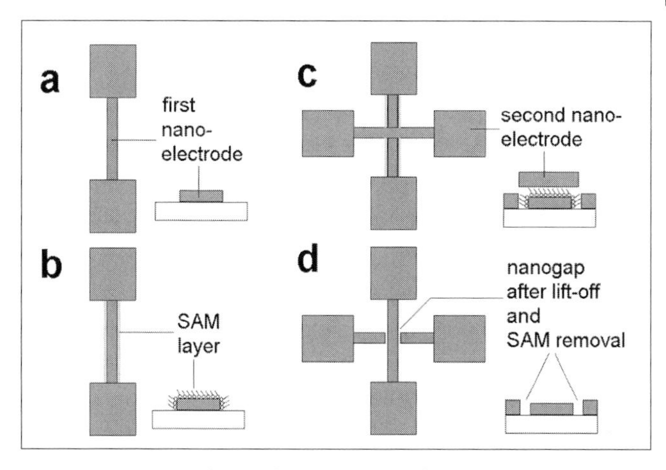

Fig. 85 Preparation of a SET device by means of molecular self-assembly: the first generated electrode (a) is covered with a SAM layer (b), then covered with a cross-electrode structure (c) before the molecular monolayer is removed in order to obtain the nanogaps (d) (R. Negishi et al. [115])

The use of self-assembled molecular monolayers to delineate functional nanostructures is a special form of the use of molecular self-organization in nanotechnology. An SET device has been prepared by a combination of EBL and SAM techniques. For this, a molecular monolayer was deposited as a first layer at a nanoelectrode. Material deposited in a second deposition step could not encroach into the space occupied by the molecular film. As a result, an electrode gap with a gap width of the thickness of the molecular monolayer was obtained (Fig. 85). Electrical characterization was indicative of a "coulomb staircase", reflecting the single-electron tunneling behavior of the device [115].

4.4 Nanofabrication by Scanning Probe Techniques

Scanning probe techniques, such as scanning force or scanning tunneling microscopy, are important ultramicroscopic methods due to their ease of application and extremely high resolution (especially perpendicular to the surface plane). This high resolution made them attractive for nanotechnology [116]. Besides the microscopic aspect (measurement and characterization of small structures), these techniques are also interesting as fabrication methods.

4.4.1
Mechanical Surface Modifications based on Scanning Force Microscopy (SFM)

Scanning force microscopy (SFM), also known as atomic force microscopy (AFM), generates a signal by interaction of the scanning tip with the object, which is used for the generation of a map of surface topography or surface interactions. The force between the tip and substrate can be used for visualization of the substrate topography, but also for the induced movement of weakly bound particles on the surface or for scratching into the upper substrate layer.

Structure generation takes place in analogy with imaging: the probe tip and substrate approach towards each other in the *z*-direction, while the distance or interaction force is regulated usually based on a light pointer (Fig. 86). Movement of the substrate in the *x*- and *y*-directions results in structures that are generated at the contact point of the tip with the surface.

An ultrathin resist layer can be utilized as a mask for SFM-based structure generation. A highly diluted commercial photoresist was spun on an Si/SiO_2 substrate resulting in a thickness of 5.4 nm, prior to writing a pattern using the probe tip and a wet etching-based transfer of this pattern in the substrate, yielding line widths of between 25 and 40 nm [117].

In addition to mechanical manipulation of the layer material, individual cluster and nanoparticle are also addressed by SFM operations. Thus 50 nm gold nanoparticles were pushed, by directed movement of the scanning tip, into a gap between two microstructured electrodes, so that the arrangement exhibited single electron tunneling effects in a subsequent electrical characterization [118]. Besides metal nanoparticles, nanorods and nanotubes are other examples of scanning probe manipulations, such as is demonstrated in the probe-based manipulation of single wall carbon nanotubes (SWNT) between nanoelectrode structures [119].

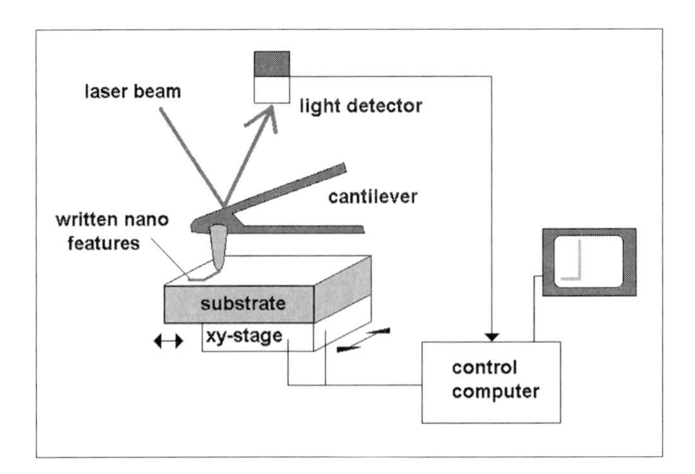

Fig. 86 Lithography with the scanning force microscope (R. Wiesendanger 1995)

Multi-step approaches are also possible, as exemplified by the positioning of latex nanospheres prior to thermal processing in order to create a sintered structure [120].

SFM is also able to dissect very small individual objects adsorbed onto surfaces. So a single DNA-plasmid could be cut into pieces down to dimensions of 20 nm [121].

Using an SFM tip, mechanical stress can be applied in a nano-localized manner in order to induce, for example, changes in the properties of certain materials. Using an SFM to apply mechanical stress in conjunction with an electric field to a local region of a thin layer of barium titanate resulted in a novel switching effect in this material (M. Abplanalp et al. [122]). At low mechanical stress, a voltage pulse polarized the ferroelectric in the direction of the applied field, as expected. However, simultaneous application of a large mechanical force (0.9 μN) during the voltage pulse created a region that was polarized in the direction opposite to the applied field. This ferroelastoelectric response may prove useful for developing new ferroelectric devices.

AFM is not only able to image individual molecules, but is also used to characterize and manipulate on the single molecule scale. It is a unique tool for studying the mechanical properties of extended molecules or the bonds between individual molecules (single-molecule force spectroscopy [123]. In combination with optical methods, for example, it can also be used to apply precisely defined mechanical forces to molecules with a high lateral resolution. An example is the development of a scanning force/fluorescence microscopy technique to study the mechanical properties of single fibrin fibers [124]. Here, the tip was used to stretch fibers that were suspended across 12 μm wide channels; the fluorescence microscope was used to image this stretching process in order to study the extendibility and elastic limit of fibers formed in the presence and absence of factor XIIIa (FXIIIa).

A scanning force microscope (SFM) is also able to mechanically manipulate selected single atoms by soft nano-indentation. A near-contact SFM operated at low temperature has been used for the vertical manipulation of selected single atoms on an Si(111)–(7 × 7) surface [125]. The strong repulsive short-range chemical force interaction between the closest atoms of the tip apex and the surface during soft nano-indentation allowed the displacement of a selected silicon atom from its equilibrium position at the surface without additional perturbation of the (7 × 7) unit cell. Deposition of a single atom at a vacancy created on the surface was also achieved. These manipulation processes were purely mechanical, since neither a bias voltage nor a voltage pulse was applied between the probe and the sample.

4.4.2 Manipulation by a Scanning Tunneling Microscopy (STM)

STM is not only a microscopic method for highest (e. g., atomic) resolution, but also a tool for manipulations of very small structures, which had already been applied before the introduction of the SFM [116][126].

The manipulation of single atoms and small molecules with a scanning tunneling probe is achieved by tuning the interaction between the tip and the atom/molecule on the one hand and between the atom/molecule and the surface on the other. Prefer-

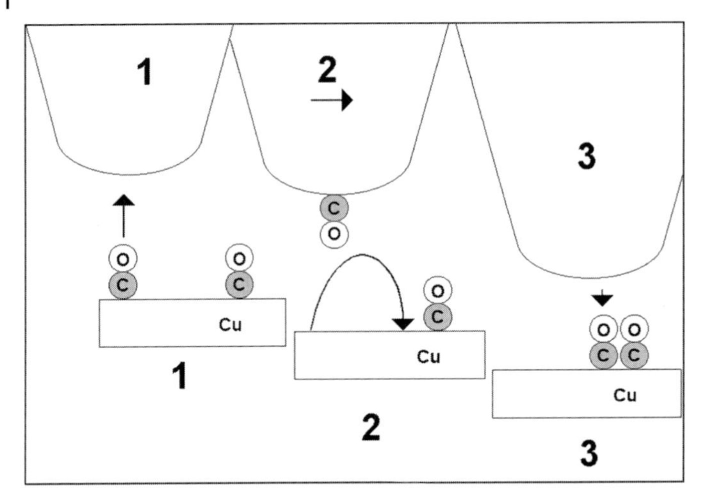

Fig. 87 Scanning probe-based manipulation of single molecules on solid substrates: transfer of CO onto a Cu substrate [127]

ential treatment of one or the other interaction is achieved by adjustment of the STM potential in combination with the distance between the tip and surface. Thus single Cu atoms were moved or CO molecules were picked up and transferred by the STM tip on a crystalline Cu surface (Fig. 87) [127]). In an analogous experiment, Si atoms were moved on an Si(001) surface and arranged into lines, resulting in trench structures of 3 nm width and 2–3 nm depth [128].

Experiments at 4 K showed the feasibility of manipulating individual atoms and arranging them into structures. On crystalline Ni, adsorbed Xe atoms were assembled by the STM tip into an IBM logo in an area of 5 nm × 16 nm [129]. The information density of such structures is about 1 bit nm^{-2}, which corresponds to 100 Tbit cm^{-2}. However, the deposition and manipulation of individual atoms and small molecules, and therefore also the generation of structures with atomic resolution is still limited to isolated demonstrations in basic research. Therefore, the patterns created by individual noble gas atoms on crystalline surfaces at low temperatures represent milestones in the exploitation of the ultimate limits of nanostructure technology, but they do not exhibit direct practical relevance for the fabrication of nanotechnical devices.

STM allows manipulation and characterization at the molecular scale at the same time. In order to study metal–molecule contacts, the interaction of a gold atom with a pentacene molecule, both adsorbed on a thin NaCl film grown on a metal substrate, was investigated using STM [130]. An STM tip was used to bring the Au atom into close contact with the molecule. Resonant inelastic electron tunneling (IET) through the lowest unoccupied orbital of pentacene led to bond formation, and the resulting changes in bond hybridization could be imaged. The bond could be broken by IET through the molecular complex. The resulting changes were rationalized by comparison with density functional calculations.

The tip of the STM can thus be used to pick up atoms and move them on surfaces. However, it may also be used to induce motion through electronic excitations produced by the tunneling electrons. Short chains of Cu atoms terminated by a Co atom were assembled on a Cu(111) surface, and the hopping induced by tunneling electrons of the Co atom between different sites at the end of the chain (which manifested itself as low-frequency "telegraph" noise) was analyzed [131]. Density functional calculations were used to rationalize the facts that the tip location that maximized this hopping was not directly over the Co atom and that the barrier to motion increased with increasing Cu chain length.

Using a low-temperature scanning tunneling microscope (LT-STM), the controlled lateral manipulation in constant height mode of a Lander molecule (a polyaromatic molecular board supported by four legs) adsorbed on a Cu(2 1 1) surface has been demonstrated [132].

Controlled manipulation of a single molecular wire along a copper atomic nanostructure by means of low-temperature scanning tunneling microscopy (STM) has been demonstrated using a Lander molecule [133]. Irrespective of its position along the copper nanostructure, the central molecular wire maintained electronic contact with the atomic wire underneath. This effect was manifested in the STM images depending on the orientation of the legs. By STM manipulation, the molecular wire could be precisely positioned in an electronic contact conformation at the end of the atomic wire.

A novel STM manipulation scheme for controlled molecular transport of weakly adsorbed molecules has been demonstrated [134]. Single sexiphenyl molecules adsorbed on an Ag(111) surface at 6 K were shot towards single silver atoms by excitation with the tip. To achieve atomically straight trajectories, an electron resonator consisting of linear standing-wave fronts was constructed.

4.4.3
Thermo-mechanical Writing of Nanostructures

Scanning force microscopy is also used for thermally based shaping of materials. Thus the advantage of local heat transfer over small dimensions is combined with the small probe diameter. High resolution is realized for systems of a deformable material with fairly low heat transfer of a substrate that is rather heat conductive, e. g., a thin PMMA layer on an Si wafer. The energy of a heated tip in contact with the PMMA layer is transferred and leads to a local softening of the material, so that the geometry of the tip is transferred into the surface topography of the resist layer. The adjacent regions of the resist are not affected, because the heat is efficiently carried off by the Si substrate, so that a steep lateral temperature gradient is realized. The pixel distances achieved of 40 nm in a PMMA layer of 40 nm thickness result in pixel densities with an interesting potential for future memory applications in the 400 Gbit cm^{-2} range [135][136].

Amplitude-modulated electrostatic lithography on 20–50 nm thin polymer films has been used to fabricate arrays of nanodots, as small as 10–50 nm wide by 1–10 nm high, through localized Joule heating of a small fraction of polymer above the glass transition temperature [137]. This approach is based on an electric bias of the AFM tip that increases the distance over which the surface influences the oscillation amplitude of an AFM cantilever, providing a process window in which the tip-film separation can be controlled.

A related method uses the tunneling current of an STM tip to locally melt glassy metal alloy material with spot diameters between 10 and 40 nm in $Co_{35}Tb_{65}$ and $Rh_{25}Tr_{75}$ [116][138].

4.4.4
Electrically Induced Structure Generation by Scanning Probe Techniques

Electrically induced structure generation utilizes an electrical current to change the local surface state of the material. Often metals or semiconductors are transformed into their oxides, using the surface humidity layer or the ambient oxygen as a reaction partner (Fig. 88). The scanning tip positioning is realized by either force sensing (SFM) or is based on the tunneling current (STM). The latter uses the tunneling current for both distance control and local electrochemical transformation of the surface.

Electrical Structure Generation by SFM

Nanostructure generation by anodic oxidation with an SFM tip prefers materials with a tendency towards stable oxide surface layers. This is the case for Si, Al, and several transition metals. Thus the substrate is used as the anode. Voltages in the range between 2 and 30 V are used; this is because they have to overcome the oxidation potential of the substrate, and often voltage drops between the substrate and tip or as a result of resistive surface layers on the scanning tip.

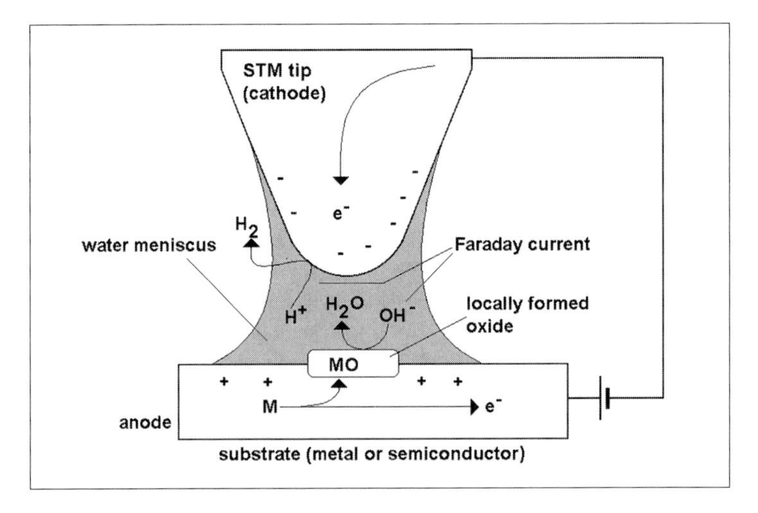

Fig. 88 Electrochemical processes during the anodic manipulation of metals and semiconductors with the scanning tunneling microscope

Conductive SFM has been applied to form thin layers of Ti into line electrode structures of 75 nm width in isolated surroundings [139]. A GaSb/AlGaSb/InAs structure for single electron tunneling processes was realized by local oxidation of GaSb/AlGaSb at 30 V prior to the selective removal of the oxidized regions in an acetic acid etching bath [140]. Structures in InAs/AlGaSb were fabricated at 3–8 V [141]. The properties of certain metals for the excellent local oxidation by SFM are utilized for SFM mask processes. Thus sub-micrometer structures in Si have been generated by an Al mask that was patterned by the local oxidation of a thin Al mask layer prior to anisotropic pattern transfer by an RIE process, using a combination of laser and SFM lithography [142][143]. Scanning force microscope (SFM)-induced oxide features can be reproducibly formed on both Zr and ZrN surfaces [144]. With increasing anodization time, only the thinnest (6 nm) films show a large enhancement in oxide feature height, demonstrating the role of the film/substrate interface. Under the same conditions, the height of features grown on ZrN films is greater than that of those grown on Zr films, indicating that nitrogen plays a role in the oxidation process.

Also thin dielectric layers are accessible for conductive SFM structure generation. Therefore, the local current transforms the material into a state that is selectively removable. So dielectric layers are usable as resist material for nanolithography, when the underlying material is electrically conductive and used as the anode. A 100 nm thick spin-on glass layer (polysiloxane) has been structured at 70–100 V into line width of 100 nm [145]. Also organosilane monolayers are used as high-resolution resists in SFM-tip induced local oxidation. Trimethylsilyl layers yielded resolutions down to 20 nm and maximum writing speeds of up to 5 mm s^{-1} [146][147].

As well as inorganic or Si–organic layers, pure organic layers are also applicable. Even a monolayer can act as a resist, when the subsequent process has a high selectivity [148]. Trenches of 100 nm width were fabricated in Si monocrystals using a multilayer resist process by local anodic oxidation of an octadecylsilyl monolayer using an SFM tip, and a subsequent transfer of this pattern into a thin SiO_2 layer, which acts as a transfer mask for an anisotropic wet etching of the Si substrate by a TMAH (tetramethylammonium hydroxide) solution [149].

Scanning probe lithography is seriously hampered by the low writing speed. The throughput can be increased by electronic control of the current during the scan and by the application of a parallel arrangement of several tips [150]. However, highly productive systems are not yet available. The development of high-speed SFMs [151] offers a solution to the need for an increase in sample throughput. In addition to relieving the tedium of waiting for images, commercial high-speed SFMs will also enable researchers to study fast processes such as protein motion [152] and crystal growth (for a review, see [153]), as well as to carry out faster force spectroscopy [154], which has hitherto only been possible in a few laboratories with in-house constructed equipment. High-speed SFM also offers enormous promise for the increased use of this technique for industrial measurements (and ultimately for possible storage or fabrication applications), where metrology is often considered by the cost per measurement site [155]. In the case of an AFM that can operate 30 times faster, this can translate into a substantially lower cost per measurement or even enable novel nanofabrication applications.

Electrical Structure Generation by STM

Nanofabrication by STM relies on several mechanisms, depending on parameters such as material, potential, current, continuous or pulsed mode. Besides a local increase in temperature, they also include activation of surface diffusion for adsorbed particles, electromigration, field ionization and electrochemical reactions [156].

STM allows the modification of a mono molecular or even mono atomic adsorbate layers on the substrates. These processes are particularly efficient when the adsorbates readily transfer into the gas phase. So special Si–H layers on monocrystalline Si are accessible for ultimate lithography [157][158][159]. At negative tip bias, a passivating H layer was removed under UHV conditions from a monocrystalline Si surface. In this way, areas of 20 nm squares as well as line gratings with a period of only 3 nm were achieved (1 nm lines with 2 nm distances). These adsorbate-free areas oxidize in the presence of oxygen and create an oxide layer, so that the structures could be transferred in a more robust nanostructured surface by incubation with ammonia [160]. Nanostructured H layers on Si surfaces can be also used to inhibit surface selective reactive deposition processes. Thus a CVD deposition of Al layers from a dimethylaluminum hydride occurs on H-free surfaces only, and STM-written structures can be additively transformed into alumina structures with line widths below 3 nm [161]. STM-generated areas with low H concentration in H-rich and less reactive surroundings can be applied for the selective deposition of organic molecules in nanostructures [162].

In analogy with conductive SFM, STM is also able to structure easily oxidizable metals and semiconductors in an anodic process [163]. Thus lines of 100 nm width were structured in 30–50 nm thick chromium layers (Fig. 89) [164]. Also for STM-induced local oxidation, a simple electrochemical model includes a thin aqueous adsorbate layer as the electrolyte, the tip as the cathode and the substrate as the anode [146].

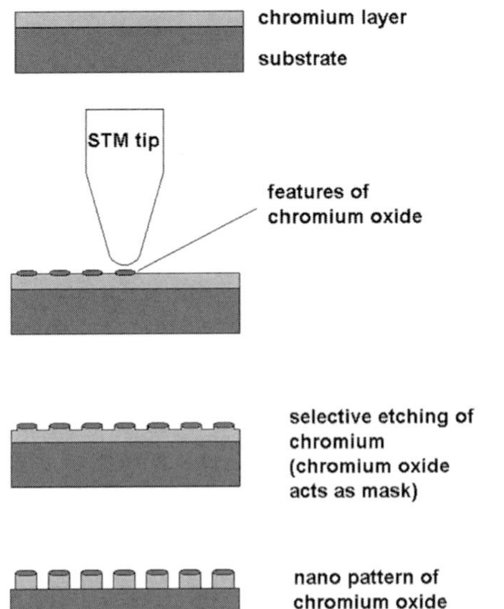

Fig. 89 Fabrication of chromium nanostructures by a combination of local STM oxidation and a subsequent selective etching process (H.J. Song et al. 1994)

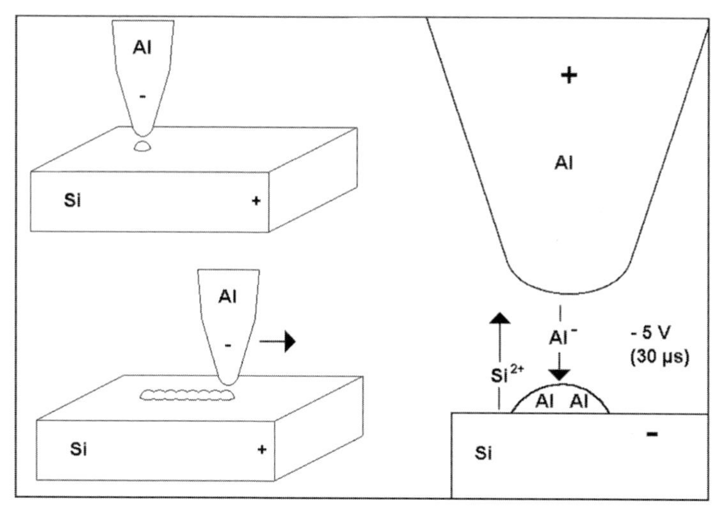

Fig. 90 Preparation of aluminum nanostructures by transfer from a scanning probe tip using an electrical pulse

Anodic oxidation of silicon yields SiO_x lines and meanders of high quality with line widths below 100 nm.

The tunneling tip also allows for a potential-controlled manipulation of individual atoms and small clusters, e. g., the extraction, transfer and deposition of Si on a substrate surface [165][166] or a conformational switching of individual organic molecules adsorbed on a surface (R. Otero et al. 2004). Extremely plane surfaces are required for reproducible manipulations of atoms and molecules. This requirement is met by crystalline surfaces, such as monocrystalline Cu.

Electrical pulses allow the transfer of groups of atoms from the probe tip onto the substrate surface (Fig. 90). Larger clusters of about 100 nm diameter are created in the case of tungsten or aluminum tips. Line gratings with a periodicity of 100 nm were fabricated by an equidistant arrangement of deposited Al clusters [167]. Voltage pulse-induced material transfer from a gold tunneling tip creates gold islands of 10–40 nm diameter [168][169][170]. A W-tip wrote lines of only 0.8 nm width on a Ge monocrystalline surface [171].

An additive writing process of metal structures can be realized based on the supply of easily cleavable gases. To create noble metal structures, an electrically induced local decomposition of organometallic gases can be applied (Fig. 91). Dots of only 30 nm diameter were so realized with gold [dimethyl-gold(III)-trifluoroacetylacetonate] and platinum [cyclopentadienyl-trimethylplatin(IV)] compounds in an STM CVD process [172]. A local STM-induced deposition from a dimethylcadmium atmosphere resulted in 20 nm cadmium structures [173][174]. Another process for the generation of nanostructures relies on the local chemical reactions of molecular layers induced by the tunneling current [175]. The decomposition of an adsorbed layer of ferrocene by the STM tip resulted in line structures with a width of 3 nm [176]. Layers of trimethyl

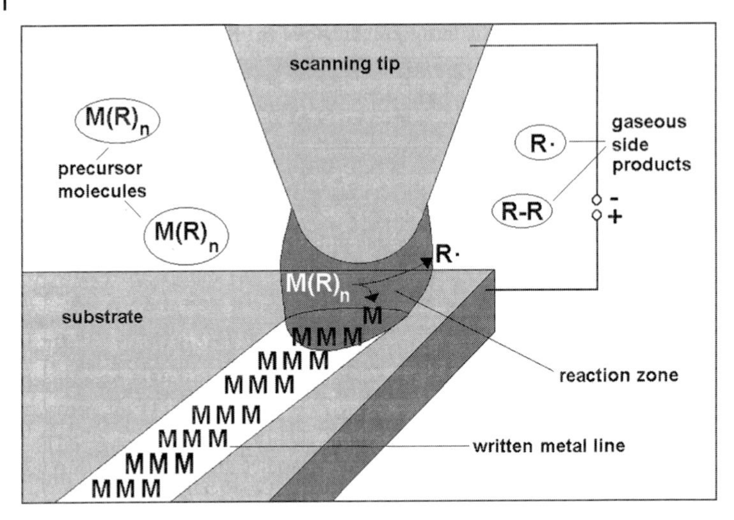

Fig. 91 Direct-writing anodic preparation of metal nanostructures by scanning probe-induced decomposition of metal organic compounds (STM-CVD) [172]

aluminum on graphite substrates yielded tip-induced structures of 20 nm width, sometimes even holes of 2.7 nm diameter [177].

Local tip-based oxidation of monomolecular alkyl layers results in areas with high densities of reactive OH groups in a matrix of chemically fairly inert alkyl residues. Using the OH groups, secondary coupling reactions are applicable in order to generate more complex molecular layers or to build up supramolecular architectures, e.g., using a siloxane linker (Fig. 92) [178].

An STM can be used for the local generation of electron beams. Because the electrons are adequately generated and accelerated only in the small gap between the tip and substrate, the effect is fairly localized. With respect to its effect on the substrate (e.g., thin layers of radiation-sensitive material), this process is related to the electron beam lithographical techniques and can be used for the exposition of typical electron beam resists with structure widths of about 15 nm being achieved [179][180]. Direct structure generation with a low energy electron beam (LEEB) uses a beam extracted from the STM tip (cathode) with electron energies of 30–160 eV and beam currents of 1 nA–1 μA. At substrate temperatures of 600–700 °C, oxide films on Si substrates were locally decomposed, yielding structure dimensions of below 50 nm [111][181]. In addition to the ultrathin layers of standard electron beam resists, also Langmuir–Blodgett-films (LB-films) are applicable as resist layers [179]. Structures written with STM in LB-films were transferred into aluminum by wet etching [182].

All procedures regarding the application of STM for structure generation are still in development and are not yet comparable with traditional approaches in lithography. They are usually applied to individual or a small number of structures. Writing times for larger areas are prohibitive. Considering the technical development and the technical maturity, electron beam lithography and focused ion beam lithography are much more advanced compared with STM lithography. On the other hand, STM accesses the

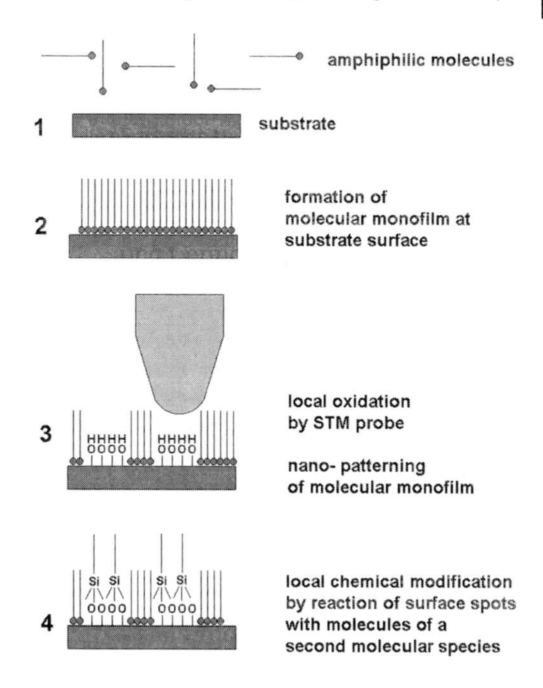

Fig. 92 Preparation of a lateral nanostructured molecular monolayer by local oxidation using a scanning probe tip and subsequent local surface modification [178]

ultimate limit of single atom manipulations, which is not possible for the conventional lithographical techniques. Moreover, STM is cost-efficient compared with electron and ion beam systems. Therefore, STM structure generation processes are interesting and important tools for nanostructure research.

4.4.5 Chemical Induced Scanning Probe Structure Generation

Changes in binding states on surfaces are not only induced by electrical processes. Near-field probes are able to induce chemical processes without electron transfers, or they can initiate open-circuit redox processes. One possibility is the structure generation by a scanning tip of a catalytically active material in thin films of reactive molecules or functional groups. In the absence of the probe material, the surface layer is stable. The approach of the tip induces a local transformation or decomposition of the surface layer. An example is the catalytic transformation of an alkyl azide surface layer into an alkyl amine layer by a Pt probe in a hydrogen-saturated atmosphere of isopropanol [183].

Another approach to nanofabrication generations is the local incubation of the surface with a corrosive solution using a nanopipette. This procedure requires a precise alignment between the pipette and substrate, which can be provided by scanning probe technology. Using quartz pipettes of 10 nm outer and 3 nm inner diameter, respectively, a solution of hexancyanoferrate(III) induced lines of 100 nm width in a chromium layer [184].

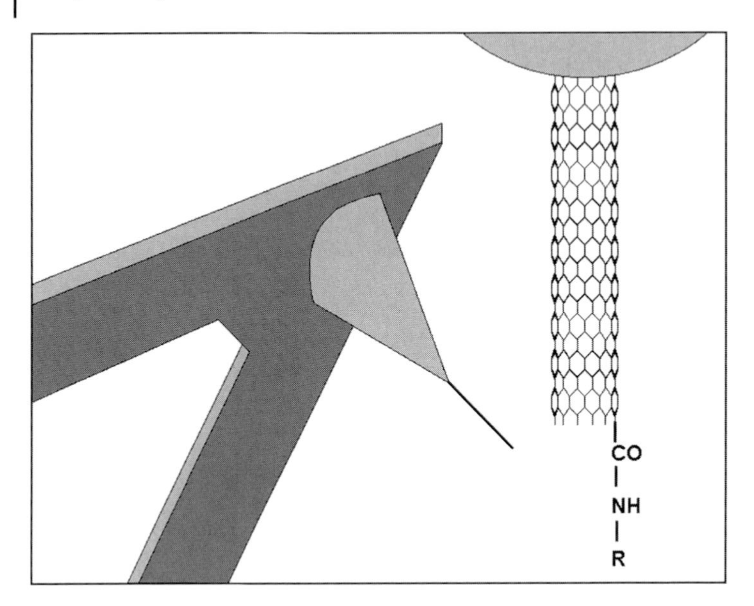

Fig. 93 Functionalized carbon nanotubes as chemically selective probes with a high aspect ratio for chemical force microscopy [186]

In another example, silicon nitride probe tips capped with palladium nanoparticles were used to catalyze the Suzuki coupling of aryl boronic acids to a layer of aryl bromides that were bound through sulfide linkages to a gold surface (J. J. Davis et al. [185]). After immersing the film in a methanolic solution of the boronic acid and a base, the probe was manoeuvered to the desired reaction site and coupling was induced by applying a force of 20–25 nN between the tip and the surface. Reducing the force to 1–5 nN allowed imaging of the patterned surface without further catalysis. To verify the spatial selectivity, coupling was performed with amine-substituted boronic acid substrates, which were subsequently labeled with a fluorescent dye.

An interesting approach to chemically based probe-induced structure generation, which also tolerates higher aspect ratios, includes the application of a carbon nanotube, which acts as a tip on a tip between a standard tip and a chemically functionalized upper region (Fig. 93). Therefore, the carbon nanotubes are functionalized at the ends. In contrast to the carbon atoms along the tube, which are saturated due to the aromatic bonds in the rings, the carbon atoms at the ends exhibit free valences. They can be addressed for oxidizing processes leading to hydroxyl, carboxyl or peptide groups, which are used for the coupling of different molecules [186]. This process results in chemically selective acting scanning tips with the dimensions of a small single molecule.

4.4.6
Nanostructure Generation by Optical Near-field Probes

Light can be applied for the generation of nanostructures by scanning probe techniques, because these techniques work in the near field, so that the resolution of optically induced structures is not diffraction-limited. Such processes require that the influence of the light quanta occurs at a small distance between a quantum-emitting probe and a light-sensitive surface in a local process.

The structure generation can be based on photochemistry or on simple ablative processes (non-reactive transfer of layer material into the gas phase). Photochemical scanning probe lithography in the optical near field can be based on photo resists similar to those in standard optical lithography. The only important difference is the limited thickness of the resist layer, to ensure that the whole layer is in the optical near field (below 100 nm). Therefore, the resist viscosity has to be low, and substrate topography should be minimal to achieve a spin-on coating without significant thickness inhomogenities. This low resist thickness demands a high absorption by the photo active component, enabling work to be performed with high dye concentrations in the layer and with dyes of very high extinction coefficients. As in standard photolithography, surface regions to be developed require an adequate dose, either by short exposure with high intensity or longer exposure with lower intensity. For the dose determination, the part of the dose that is diffracted by adjacent regions has to be taken into account. Because of the serial character of the probe technique, the exposure times for an area should be not too long in order to achieve reasonable process times.

For a physical ablation, a high local energy density is required to achieve evaporation of the material. Hence the locally deposited energy density per time interval has to be sufficient, and thus the intensity of the optical excitation. On the other hand, the mean intensity has to be limited to avoid an extensive material decomposition beyond the desired area, resulting in line broadening. Therefore pulsed lasers are applied. Using a metal-coated hollow fiber as a near-field probe and laser pulses, holes were structured

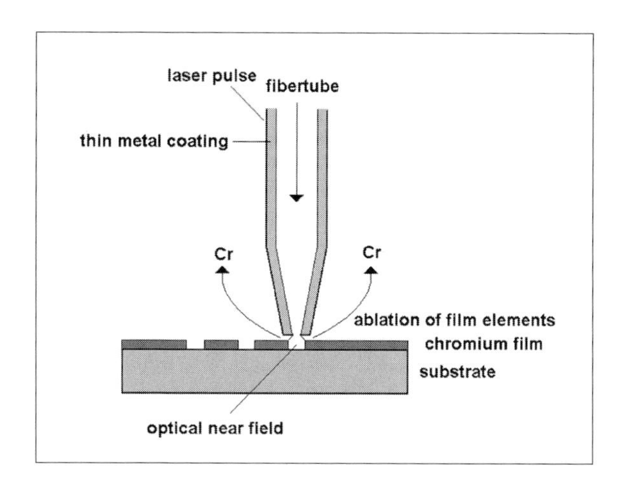

Fig. 94 Coated fiber probe for near-field optical structure generation in metal layers by laser ablation [187]

by ablation into thin chromium oxide layers on chromium substrates (Fig. 94), prior to a pattern transfer of the created mask into the chromium [187].

4.4.7. Scanning Probe Methods for Nanoscale Transfer

The family of SFM-based structure generation techniques also includes dip-pen nanolithography (DPN) [188]. Here, molecules are transferred from the scanning tip to the surface during the scanning process.

Applications include the use of alkane thiols, DNA [189], proteins (K. B. Lee et al. [190]), and conducting polymers [191].

Additionally, using multiple steps, nanopatterning for a broad range of applications is possible by means of ing DPN, as, for example, in magnetic nanostructures [192] or DNA-directed positioning of metal nanoparticles [193]. Beside capillary forces, other means can also be utilized to control the transfer of molecules from the tip to the surface, such as an electrical field, as demonstrated in the case of short DNA molecules [194]. Thereby, one can either switch between writing and reading modes, or even incorporate an electrophoretic separation of, for example, DNA into the transfer process.

4.5 Reduction of Feature Sizes by Post-Lithographic Processing

4.5.1. Narrowing of Nanogaps by Material Deposition

Nanostructures can be realized by reducing the size of lithographic features. Post-lithographic processing is not able to enhance the density of a pattern, because the principal topology of the transferred structures is conserved. However, the individual structures can be modified. Spaces can be made smaller by the enlargement of neighboring lines and vice versa.

One way of reducing the size of gaps in metal films is the cathodic reduction of metals [195]. The resultant increase in film thickness over the whole surface leads to a decrease in gap width. In this case, the gap width can be controlled electronically (Fig. 95). A similar process can be performed without the application of an external current by metal-catalyzed surface deposition through a chemical redox reaction. Thus, the thickness of patterned Au films can be enhanced and the gap can be reduced if salt solutions of noble metals (Au^{3+} or Ag^{+}, for example) are deposited by the application of a reducing agent. In this case, the process cannot be controlled electronically. Nevertheless, a reduction of nanogaps of 18–52 nm down to just a few nm has been successfully accomplished, for example, by the reduction of a gold salt solution with hydroxylamine [196]. The authors claimed to have realized gap widths of around 3.3 ± 1.4 nm, indicating that gaps of the order of magnitude

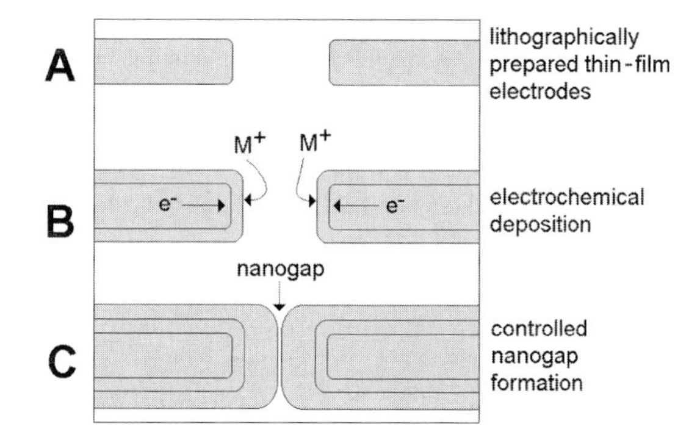

Fig. 95 Formation of nanogaps between microelectrodes by controlled electrochemical metal deposition: A) lithographically prepared thin-film electrodes, B) reduction of gap width by electrochemical metal deposition on the electrode surface, C) electronically controlled formation of a nanogap [195]

of tunneling distances are accessible. Indeed, nonlinear current-voltage characteristics of the prepared narrow nanogaps were seemingly indicative of electron tunneling.

Particularly small structures can be generated by the deposition of material at side walls. Nanostructures of high aspect ratio can be produced by the selective removal of originally deposited material. So-called "sub-lithographical nanofabrication" yields features down to the sub-10-nm level [197].

4.5.2
Size Reduction by Thermally Induced Reshaping

Lithographic structures in polymer materials can be easily modified by thermal treatment after development. The response of isolated dots of high aspect ratio to thermal treatment is a reduction in height and an increase in diameter [1]. This change in shape is caused by surface-tension-induced local material flow due to the thermally reduced viscosity of the material. The material transport results in a reduction of the surface curvature.

The same effect is observed in the case of isolated holes [198]. Here, it leads to a reduction of feature size due to the surface-tension-induced transport of material into the hole. As a result, a narrowing of the gap between the walls is achieved (Fig. 96). The line width of isolated trenches and the diameter of holes can be considerably reduced in this way.

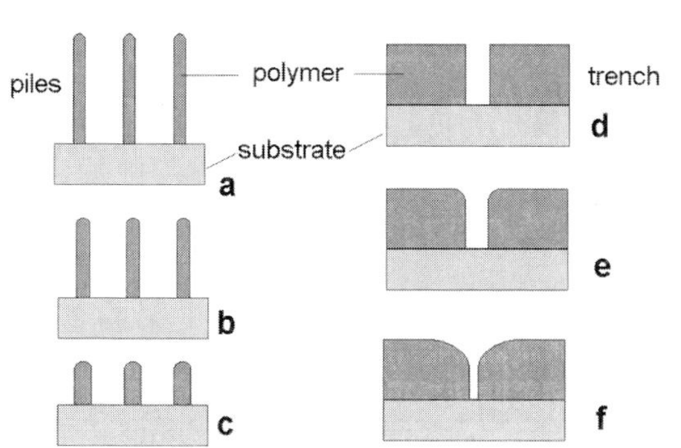

Fig. 96 Generation of nanoholes by thermal flow: reduction of height and increase of diameter during thermal treatment of pillar structures (a – initial state, b – at the start of thermal flow, c – final state after thermal flow); reduction of diameter and height of hole structures (d – initial state, e – at the start of thermal flow, f – final state after thermal flow) [198]

4.5.3
Size Reduction by Sidewall Transfer

The line width of lithographic features can be drastically reduced if side wall depositions at perpendicularly etched micro- or nanostructures are used as masks in subsequent anisotropic etching or deposition processes. This technique transfers the side-wall topology rather than the original structure to the substrate. It is therefore named "sidewall transfer lithography" (STL).

The application of STL to silicon leads to the formation of sub-10-nm structures. The process involves the deposition of a layer of silicon dioxide by a CVD process at the steep sidewalls of preformed anisotropic trenches in the silicon substrate [197].

4.5.4
Formation of Nanodots by Dewetting

It is known that macroscopic droplets can split into smaller droplets due to a reduction of adhesive forces. A subdivision of larger droplets or dots into smaller ones at surfaces can also be induced by lowering the viscosity or the melting of materials. This principle can be extended into the submicron and nanometer ranges and can be used for the preparation of small structures.

The local formation of small metal dots can be achieved by the activation of metal lines that were previously prepared by lithographic means. The local transport of gold and platinum and the rearrangement into small dots by dewetting can be induced by irradiation with high-energy ions [199][200]. For this, 1 MeV Kr^+ ions and 30 keV Ga^+ ions were used. The high local power density activation was achieved by the application of focussed ion beams (FIB). A certain amount of the line material is removed by the sputter effect. The remaining material resolves into droplets below a certain residual line width. The critical line widths for dewetting and droplet formation are in the region of 20–35 nm, depending on the original width of the metal lines.

5
Nanotechnical Structures

5.1.
Nanostructures and Nanomaterials

Nanostructured materials have been known for a long time and have been used for specific applications without any knowledge of the tiny size of the particles involved. For example, colloidal gold (which can be considered as belonging to this broad class of materials) has been used in colored glass since ancient times, but its particular character was only explained in the middle of the 19th century by Michael Faraday.

The term "nanomaterial" is nowadays applied to very different classes of materials, including:

- dry solid nanoparticles and clusters
- dispersions of nanoparticles in liquids ("nanosuspensions" and "colloidal solutions")
- nanocrystalline materials held together by glassy material or embedded in a glassy matrix, such as ceramics or glass ceramics with nanosized phases
- nanocomposite materials including organic as well as inorganic components
- stiff macromolecular or supermolecular aggregates composed of fullerenes, nanorods, nanotubes, or similar rigid molecular objects

Normally not included in the term "nanomaterials" are nanopolycrystalline materials, partially crystalline polymers, and emulsions with nanodroplets. However, such materials fall within the focus of nanotechnology when they are to be used for functional nanostructures or as materials with special properties in the framework of planar technology.

In this book, nanomaterials are only considered as far as they relate to the preparation of particular functional geometries or nanodevices. The specific material properties and selected material combinations are very important for the realization of special device functions. In addition, nanostructuring technologies are strongly dependent on material properties. Therefore, the nanostructuration of different classes of materials is discussed in the following.

Nanotechnology. M. Köhler and W. Fritzsche

ISBN: 978-3-527-31871-1

5.2
Inorganic Solids

5.2.1
Influence of Material Morphology on Nanoscale Pattern Processes

Material properties are determined by the elementary composition and the morphology. The size of standard nanotechnical devices is above that of the atomic scale, so that the elementary composition appears as an integral property of the material.

This is in contrast to the morphology, where for example larger inhomogeneities lead to problems even on the macroscopic scale. Apart from such failures, most bulk materials exhibit inherent morphological structures with dimensions between a few nanometers and several micrometers. These values cover the range relevant to nanotechnology. Crystallites and grains, phase and grain boundaries appear in macroscopic structures as material properties. However, in nano- and sometimes in microstructures, they become discrete features that do not act in an integral manner on the function of the devices, but which influence as individual structures the individual function of an individual element.

The relationship of the material-inherent morphological dimensions to the dimensions of nanotechnical structures is a basic requirement in nanotechnology. From the standpoint of material science, only monocrystalline and ideal glassy materials are not problematic. In other cases, including nano- and partially crystalline, and also crystalline domains, grain and column structures, the effects of subdomains on the individual device have to be considered.

5.2.2
Inorganic Dielectrics

Microtechnology utilizes a wide range of inorganic materials that includes the majority of elements in quite different compositions. Although nanotechnology relies more on molecular construction than microtechnology, organic compounds are thus important, and also the number of applied inorganic compounds increases.

In addition to conductive materials, which are particularly important for nanoelectronics, crystalline and glassy solid materials are applied as dielectric layers and structural elements. Of specific interest is the application of slat-like materials such as AlF_3 and CaF_2 as mask material in the lower nanometer range. Aluminum fluoride is especially well suited for pattern generation by ion and electron beams [1][2]. Therefore, thin layers of AlF_3 are thermally deposited on a plane substrate. After exposure, the structures are developed by simple rinsing with water into topography. As a result of high resistance towards beam etching processes, such AlF_3 nanostructures are well suited as high-resolution ion beam resists. For very high doses, the material reacts as a self-developing positive resist, which means that the exposed regions are removed by local sputter and evaporation processes. Low exposure by Ga^+ results in a decrease in solubility and therefore in a negative contrast after development with water. Particularly narrow structures are possible through a combination of both effects. It is possible to project edges as lines, similar to solarization in classical photography. Therefore, a broader structure is exposed to high doses, so that the structure is removed due to self-development; but the edges are preserved because they were subjected to lower doses. This region is doted with Ga ions, so that it is preserved in a subsequent selective wet-etching development. This development removes the AlF_3 in the surrounding, unexposed areas, and structures with dimension down to 10–20 nm are preserved [3].

Two-dimensional arrays of ZnO nanostructures have been generated by gas-phase transport deposition (VTD) at 860°C [4]. ZnO piles were generated by the deposition of ZnO catalyzed by gold spots. The gold pattern was generated in situ by the direct deposition of a patterned gold film through the gaps in an ordered array of surface-adsorbed microspheres (Fig. 97). Very long ZnO fibers can be obtained from regular array structures of gold-capped ZnO piles with aspect ratios above 20 and pile diameters in the range 50–140 nm. The formation of c-axial wires or of a-axial belts can be controlled by indium doping. J. S. Lee et al. [5] demonstrated the formation of nanobridges of ZnO over lithographically formed trenches of gap width 4–6 μm.

Particular structures of semiconductor nanomaterials can be achieved by special synthetic processes. The formation of nanosprings in a remarkably high yield has been accomplished in gold-catalyzed synthesis of silica at 350°C. These nanosprings were quite regular in structure. The outer diameter and the repeat length ranged from about 50 to 200 nm [6].

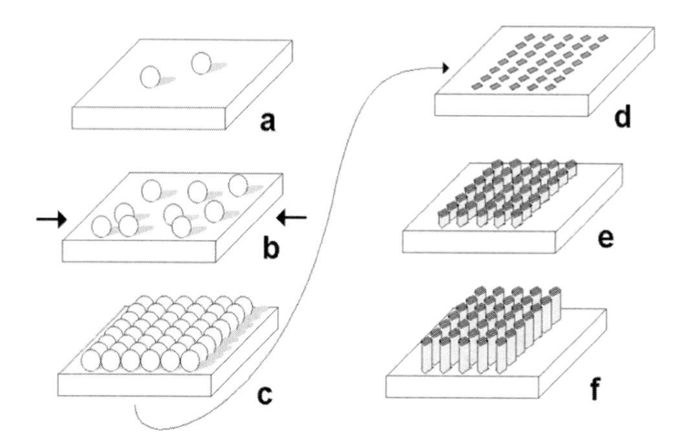

Fig. 97 Formation of arrays of high aspect ratio ZnO nanostructures by means of microsphere assembly: a) adsorption of microspheres at a clean, smooth substrate surface, b) increasing microsphere density by increasing particle adsorption and lateral compression, c) regular microsphere array formed after compression at substrate surface, d) gold deposition from gas phase using the microsphere array as mask; obtained gold spot array after removal of microspheres, e) gold-catalyzed growth of ZnO piles, f) obtained array with high aspect ratio ZnO nanostructures and gold capping [4]

Self-organized local deposition of material on single-crystalline substrates yields structures in the lower nanometer range with the following requirements. The substrate has to present terraces on its surface, and the lattice parameter of substrate and deposited material has to be similar. Silicon substrates etched at a low angle with respect to the (111) plane are thus patterned with regular parallel stripes of CaF_2. Line gratings with widths of 1–15 nm and periodicities of, for example, 20 nm were prepared according to this approach and applied as a mask for the deposition of one-dimensional molecular electronic confinement [7][8].

Also inorganic materials with a tendency towards spontaneous creation of structures at the nano- to micrometer scale are taken into consideration in nanotechnology. Morphologically interesting dielectric materials are layered silicates with helical structures that are created from 0.5–1 μm crystal block screw structures with ca. 4 μm diameter and 5 μm pitch height [9].

5.2.3 Metals

High-resolution structures in metals are usually fabricated through the combination of electron beam techniques and high-resolution etching or lift-off techniques. This lift-off process applies a high-resolution resist – usually PMMA or a related material – onto the substrate prior to exposure and development. Then the functional metal layer is deposited on the resist mask, prior to removal of the mask which includes the covering metal. By 1981, a similar approach had yielded Pd structures of 16 nm width with distances of 53 nm [10]. The combination of electron beam exposure and the lift-off technique is also applicable for the fabrication of freestanding membranes (Fig. 98). Using the lift-off technique, nickel structures of only 20 nm width and a thickness of 35 nm were prepared, resulting in an aspect ratio larger than 1 [11].

Interdigitated structures of titanium electrodes with a line width of 50 nm have been produced by focussed ion beam etching (FIB). For this, titanium was deposited (at a thickness of 40 nm) on a silicon wafer and patterned using Ga^+ ions (30 kV). The electrode structures thus obtained showed very high resistances, reflecting the efficiency of the etching process [12]. Gaps in titanium films of width less than 50 nm could be achieved by anodic oxidation of the side walls of microlithographically patterned titanium after RIE. After application of DNA solution, these structures showed electrical conductivity behavior of Schottky contacts, indicating the potential use of these gaps for the detection of biomolecules [13].

Arrays of Ti structures with 35 nm diameter and 75 nm periodicity were fabricated using a lift-off technique and a fullerene-doped electron beam resist [14]. Weak-link structures of 10 nm broad layer elements of Nb were realized by a combination of electron beam lithography, RIE and lift-off technique for the fabrication of niob S-N-S Josephson contacts [15]. Deposition of copper in prestructured trenches prior to chemical–mechanical polishing (CMP) resulted in 50 nm broad Cu contact pads in an SiO_2 layer [16]. A nanolithographic process (NIL) was utilized for the fabrication of nanostructures in nickel and cobalt, which are interesting as structures for high-

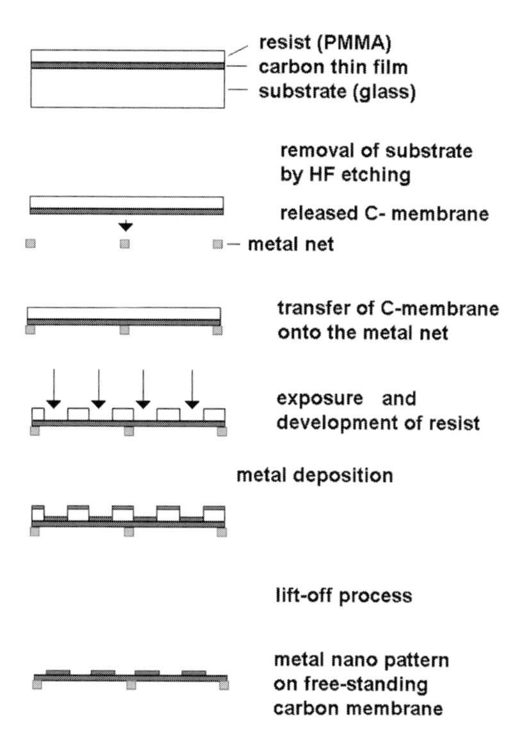

Fig. 98 Fabrication of metal nanostructures on a free-standing carbon membrane (S.P. Beaumont et al. 1981)

density magnetic storage. Using a molded PMMA-mask, holes were structured in SiO_2 by reactive ion etching prior to deposition of the metal either galvanically or by vacuum evaporation. A memory density of up to 30 Gbit in^{-2} (about 5 Gbit cm^{-2}) could be achieved with Co structures of 25 nm $\times$ 75 nm [17].

A line space pattern with a half pitch of 93 nm has been created in iron without the use of a lithographic mask by focussing an atomic beam of iron by radiation with interfering laser light. The deposited line pattern was formed directly by the standing wave pattern projected onto a glass-ceramic substrate [18].

Arrays of parallel nanochannels have been formed by electrochemically supported etching of aluminum under the formation of aluminum oxide at the surface. The nanochannels in alumina thus obtained can be used for the deposition of other materials. In this way, the structure of the alumina channels can be used as a template and transferred to other materials. Arrays of gold structures, for example nanotubes, have been prepared by such application of alumina nanochannels[19][20].

Free-standing nanowires and arrays thereof can be generated by a special pattern-transfer strategy [21]. For this, a semiconductor substrate is first patterned. A substrate with arrays of trenches is then tilted and the wire material is deposited at one sidewall of the lithographically prepared trenches. The material is then transferred onto a silicon substrate by the use of an auxiliary adhesive film. The previously patterned semiconductor material is then removed by selective etching. Air bridges down to less than 10 nm have been obtained after plasma removal of the adhesive film on the silicon substrate.

Additive structure generation in the nanometer range also leads to high aspect ratios. For metals, such structures can be achieved by local galvanic deposition in the windows of a nanolithographic mask. This approach yielded, e. g., nickel columns of 75 nm diameter and 700 nm height, leading to an aspect ratio of 9.3 [22].

Exceptionally small structures are made by scanning tunneling lithography (cf. Section 4.4), such as lines of only 3 nm width and a depth of 0.3 nm in Ag_xSe films [23]. The controlled transfer of Au from a scanning tip onto a gold or platinum substrate yields gold spots of only 3 nm diameter in various patterns [24][25]. Interactions of a W tip with germanium single crystals resulted in particularly small structures of 0.8 nm for both height and width [26]. Structures of Cd with 20 nm width were written by an STM tip in an adsorbate layer of dimethyl cadmium [27][28]. Thin Cr structures down to 20 nm thickness were prepared by anodic oxidation using an AFM tip [29]. These examples show the possibilities for the preparation of nanometer structures in metals. In particular, the fine structures possible through STM are still at an early stage of laboratory development, so that the production of highly integrated metal devices in the lower and medium nanometer range has not yet been achieved.

A coaxial metal nanowire may be fabricated by taking advantage of the molecular self-organization of peptides. For this, a nanotube assembled from peptide molecules was used as a scaffold for the deposition of metal nanoparticles on both the outer and inner parts of the tube [30].

5.2.4
Semiconductors

Because of the significance of nanotechnology in optoelectronics and the fabrication of quantum devices, nanolithography of semiconductors (especially of compound semiconductors) and stacks of the layers is of great importance. Very narrow structures have been fabricated through a combination of electron beam lithography with reactive ion etching [31] and ion beam etching or lift-off techniques, yielding, for example, GaAs lines of 10 nm [32]. Silicon structures with high aspect ratios (up to 7) were fabricated by masking Si surfaces with gold nanoparticles and subsequent reactive ion etching in an $SiCl_4$ plasma. Using 15 nm diameter particles, an Si column with a height of up to 100 nm could be achieved. Particles as small as 2 nm were still efficient etching masks [33].

Si needles with extremely high aspect ratios and very small tip diameters were fabricated by a combination of the so-called vapor liquid solid (VLS) techniques with ion beam etching. Initially, silicon needles several micrometers in length but only a few micrometers in diameter were created by catalytic deposition on Si from an $SiCl_4/H_2$ gas mixture. In a second step, these needles were sharpened using Ar or N ions at 30 keV, resulting in tip radii as low as 2 nm [34]. Comparable values were realized by deposition of gold particles on single silicon crystals prior to catalytic silicon wafer growth [35].

Silicon has been subjected to a variety of scanning probe nanolithography experiments. The preferred approach was the oxidation of Si–H surfaces or the direct oxidation of Si into the oxide for local structure fabrication [36]. Another approach applies the direct extraction or deposition of silicon atoms [37]. The width of STM-based, anodically oxidized lines in silicon increases with voltage, tip current or exposure time [38].

Small line widths in silicon and nickel silicide have been achieved by so-called sidewall transfer lithography (STL). The initially formed silicon structures obtained after RIE lithography had line widths of about 20 nm. These structures were covered by a thin film of nickel. Thermally induced silicidation led to the formation of Ni_2Si structures with line widths between 25 and 37 nm [39].

A gate length in silicon of only 12 nm has been achieved by using EBL in the fabrication of a MOSFET. The beam was used to expose a silicon polymer as a resist material [40]. TMAH was used as a developer for the silicon-containing resist material. The precise pattern transfer was accomplished by plasma processes using selected compositions of halogen-containing etch gases.

An aspect ratio of about ten was realized in the case of polysilicon nanostructures with feature sizes of just 20 nm. The structures were obtained after EBL and plasma etching at 2 mTorr using an optimized mixture of fluorinated gases [41].

Similar to metals, nanostructured semiconductor materials are of interest for the fabrication of single electron tunneling devices. Thus free-standing nanobars of highly-doped silicon with a length of 800 nm and a cross-section of 24 nm × 80 nm have been fabricated [42]. A combination of EBL, RIE and lift-off resulted in 7 nm × 7 nm gate structures. These dimensions are of interest to single electron memory devices at room temperature [43].

Exceptionally fine structures were fabricated in GaAs through implantation using Ga^+ ions. Highly focused exposure produced lines of only 2–3 nm width in a GaAs surface, prior to STM visualization of these lines based on their increased electrical conductivity [44]. For the fabrication of nanometer field effect transistors, a WSi_x/Si system with an extremely thin SiO_2 gate film was used in a combination of electron beam direct exposure with etching in a high-density plasma, resulting in a gate stack with dimensions of 60 nm and 170 nm for lateral extension and height, respectively [45].

Sub-10-nm structures of GaAs have been generated by NIL. The mold was fabricated by EBL and RIE in SiO_2 using a polystyrene mask layer. Imprinting was executed by pressing the SiO_2 mask into a resist on a PMMA sublayer on the GaAs substrate. In this way, structures down to 6 nm line width could be transferred [46]. High aspect ratios of GaP structures have been achieved in pillars deposited by laser ablation and epitaxial growth on silicon [47].

Tin dioxide films are of particular interest due to the high sensitivity of their electrical conductivity to interactions with various gases at elevated temperatures. Sub-100-nm line-space patterns have been fabricated by a standard photolithographic RIE process with a lift-off step. The structures obtained showed concentration-dependent resistances for various gases and vapors (CO, ethanol, acetone) in the 10–100 ppm range [48].

Nanometer structures of semiconductor and metal materials that exhibit different etching behavior to the pure elements can be achieved by in situ lithography (which means without resists) for compound formation with local activation. This can be achieved, for example, by interdiffusion and reaction of platinum and silicon by local activation utilizing a focused ion beam. So platinum silicide structures of 50 nm width were prepared by applying selective plasma etching or wet etching to the separation from the substrate material [49].

Very small Si structures embedded in silicon oxide can be fabricated on chips starting with small Si structures such as stamps, etc., and their oxidation. The size of the remaining Si-core is controlled by the extent and the parameters of the oxidation. Based on starting structures of 10–45 nm diameter, embedded Si cores of 5–30 nm have been demonstrated [50].

5.3 Carbon Nanostructures

The high variability in the topology of interconnections between carbon atoms in larger molecules makes such systems interesting for the generation of nanostructures of lithographic quality, if the structures are rigid and possess a well defined geometry. They include Buckminster-fullerenes and carbon nanotubes. While the regular, spherical Buckminster-fullerenes (e. g., C_{60}) exhibit properties of individual single molecules with dimensions in the lower nanometer range, the carbon nanotubes represent a class of nanoobjects that combine properties of molecules with those of solid substrates, and cover a range of dimensions from several micrometers (length) down to a few nanometers (diameter). Therefore, carbon nanotubes exhibit properties of both the mesoscopic world, such as controlled single electron transport [51], with access to manipulations typical for micro objects. The fabrication of larger quantities of carbon nanotubes has been demonstrated [52].

Carbon nanotubes (CNTs) can be deposited on metal films and used as masks for metal-film nanopatterning by sputter or ion etching. In this way, metal nanowires with line widths of less than 20 nm can be generated without the application of any resist technology [53].

Carbon nanotubes are often complex structures. Multi-walled tubes reach diameters above 100 nm with hierarchies of several orders of helical structures. Beside continuous walls, spiral cross-sections are also observed [54]. Such tubes with a certain rigidity and length are interesting objects for nanomechanics. One application is their use as ultrathin tips for scanning force microscopy that can be enhanced by subsequent chemical modification for chemical affinity scanning force microscopy [55][56].

Carbon nanotubes are under investigation for possible nanoelectronic applications, where the key factor is the mobility of electrons in the planes of conjugated π-bonds (from C sp^2 orbitals). The nanotubes themselves act as “quantum wires” if single charges are introduced into them. Links between nanotubes, kinks, substitutions, and other local changes in the molecular bond network lead to modifications in local electron mobility. The transport of charges can thus be controlled by the topology of π-bonds, substituents, tunneling barriers, and so on, and specific nanoelectronic networks can be constructed. Changes in the local conductivity in CNTs have been demonstrated, for example, by P. G. Collins et al. [57] by means of STM measurements.

Carbon nanotubes can be arranged in different positions at plane substrates. Densely packed, highly-ordered parallel CNTs have been arranged by means of acetylene pyrolysis on cobalt within the nanochannels of nanochannel alumina (NCA) at 650°C [58]. A high-quality array of densely packed MWCNTs has been achieved on silicon surfaces using photolithographically patterned iron catalyst structures for the generation of CNT bundles [59]. Bundles of CNTs can also be selectively generated at the tips of anisotropically etched single-crystalline silicon structures [60]. A regular lattice-like arrangement of single CNTs oriented perpendicularly to the substrate plane in extended two-dimensional arrays has been obtained by means of NIL (S. M. C. Vieira et al. [61]). Such arrays can serve as field emission devices with high electron density.

A precise electrical characterization of single carbon nanotubes (CNTs) has been achieved following their deposition on four lithographically patterned thin-film electrodes. T.-Y. Choi et al. [62] determined the electrical resistance of a single MWCNT and the temperature dependence thereof by means of a four-point measurement. The measured resistance was of the order of 9.3 kΩ and the T_K amounted to about $1.6 \times 10^{-3}\ K^{-1}$.

An important consideration in CNT applications in planar technology is the exact positioning of the nanotubes. The arrangement of CNTs at planar surfaces can be directed by classical lithographic techniques as well as by soft lithography [63]. The particular conditions for formation can be used as a direct means for the positioning of CNTs. CNTs are normally formed by a high-temperature CVD process from organic vapors using transition metals such as iron, cobalt, and/or molybdenum as seed catalysts. In this technique, the well localized generation of metallic catalyst dots defines the positions of the subsequently generated CNTs. The placement of CNTs can also be supported by lithographically defined affinity spots. Metal contact areas preferentially bind functionalized CNTs, whereas the same CNTs have no affinity for SiO_2 surfaces [64].

A. Nojeh et al. [65] were able to show that CNTs can be assembled to form nanoconnections between lithographically prepared thin-film molybdenum electrodes by electric-field-directed lateral growth. Without the application of a voltage, circularly grown CNTs give rise to a quadrupolar thin-film arrangement. However, CNT connections between the thin-film electrodes were found after CNT formation in a metal-catalyzed synthesis from methane and ethylene at a maximum field strength of 1 V μm^{-1} (voltages of 1–5 V).

The integration of metal nanostructures, inorganic semiconductor nanostructures, and single CNTs is of particular interest in relation to nanoelectronic and nanooptoelectronic devices. Carbon nanotubes can be directly coupled with metals by combining the CVD synthesis of CNTs with electrochemical deposition. In this way, Ag/Si and Pt_6Si_5/Si nanostructures have been directly coupled with CNTs. These heterojunctions can be formed in perpendicular orientation to the substrate plane and arranged into arrays [66].

The conversion of discrete CNTs into complex networks of nanotubes is of interest with regard to the formation of three-dimensional wires and of nanoscale, complex topologies of coupled nanoelectronic functions. In principle, the impact of high-energy particles (40 keV Si ion beams, for example) on assemblies of CNTs can cause the formation of tube connections and the appearance of nanotube networks [67].

Fullerenes are not only interesting as individual nanoobjects, but also as material for structures in the medium nanometer range. So regular structures in this dimension can be realized by a self-organization process during the thermal activation of fullerene crystals. Single crystals of C_{60} exhibit a photon-induced surface reconstruction that leads to periodic structures with dimensions of 30–40 nm [68].

5.4
Organic Solids and Layer Structures

5.4.1
Solids Composed of Smaller Molecules

Solids that are made from smaller molecules usually exhibit a regular internal arrangement and a fairly low melting point. Size and structure of the molecule, but also purity of the material and the preparation method, and in particular the speed of transformation from the mobile into the solid phase determine whether the material is fairly amorphous or crystalline. Pure materials of small molecules that slowly solidify from the melt or are gradually deposited from solution often result in crystals. On the other hand, larger molecules, mixtures of materials and fast cooling from the melt usually yield amorphous solids.

Organic molecules with low vapor pressure can also be transferred into glassy layers, e. g., by sublimation or deposition from solution. Mixtures of small molecules and polymers can be used to fabricate stable amorphous layers from solutions. Thin calixaren layers were applied to generate lines with widths of 10 nm using focused electron beam writing [69].

5.4.2
Organic Monolayer and Multilayer Stacks

Organic molecular monolayers and multilayer stacks on solid substrates are accessible lithographic methods for the generation of lateral nanostructures. SAMs as well as LB films are suitable substrates (cf. Sections 5.3.1 and 5.3.2). Nanostructures are fabricated either by direct writing (in the case of layers susceptible to electron, ion and X-ray radiation or where they can be manipulated by scanning probe techniques) or by the use of a transfer mask. Monolayers of alkyl thiols on gold substrates are especially suitable for the direct fabrication by scanning probes. STM-based nanofabrication has been applied to prepare holes of 10 nm diameter [70].

The use of LB films as high-resolution electron beam resists has been investigated. Adequate sensitivity and high resolution were found. However, the high defect density of such layers hampers routine application. SAMs exhibit fewer defects. They are applicable as resists when the functional layer supports the creation of a defect-free monolayer, such

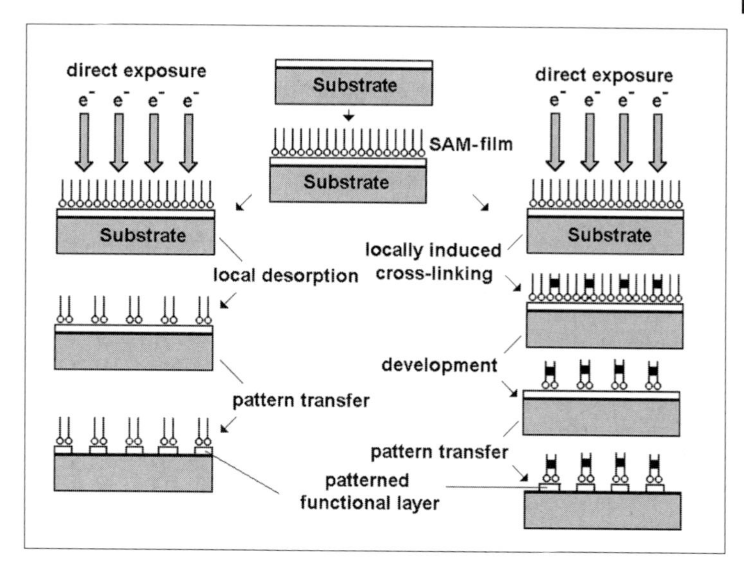

Fig. 99 Application of molecular monolayer as an electron beam resist. Positive (left) and negative (right) process

as in the case of octadecyl monolayers on GaAs substrates. Exposure of these monolayers to an electron beam (50 keV, positive resist mode) prior to transfer of the written structure by wet-etching in an $NH_4OH/H_2O_2/H_2O$ mixture into the semiconductor material yielded GaAs structures of 50 nm width [71]. For lower electron doses, an SAM can be applied as a negative resist. The electron beam induces a cross-linking of molecules inside the monolayer in the case of 1,1-biphenyl-4-thiol with 50 eV [72].

Electron beam exposure of resist monolayers allows both modes of operations: a positive process occurs in the case of direct desorption of the resist material under the electron beam (Fig. 99, left). A negative process results when adjacent molecules of the monolayer cross-link under the influence of the electron beam. The exposed regions of the layer are thereby stabilized and stay on the substrate, even during the subsequent removal of the unexposed regions of the layer in an etching step (Fig. 99, right).

LB as well as SAM layers are susceptible to interactions with electrons from a tunneling tip, and can be locally modified or desorbed. This procedure yielded lines of 20 nm width in an SAM of the benzyl chloride based material CMPTS [73]. Structures of 10 nm were the result of STM-based modification of docosanic acid. For a molecular length of about 3 nm in such an LB layer, the structure constructed represents an aspect ratio of 0.3 [74].

5.4.3
Synthetic Organic Polymers

Polymers represent an important technical construction material, and also play a significant role in microtechnology. They are a key matrix material for lithographic resists in micro- and nanotechnology. Radiation-sensitive polymers on a polyacrylate base are the preferred resists in electron beam, ion beam and X-ray lithography. The Novolaks, which are applied as resins in positive photoresists, consist of relatively short molecular chains and exhibit only slight swelling during development. This fact and the susceptibility to electron beams means they are suitable for nanostructure fabrication. The minimum structure size correlates with the molecular weight of the polymeric resist. For a polystyrene resist, weights of 17 500 Da yielded 21 nm structures, but with short-chain polymers with molecular weights of about 1100 Da, structures down to 11 nm were feasible. The structure width increase due to larger doses correlate with the molecular weights of the resists [75].

In electron beam lithography, the best lithographic resolution is achieved with PMMA or other such beam resists. Structure widths between 50 and 100 nm are standard, but dimensions below 10 nm are accessible. Scanning probe techniques can also be applied for the nanofabrication of organic polymers. Thus low-energy electrons applied by a scanning tip resulted in line structures of 20 nm width in a urethane diacetylene polymer [76].

Structures in the sub-micrometer and the medium nanometer range can be transferred by oxygen plasmas in virtually all types of organic polymer layers by using a structured transfer layer that is resistant to an oxygen plasma. A variety of organic polymers were studied with respect to their suitability in a three layer resist system when applying transfer layers of Si, SiO_2, Si_3N_4 or metals (preferably Cr and Ti) [77][78][79].

Even molecular monolayers of polymers can be applied as resists. To demonstrate this alkyl and oligoglucol-substituted polythiophenes were spread in a Langmuir trough, compressed and collapsed. The resulting polymer filaments were transferred onto a solid substrate. They were electrically conductive due to their conjugated π-electron systems. Dimensions of 60 nm and 15 nm for width and height, respectively, were measured using AFM [80].

Self-assembly processes in polymers can also probably be used for the generation of nanostructures. Block polymers are suitable candidates for this approach. Of special interest are highly ordered structures with electron-conduction islands embedded into a non-conductive matrix. Such structures of 12 nm length could be demonstrated in a PMMA-polyparaphenylene (PPP) block polymer [81].

Nanoobjects of organic polymers are also possible by a so-called template synthesis. This approach uses existing cavities with nanometer dimensions for polymerization. Tube-like objects are accessible through polymerization inside the cylindrical pores. Hence nuclear track membranes were utilized as a polymerization matrix in order to fabricate nanotubes of polyheterocycles with about 0.5 μm diameter and wall thicknesses of about 0.1 μm [82].

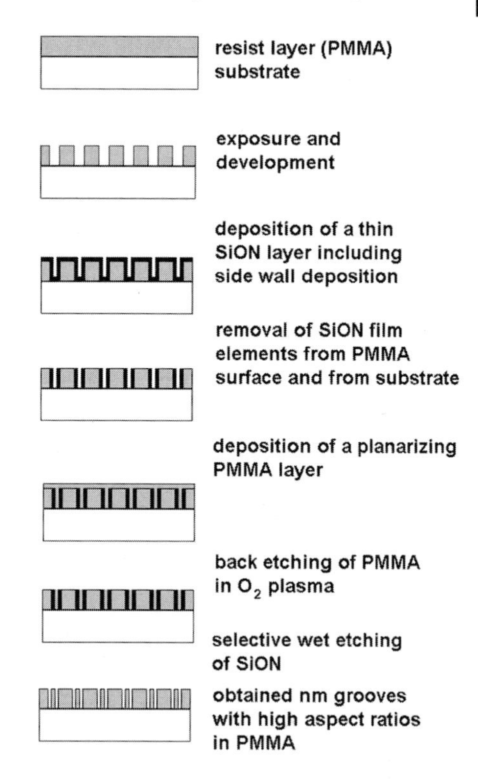

Fig. 100 Fabrication of nanostructures with extremely high aspect ratios by side-wall deposition, planarization, back-etching and selective etching

Inorganic assisting materials are also applied to nanostructure fabrication in planar technology. Thus ultrathin inorganic wall structures in microstructured thin layers were transferred by overlay techniques, back etching and subsequent selective etching into trench structures with dimensions in the lower nanometer range (Fig. 100) [83].

Recent years have witnessed a growing interest of micro- and nanotechnology into the physical–functional properties of polymer molecules. So polymers with conjugated double bonds provide the basis for organic conductors, semiconductors or even nanowire applications. Optoelectronic applications are envisioned for, e. g., polymers exhibiting electroluminescence behavior. Electrooptically active polymers can be applied in organic photodiodes or luminescence diodes [84][85][86][87][88].

5.4.4 Biopolymers

Besides synthetic organic polymers, films of biopolymers are also in the focus of interest of biosensoric applications, molecular nanotechnology and biochips [89][90]. In principle, such biogenic materials can be manipulated (in analogy with synthetic materials) with scanning probes or beam lithography. Biopolymers are usually more sensitive with respect to their composition and structure, compared with synthetic polymers. So the removal rates in reactive plasma processes for thin carbohydrate layers (e. g., cellulose, starch) are higher than the rates for aliphatic polymers containing less oxygen, and clearly higher than the rates for aromatic polymers. Protein layer in particular are highly sensitive with respect to thermal and chemical effects, resulting in damage in lithographic processes. Therefore, protein structures are preferably prepared by patterning a binding layer on the surface prior to adsorption of the protein on this structure. Hence the adverse effects of lithographic processes on the protein layer are avoided.

Biopolymers and analogous synthetic structures, such as polypeptides, exhibit three advantages which make them very interesting for future technical applications: such molecules are created by nature for the interaction with functional supramolecular systems, so that they often exhibit the capacity for the spontaneous generation of predetermined three-dimensional geometries [91][92]. This capacity is encoded in the binding topology of such molecules. Molecules with an inherent self-organization potential will play an increasing role in future technical solutions.

On the other hand, a whole toolbox of highly specific working enzymes for the manipulation of these macromolecules exists. An additional factor is the possibility of degrading such materials biologically and without toxic byproducts, so that their application would ease the growing waste disposal problem.

5.5 Molecular Monolayer and Layer Architectures

5.5.1 Langmuir–Blodgett Films

Amphiphilic molecules can be enriched, oriented and compressed on a liquid–gas interface (cf. Section 5.4.6). This can be achieved in a Langmuir trough by spreading the molecules on a fluid surface and compressing the surface area with a movable mechanical barrier (Fig. 101). A compression of the molecular film results in a measurable pressure. This pressure increases with decreasing area until a condensed monolayer is generated. Then, only the onset of multilayer generation leads to a pressure jump [93].

In the state of the compressed monolayer, a lipophilic interaction between aliphatic residues leads to a stabilizing influence. Thus the generated film can be transferred onto a solid substrate. Therefore, the substrate has to be inserted before film generation in the trough, and must be carefully removed afterwards. For a reversed orientation of the molecules with respect to the substrate, it must be inserted in the trough after the monolayer formation.

This simple approach allows the transfer of molecular monolayers of nearly all amphiphilic substances onto solid substrates. Repeated insertion and retraction of the substrate yield multilayer arrangements (Fig. 102).

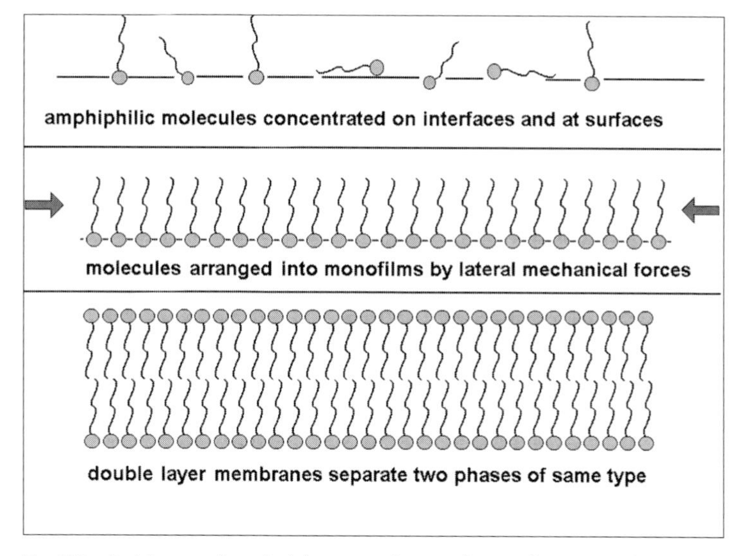

Fig. 101 Enrichment of amphiphiles in interfaces and on surfaces (top), formation of continuous monolayer of amphiphilic molecules on fluid surfaces (below) and separation of similar phases by double layer formation (bottom)

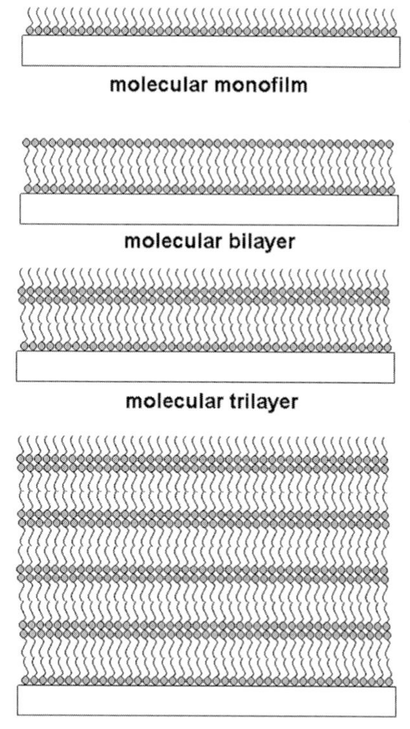

Fig. 102 Langmuir–Blodgett films on solid substrates. A single passage of the substrate through the phase interface generates a monolayer (top); repeated passages result in multilayer formation with alternating orientations

A major problem of Langmuir–Blodgett films is the existence of a significant number of failures (holes). So these films are suited for proof of principle studies, but not yet for integrated systems.

5.5.2
Self-assembled Surface Films

Several amphiphilic substances also form ordered layers on solid substrates. A prerequisite for such behavior is an adsorption of these molecules onto the surface, an intramolecular mobility and intermolecular stabilizing interactions. Typical examples are the n-aliphatic tail-groups with medium chain length (8–30 carbon atoms) connected to a hydrophilic polar – or easily polarizable – head group that can react with the substrate surface, and with end-groups that do not react with the substrate. In the case of head-groups bound on the surface in a high density, the end-groups stabilize each other by van der Waals interactions. So a cooperative effect of layer stabilization can be observed, which results in two-dimensional highly ordered monolayers. These arrangements are denoted as self-assembled monolayers (SAM). In addition to aliphatic SAMs, SAMs with aryl groups have also been described (Fig. 103).

The formation of a compact monolayer occurs spontaneously. This is different from LB films on liquid surfaces, which require compression for the generation of a monolayer. Organic thiols, sulfides and disulfides, phosphonic acid, phosphines and isocyanates exhibit the ability to generate SAMs. Suitable substrates are relatively noble metals with easily polarizable shells (metals) that form poorly dissolvable sulfides, such as Au, Ag, Cu, Pd, GaAs, InP, or Pt [94].

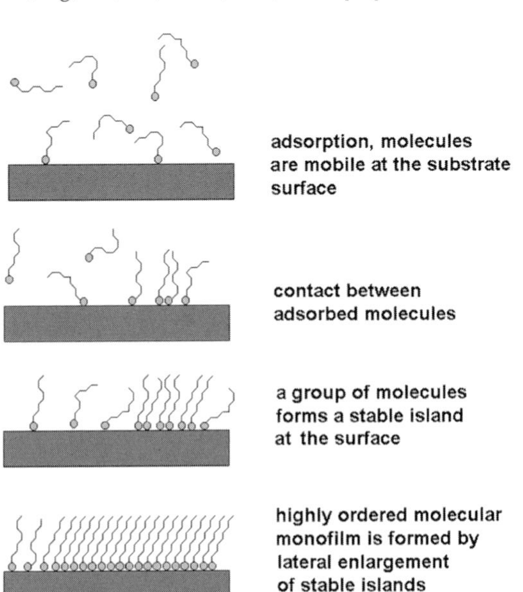

Fig. 103 Formation of self-assembling monolayers. Continuous monomolecular layers are generated by chemisorption of molecules that are mobile on the surface but exhibit intermolecular stabilizing properties

The property of self-assembly is not limited to gold (or other metal) surfaces and thiols. However, noble metals are preferable, because under ambient conditions they exhibit clean metal surfaces. This is in contrast to less noble metals that are usually covered with oxide films and therefore prohibit the direct interaction of the head-group with the metal surface. For such cases, reactions of thiol vapor with a cleaned surface in a water-free atmosphere can be utilized. In principle, other functional groups are also able to form SAMs, when the head-group reacts with the solid substrates, and the adjacent chains interact stabilizingly, but not immobilizingly. Direct covalent coupling does not provide the required remaining mobility in the case of hydroxyl or amino groups, but in the case of hydrogen binding.

Although ultrathin, SAM films are usually sufficiently dense to act as etch masks for wet etching processes of underlying metal layers. A straightforward approach is the formation of such a layer on gold using alkyl thiols with chain lengths of between a few carbon atoms (propane thiol) and several tens carbon atoms. The C-18 thiol (octadecyl thiol) is often preferred. Besides the n-alkyl thiols, also branched thiols or multiple functionalized thiols are utilized.

5.5.3 Binding of Molecules on Solid Substrate Surfaces

The fabrication of ordered dense molecular films on substrate surfaces requires self-assembly principles, such as in the case of SAMs or LB films. Many nanotechnical applications need only a functionalized layer with a certain density of functional groups, with the requirement of high order or exact thickness. The layers should be free from larger holes, so that the level of statistical failures in the distribution of the molecules should be significantly below the smallest lateral dimension of the layer.

When there is a sufficient density of binding functional groups on the surface, dense molecular films form without ordered layer generation. The density depends on the density of binding surface groups on the one hand and the reaction probability on the other. The latter is not only influenced by the reactivity of the reaction partners in their original state, but is strongly affected by neighborhood effects such as screening. Similar to SAM, additional lateral interactions between the immobilized molecules result in additional stabilization of the layer.

The formation of a monolayer or sub-monolayer requires that the first layer saturates all free valences on the substrate surface and generates no additional free valences for binding of molecules from the liquid phase. Therefore, the film-forming molecules should not react with each other, otherwise films of several molecular layers with local inhomogeneous thickness distribution occur. Similar to solid phase coupled synthesizes, a defined multi-step formation of a molecular monolayer requires a protection of secondary functional groups in the first assembly step prior to a subsequent activation.

In a wide variety of applied materials, Si, SiO_2 and glass play the leading roles as substrates and functional materials in micro- and nanotechnology. Apart from their availability and their well-established technology, in particular their outstanding physical and chemical properties contribute to this position. Therefore, binding of molecules on these substrates is of particular interest.

The Si atoms are in a tetrahedral arrangement in both SiO_2 and glass. This geometry is disturbed on the surface, and can be rearranged either by double bonds between Si and O, or the second free valence of an Si–O bond is occupied by another element. Under ambient conditions and after reactions in aqueous solutions, the surfaces usually exhibit a certain level of hydrogen, resulting in OH groups. Additionally, metal atoms can be found, originating from the glass or from the process media.

For elementary silicon, the surface is always covered by a thin film of oxide and so (under humid conditions) O= and OH groups are predominant. After wet etching steps and plasma treatments, the surface valences are also filled by hydrogen (Si–H) or other elements from the process media (e. g., Si–F, Si–Cl).

The Si–OH groups represent a suitable platform for covalent surface coupling approaches, because the Si–O bond exhibits a stable interaction with the solid and – at the same time – the –OH group represents a reactive binding group. When there is an insufficient density of Si–OH groups on the surface, an activation step has to be applied. Also Si surfaces can be functionalized to contain a high density of binding groups by such an activation procedure. The efficiency of this procedure can be monitored by measurement of the wetting angle of water using contact angle measurements. A small angle points to a high density of OH groups.

There are several approaches to the coupling of molecules on Si–OH groups on the surface. It is favorable to include the formation of Si–O bonds in the coupling scheme, because such binding reactions show a high efficiency and yield a stable bond. Alkylhalogenide silanes or alkylalkoxy silanes (in particular the three-fold functionalized silanes) are preferred:

$$\text{R-Si}(-\text{X})_3 + 3(\equiv\text{Si}-\text{OH}) \rightarrow \text{R}-\text{Si}(-\text{O}-\text{Si}\equiv)_3 + \text{HX} \qquad (5.1)$$

Or

$$\text{R}-\text{Si}(-\text{OR}')_3 + 3(\equiv\text{Si}-\text{OH}) \rightarrow \text{R}-\text{Si}(-\text{O}-\text{Si}\equiv)_3 + 3\text{R}'\text{OH} \qquad (5.2)$$

In both cases water acts cases as an efficient competitor of the surface OH groups, this therefore demands that water and air are excluded for high binding group densities.

In analogy to Si, SiO_2 and water, also metals that form stable bonds to oxygen (such as highly passivating metals) can be modified via surface OH groups with organic monolayers. This applies especially to metals that exhibit passivating layers under ambient or aqueous conditions, e. g., the following metals that are important for micro- and nanotechnology: Ti, Cr, Ni, Nb, Mo and Al.

Sometimes, metal oxide surfaces act as substrates for SAMs, e. g., by binding of carboxylic acids, hydroxyl amides or phosphonic acids on zirconium oxide or indium tin oxide (ITO) surfaces [94].

Surfaces of organic polymers show the whole repertoire of organic coupling chemistry. So a variety of approaches exist for the coupling of molecular monolayers onto such surfaces. Owing to the high stability of the C–O bond, but also because of technical reasons, coupling via oxygen bridges is a convenient approach.

Some micro- and nanotechnical materials already contain inherent OH groups that are usable for the coupling: the resins of positive photo resists with phenolic OH groups, or the poly(vinyl alcohols) with aliphatic OH groups. In analogy with inorganic coupling chemistry, silicon chemistry can yield molecular monolayers. A technique that has already been applied for adhesion enhancement and in dry-developed resists replaces the hydroxyl hydrogen by alkyl silyl groups, e. g., by reaction with hexamethyldisilazane (HMDS):

$$2R{-}OH + [(CH_3)_3Si]_2NH \rightarrow 2R{-}O{-}Si(CH_3)_3 + NH_3 \qquad (5.3)$$

However, both longer and secondary functionalized organic molecules can also be addressed by this approach:

$$R{-}Si{-}OR' + 3({-}R{-}OH) \rightarrow R{-}Si({-}O{-}R{-})_3 + 3R'OH \qquad (5.4)$$

Polymer surfaces without OH groups can be activated by an oxygen plasma or by oxidizing baths, so that the coupling approach can be universally applied. In analogy to the natural importance of peptides, the peptide bond (which is the coupling of carboxyl with amino groups) plays an important role in the immobilization of organic molecules. However, a variety of other approaches also exits.

5.5.4 Secondary Coupling of Molecular Monolayers

Molecular functions required on surfaces are often not generated in just one step. Frequently further construction steps are required after the primary coupling. Such a system of reactions can be required for several reasons:

- due to the synthesis strategy, which means as a result of the compatibility of the functional groups
- the introduction of spacers
- special layer geometries
- integration of guest molecules into host layers
- control of binding density in the case of sub-monolayer coverage
- construction of local nanoarchitectures
- combinational synthesis on micro- or nanospots

In contrast to polymerization reactions, the creation of layer arrangements are conducted in analogy with solid phase synthesis. Thereby, the number and type of molecules that bind on the surface are defined, and there is no statistical growth. A disadvantage of this approach is that several steps are required to achieve thicker layers. The application of just one, bifunctional molecular unit is not possible. Either a protection group chemistry with alternative introduction of protected groups and subsequent deprotection and activation of these groups in order to realize the next layer, or at least two units with two different (and complementary in only one direction) coupling groups are necessary.

From the point of view of synthetic chemistry, the defined assembly of molecular layered architectures is rather a synthesis of bio-analogue sequence molecules than one of synthetic polymers. A combination of both approaches seems possible, e. g., by a layered immobilization of pre-synthesized oligomers with coupling groups on both ends. So layers with a fairly high density can be realized while preserving the internal architecture in a limited number of synthetic steps.

5.5.5 Categories of Molecular Layers

Monolayers can be stabilized by the formation of strong bonds between adjacent molecules. Cross-linking is possible in the case of layers of olefins or condensable groups, and is usually induced by thermal or photochemical activation after layer deposition.

Such arrangements differ from polymer layers of similar thickness generated by spin-on or casting in their significantly increased regularity of the binding topology. These architectures are of particular importance, when functional molecules are integrated in a directed fashion into layers, when layers are subjected to nanos-

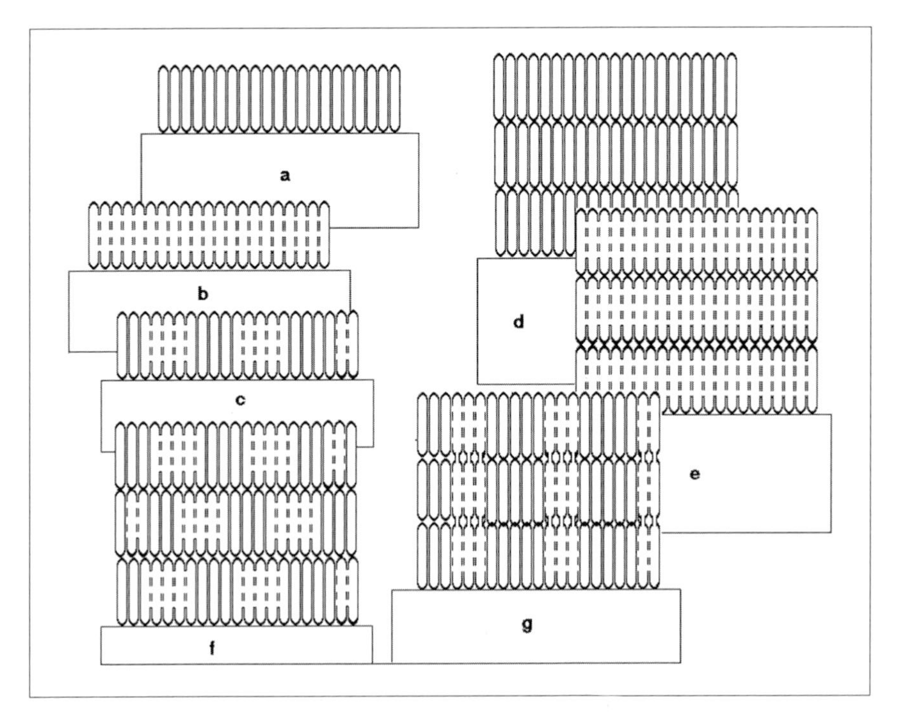

Fig. 104 Scheme for the interconnections in molecular layers

tructuring processes, or when molecular nanoarchitectures are constructed. Layer architectures are interesting options for the control of optical-anisotropic layer properties or specific optoelectric characteristics.

Independent of their chemical composition, some basic categories of molecular monolayers and layer stacks can be differentiated due to their layer geometry and binding topology. For a system based on the strength of bonds in the x-, y- and z-directions, the following categories can be distinguished (Fig. 104):

a) Assembled homogeneous monolayer
Strong, usually covalent bonds perpendicular to the substrate plane, weak bonds of the molecules in the lateral direction
b) Cross-linked homogeneous monolayer
Layers with a cross-linked structure, which implies strong bonds in all three directions
c) Assembled heterogeneous multilayer
Strong, usually covalent bonds perpendicular to the substrate; in the lateral direction, areas of both stronger and weaker bonds are found
d) Assembled homogeneous multilayer
Stack of layers with strong bonds perpendicular to the substrate plane inside the layers but only weak interactions between layers or in the lateral direction
e) Cross-linked homogeneous multilayer
Stack of layers with each layer exhibiting strong bonds in all three directions, but only weak bonds between the layers
f) Assembled heterogeneous multilayer
Stack of layers with alternating areas with weak and strong bonds of the layer generating molecules in the lateral direction, and with weak interactions between the layers
g) Partial cross-linked heterogeneous multilayer
Stack of layers with alternating areas of weaker and of stronger bonds of the layer-generating molecules in the lateral direction, and alternating areas of weaker and of stronger bonds between the individual layers

Besides the complete layers and stack of layers, layers with synthetic surface relief are also of interest in nanotechnology. A relief is formed when the surface coverage is changed in a stack of layers or in a series of coupling reactions. Quasi relief structures can be generated or removed by a change of the process media, through a certain surface mobility of parts of a molecular organized surface in combination with a fixed connection to the solid substrate.

The generation of a molecular monolayer often happens in a series of synthesis steps. Even spacers are sometimes constructed out of units instead of being introduced as pre-synthesized units. Every time bi-functional molecules are provided in sufficient density and efficiency, a layer of linear molecules is generated that covers the surface and grows with every synthesis step. The stepwise increase in film thickness by synthetic coupling of molecular units is based on protocols using protecting groups. This behavior is found in classical solid-phase protein synthesis (after Merrifield) and in biochip synthesis [95] [96].

Instead of bi-functionalized units, also molecules with multiple coupling sites are applicable for the establishment of layer architectures. In this case, a branching of the molecular structures on the surface is observed. Depending on the sequential or stochastic coupling steps, either regular layered or dendritic increasing bond topologies are the result (Fig. 105).

Besides chain-like molecules, spherical molecules can also form molecular monolayers and colloids, and also such films can be secondary functionalized. Synthetic and naturally occurring high molecular weight polymers as well as sequential biopolymers create globular shapes. Owing to the space requirement for this folding, such molecules need either to be immobilized at a lower density prior to folding or have to be presynthesized in solution in order to be immobilized only after folding.

In monolayers of globular molecules, a three-fold hierarchy of bonds is observed:

a) strong, usually covalent bonds in the backbone (primary structure) of the molecules and in the coupling to the substrate
b) weaker bonds stabilizing the secondary and tertiary structure of the folded macromolecule
c) weak interactions between the folded macromolecules

The stability of molecular monolayers is determined by the adhesion of the molecules to the substrate and the interactions between the molecules of the layer. If the adhesion is rather low, cross-links between the molecules also stabilize the adhesion due to assistance of the cooperative effect of weak bonds to the surface.

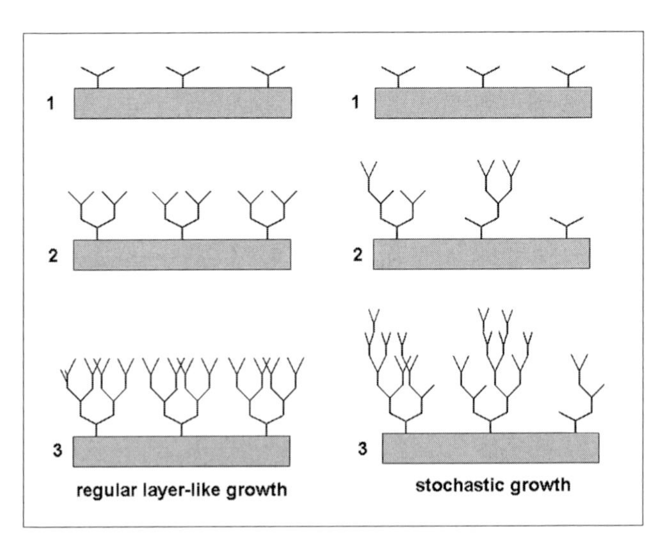

Fig. 105 Generation of immobilized dendrites: stepwise supermolecular synthesis (left), stochastic growth in analogy with polymerization processes (right)

5.5.6
Molecular Coupling Components (Linkers) and Distance Components (Spacers)

Established techniques for the generation of molecular monolayers, layer stacks and single molecule-architectures often apply principles that include molecular coupling components and distance components in addition to the primary coupling group on the substrate and a functional component (Fig. 106). The linker should provide at least two functional units, one able to bind to chemical functions on the surface, the other able to couple to the functional component directly or to a spacer.

Couplings onto OH surfaces are often mediated by methoxy silanes. In particular glycidoxypropyl trimethoxy silane (GOPS) is utilized to functionalize surfaces with epoxy groups. To realize amino-terminated surfaces, amino propyl trimethoxy silane (APTES) is used. To generate such amino surfaces on metals that bind thiols, amino propyl thiol is applied as the linker.

Linker molecules should not react with each other. This non-complementary behavior should be achieved without protection groups, so that linkers differ from the standard units of successive solid phase synthesis, as in combination chemistry. Linkers need a high affinity towards surface groups in order to achieve a complete reaction. The secondary coupling groups should not react with the surface, but with the functional element or a spacer in a highly efficient manner.

The functions of immobilized molecular units are usually determined by their direct molecular environment and by their mobility. Often simple fixation is not sufficient, but a certain density, surface distance and degree of mobility is required. These parameters can be realized by distancing components (spacers) that provide density, length and internal mobility to fulfill the requirements.

Spacers are usually of defined length. Hence polymer molecules are not particularly suitable, so monodisperse oligomers or sequential molecules are applied. They should be long enough to reach through a surface monolayer of inert molecules.

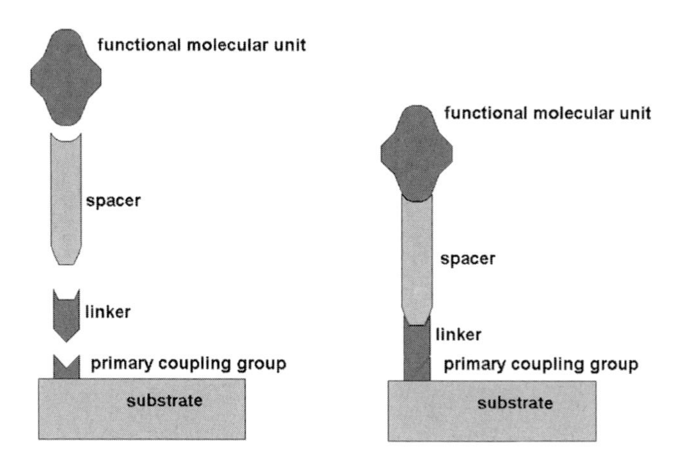

Fig. 106 Typical arrangement for the immobilization of functional molecules on surfaces

Tab. 6 Examples of spacers

Spacer	*Length*	*Internal mobility*	*Chemical stability*
Alkyl chains	up to 3 nm	high	high
Oligoethylene glycol	up to 20 nm	high	high
Oligonucleotides	up to 100 nm	considerable or low	considerable

Apart from the length, rigidity of the spacer is a second important parameter. For many applications, spacers with a certain mobility are required, usually molecules with a chain of bonds that can be freely rotated. This is the case for a simple aliphatic spacer or polyether. Such spacers are usually utilized for solvated functional groups, when the spacer mobility does not lead to unspecific immobilization.

The mobility of spacers is determined to a great extent by their tendency towards solvation. So the application of spacers has to consider the physicochemical conditions.

Oligonucleotides represent spacers of special interest, due to their high solvation in an aqueous environment, because they can be readily obtained through sequential synthesis in a monodisperse form and because of their capacity for internal stabilization due to the choice of base sequences (by loops or double strand formation).

As with a linker, the spacer also requires at least two functional groups to be connected to both the linker and the functional group. Often the coupling function is adjusted to the substrate part, by choice of linker and choice of the coupling group between the spacer and functional component. If possible, the spacer could also be directly coupled to the functional groups of the substrate. Then the spacer takes over the tasks of the linker, so that no additional linker is needed.

5.5.7 Definition of Binding Spots on Solid Substrates

Lithographic methods have not yet been successful in the generation of individual, technically usable functional groups on solid substrates with nanometer precision. However, it is possible to modify small surface areas within the lithographic resolution in order to achieve selective binding of larger molecules only in the predefined areas. The dimensions of the areas determine the precision of the binding. A higher precision is only possible by secondary relative positioning of units, which are connected to predefined positions inside the molecule. The geometric precision of such binding is determined by the specificity of the bonds and the rigidity of the molecule (Fig. 107).

Coupling groups of lithographically defined binding areas are either similar or consist of a statistical mixture of two or more types of binding groups. If only one bond is formed between the surface and the immobilized molecule, the molecule can rotate freely as long as adjacent molecules or the surface topography does not lead to steric hindrance. Two or more bonds secure the molecule against rotations; three bonds will also limit oscillations with respect to the substrate surface (Fig. 108). Thus a coupling is always associated with a break in the symmetry at right angles to the substrate surface.

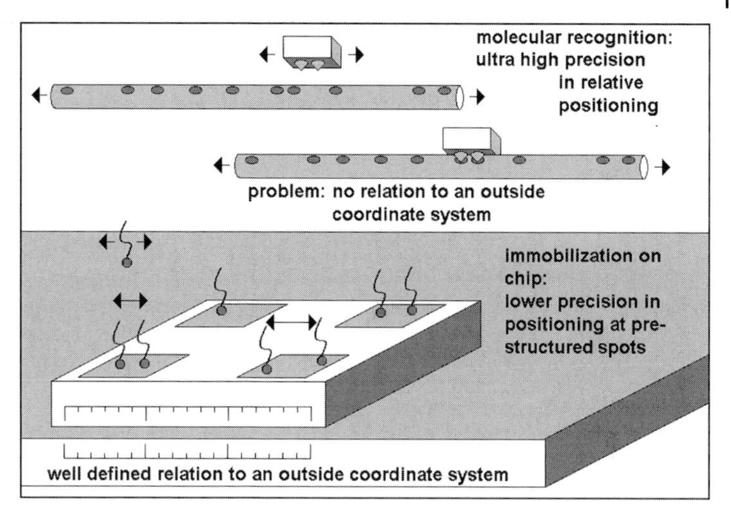

Fig. 107 Comparison of methods for fine positioning. Top: relative fine positioning by molecular recognition on macromolecules with specific sequence. Bottom: absolute positioning by lithographically defined binding spots on chip surfaces

A statistical distribution of the binding groups on the substrate surface preserves random orientations of the immobilized molecules. For regular arrangements of the binding groups on the surface and for their reference lines to be fixed in an external coordination system, only discrete orientations of the immobilized molecules are possible. The number of these orientations is determined by the interchangeability of the coupling groups of the immobilized molecule. Only for a minimum of two coupling groups from the molecule and substrate, each with high geometrical definition, or the formation of a non-rotating bond (e. g., double bond), can a well defined orientation of the coupled molecules be found.

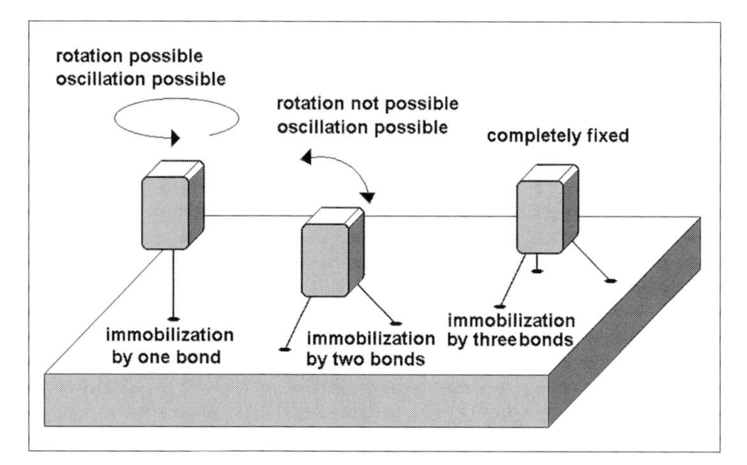

Fig. 108 Degrees of freedom for the mobility of immobilized molecules for single, double or triple coupling to the surface

5.6
Molecular Architectures

5.6.1
Single Molecules as Nanostructures

The potential for self-assembly of organic molecules can be exploited in very different strategies for the construction of functional elements at the nanolevel. In this regard, the correct application of covalent and noncovalent bonds is of particular importance. Molecular self-organization can be of use not only in the formation of new types of nanoparticles, but also in establishing connections between essentially rigid nanoparticles and molecular structures. The use of both large and small amphiphilic molecules, biomolecules, bioanalogous synthetic materials, and, in particular, amphiphilic block copolymers, allows the generation of supermolecular organized structures [97]. The spontaneous arrangement and self-organization of nanoparticles under the influence of molecular interactions can be exploited for the formation of regular grids and regular two- and three-dimensional nanopore structures (see, for example [98][99][100]).

A particular challenge is the integration of these self-organization principles and the use of "soft" matter in relation to nanolithographically produced inorganic solid-state structures. Recently, much research activity has been concerned with the development of nanolithographic techniques for the patterning of soft materials and for supporting self-organized structuring of molecular materials and nanoparticles. On the other hand, self-organizing nanoparticles and molecular materials are used to generate regular patterns similar to nanolithographic structures. For example, polymer nanoparticles may be applied for this purpose [101]. In the future, more efforts will be directed towards the development of functional nanostructures capable of guiding molecular self-organization in relation to nanolithographic structures in order to realize functional networks of nanodevices.

As the dimensions of technical structures approach molecular dimensions, a point can be reached where an individual molecule could be identical to a nanostructure. For today's nanotechnical structures, this point is not yet in the dimension of small molecules (sub-nanometer range). However, there are classes of molecules that provide a link to small lithographic structures by exhibiting dimensions in the medium and lower nanometer range. Moreover, natural chain-like molecules with lengths above 10 μm exceed, at least in this dimension, standard lithographic resolution. One example of such molecules is the DNA of the lambda phage, which has a length of 16 μm in its extended state.

The overlapping of the length of large molecules with minimal lithographic structure widths represents no real problem to nanotechnology (Fig. 109). However, a genuine problem is the precision of the integration of such molecules into lithographic structures. This problem can be considered from two aspects:

a) the precision of lithographic structures on the solid substrate surface and
b) the precision with respect to the position of molecular binding.

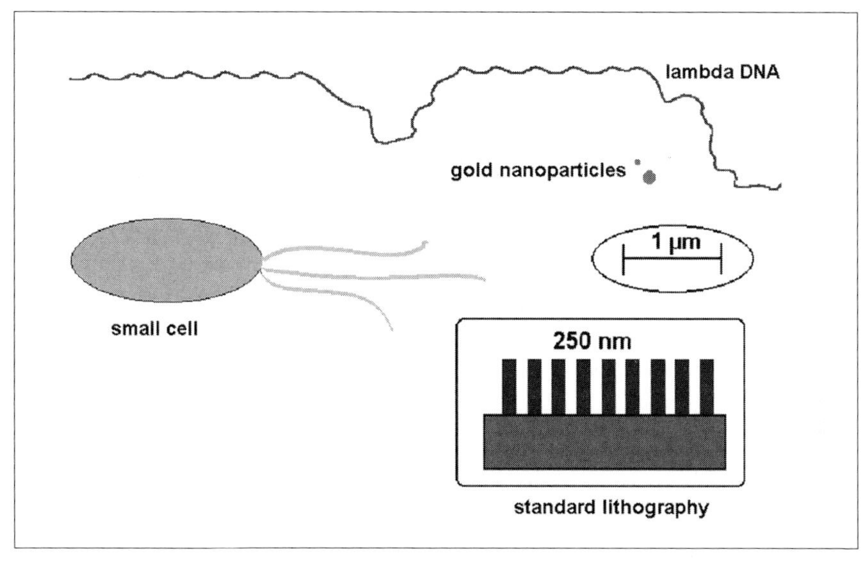

Fig. 109 Size comparison of a cell, stretched DNA molecule, gold nanoparticle and typical lithographic structures

The first aspect requires a nanolocal chemical activation of substrate surfaces. In principle, the required resolution is provided by today's lithographic techniques. However, these techniques are not yet standard procedures, e. g., the microlithographic techniques. The second aspect requires a position-selective reactivity of molecules that are sufficiently long to be integrated into microstructures. These conditions are only fulfilled by a few classes of molecules such as DNA, or are limited to a handful of locations along a large molecule.

The position-defined coupling of large molecules onto surfaces is restricted by the internal mobility of the molecules. Even when a large molecule exhibits several addressable coupling groups that are arranged in a defined pattern, this topological definition does not imply a geometrical definition. In particular, large molecules can show a variety of quite different geometries, which are only limited when in addition to the primary structure of stronger bonds, a three-dimensional geometry (e. g., through covalent or coordinative ring closures or through weaker bonds such as hydrogen bonds) is overlaid.

Long-chain molecules with a multiplicity of freely rotating bonds, such as in many synthetic polymers with high molecular weight, allow a large numbers of possible conformations that are energetically comparable. The probability of conformations that have certain positions along the molecule that are screened by other parts of the molecule (so that no binding occurs) increases with chain length.

Rigid molecules are therefore preferred for the construction of molecular architectures on solid substrates. In addition to the topological correlation of parts of the molecule due to the order of the bonds, molecules show a fairly fixed geometry. Hence not only their binding topology, but also their structural geometry is well defined. Therefore it is advisable to pre-synthesize fairly stiff units in solution prior to coupling on the surface.

An extended pre-synthesis of complex architectures is usually restricted by the decreased solubility. Flexible molecules are often significantly better solvated compared with rigid ones, and show only a low tendency towards aggregation. A prerequisite for the stabilization of larger and more rigid molecular constructs in the liquid phase is a strong interaction between the molecule and the solvent. In an aqueous phase, the following factors contribute to stabilization in solution:

- similar surface charges
- strong local dipole moments inside the molecule
- ability to form many hydrogen bonds

These factors show that the conditions in an aqueous environment depend significantly on charge distribution and ionic components. Because charged groups, local dipole moments and hydrogen bonds are usually found in combination with free electron pairs (non-binding electrons), such molecules are often interesting electron donors that can act as ligands with respect to metal cations. In addition to pH value and ionic strength, also the type of metal ion involved determines the stability of solutions of large rigid molecules, which in turn influences their reactivity.

In nature, aqueous phases are predominant. However, technical approaches are not restricted to aqueous conditions, and also the construction of nanostructures can be conducted under other conditions. The exclusion of water and air leads to a significant gain in the variability of synthetic chemical approaches to nanoconstruction. Water and oxygen represent highly reactive small molecules, which readily react with many groups, so that such groups, including water and/or oxygen, have to be excluded from synthesis strategies. Also, nature utilizes regions with decreased water contents and the exclusion of oxygen for the formation of complex living systems. Thus the hydrophobic interactions of certain protein domains or inside lipid double layers are used for the generation of membrane functions. Oxygen exclusion is essential for a whole class of organisms, the anaerobic microorganisms. Their chemical and biochemical processes are tuned to the absence of oxygen.

On increasing the size of a rigid molecule, the probability of a certain position along the molecule coming into contact with the surface prior to a reaction decreases. Moreover, with increasing molecular size, the frequency of the rotations of the whole molecule and the rate of translational movements also increase. So even in the case of soluble large units, it is not advisable to rely on a coupling involving only individual groups. More effective is the generation of larger complexes through the successive coupling of fairly small units on the solid surface.

With the coupling of a molecule on a substrate surface, the problem of solubility transforms into a problem concerned with the surface reactivity. Each molecular building unit has to have (as well as the specific coupling group) additional coupling groups for further coupling reactions. The topologically well defined assignment of coupling locations and additional reaction partners from solution is required but is not sufficient. Again, flexible parts of the molecules should not screen the coupling groups.

Under certain conditions, a long but very thin and flexible chain is sufficient to screen large parts of the surface. This is the case when the characteristic time for molecular movement (e. g., the rotational period) is smaller than the time required by a coupling molecule to penetrate, by diffusion, into the coordination sphere of coupling surface groups prior to binding. When two or more coupling groups are required, the first bond should not induce a steric hindrance for additional bonds. Finally, symmetry conditions should be fulfilled in order to generate a defined geometry.

Large molecules (macromolecules) should be optimal to bridge the gap between molecular dimensions and lithographic accessible size range. Molecular size itself is a fairly poor criterion for the classification of particles. The macromolecules constructed should include molecules consisting of a greater number of molecular units. Such macromolecules are therefore modular objects with more than 100 molecular units.

A simple class of macromolecules consists of only one type of unit (homopolymer). A linear arrangement leads to the linear homopolymers. In copolymers, two or more types of building units are used. They are denoted as block-copolymers when they consist of alternating groups of the same units. These three types of macromolecules are technically produced with a certain distribution in molecular weight and also of the order of the units (usually a statistically random distribution). For a conventional technical application, neither the exact molecular weight nor the exact positions of the units are important.

Also biopolymers such as cellulose and starch represent chain-like homopolymers with various distributions as regards the molecular weight. There are other natural macromolecules (proteins and nucleic acids), which exhibit a very high molecular weight and a chain-like primary structure, but that differ significantly from the synthetic copolymers in the following two parameters: their chain length and molecular weight are defined exactly, and the arrangement of the units in the chain is fixed.

Supramolecules in a narrow sense describe molecules with atoms or groups of atoms connected by reversible, non-covalent interactions.

Supramolecules are aggregates consisting of several molecules. The atoms inside the individual molecules are held together by strong (e. g., covalent) bonds; the molecules are connected, at least temporarily, by weaker interactions, such as dipole–dipole effects, coordinative van der Waals or hydrogen bond interactions. The interactions are often influenced by a change in solvent or an increase in ionic strength or pH [102][103].

The term nanoobject reaches far beyond supramolecules. It includes single molecules, but also metal colloids and vesicles. Supermolecular nanoobject also include objects that consist of a greater number of molecular units, defined in a way that considers their size and shape.

To distinguish the various types of macro- and supermolecular nanoobjects, two groups of parameters are important for nanoconstruction purposes (Fig. 110). The first group includes the chemical definition that covers the elementary composition, the number and arrangement of low molecular units or groups of atoms and the topology of strong directed bonds. The second group describes the geometrical definition. It includes the internal flexibility (which means the density and number of groups that can rotate and vibrate), the tendency to form very different conformers,

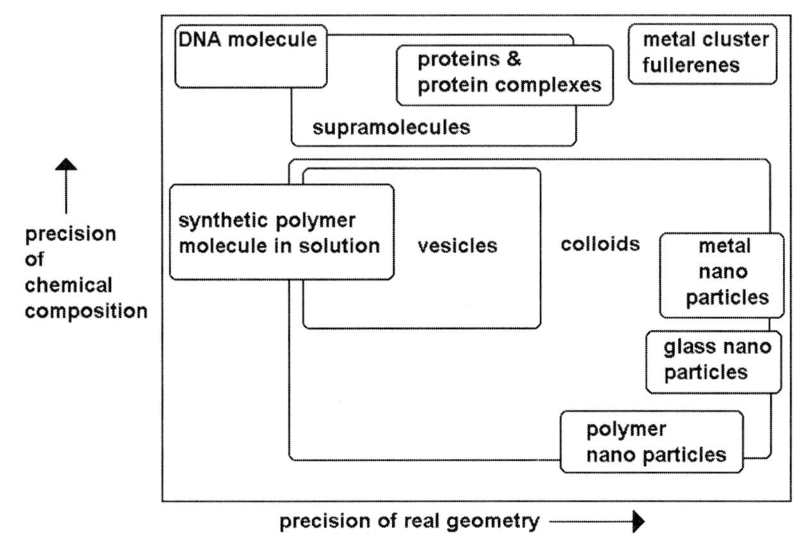

Fig. 110 Classification of nanostructures and molecules with respect to their definition of composition as well as geometry

the density of strong bonds and thereby also the extent of cross-linking. Supramolecules show a high level of specificity of the composition and binding topology. However, the term also includes particles with less defined geometries, e. g., particles able to form quite different shaped conformers. Individual molecules of linear synthetic polymers, and also copolymers and block-copolymers in solution are nanoobjects with respect to their dimensions, but are not particularly specific in their geometry and exhibit a varying composition. A higher degree of determination is observed in the case of colloidal particles, with a higher value for nanoparticles compared with vesicles.

5.6.2
Strategies of Molecular Construction

Molecular construction is a branch of synthetic chemistry. It differs in two aspects from traditional synthetic approaches. In a single molecule (and not just in the crystal), both the specified binding topology and molecular geometry should be preserved. Another point is that the macromolecular units are defined by molecular weight and structure and exhibit specific binding groups that in subsequent construction steps serve as binding sites. As long as there are no autonomous nanosystems, the aim of nanotechnical construction is an integration of the molecular constructions in planar technological environments. Surface reactions are therefore not only supporting tools, but are also required with respect to function. The synthesis strategy determines in what state the molecular units are pre-synthesized and when the subsequent construction will proceed better on the surface. Both surface processes and reactions in solution are spontaneous molecular interactions driven by the laws of molecular dynamics. The generation of structures is driven by self-assembly in combination with chemical synthesis, which is only supported on solid surfaces (as a planar substrate or as a nanoparticle). The processes are determined by molecular stochastics. The construction of molecular architectures is more based on the laws of chemical thermodynamics and kinetics than on the rules of microsystem technology and planar-technical nanolithography.

A modular approach is required to manage the synthetic efforts necessary for the construction of highly defined and stable macromolecules. The modules have to enable the application of the principles of molecular recognition in combination with long-term stable bonds to be made. These requirements include a certain contradiction. Molecular recognition requires at least intermediary weaker bonds that can open again to give way to the recognized stronger bonds and the related coupling component. Hence the bonds forming the basis of the molecular recognition should be weak enough to be replaced later, or the recognition process with weaker bonds has to be complemented by a second step that leads to more stable bonds on the recognition site.

Nature provides beautiful examples of complex arrangements through spontaneous self-organization. It is based on a hierarchical system, which requires only a fairly small number of associations of elements at every level of organization. Thus, defined monomolecular geometries are embedded in environments that act as ensembles and are independent of the specific position of an individual molecule. This dual principle includes determined complexes (such as protein complexes) but also statistically based structures such as lipid bilayers. Liquid and liquid-crystalline phases and micelles represent an outer environment for the supermolecular complexes of biological macromolecules[1]. Of particular importance is the reduction in the degrees of freedom of mobility of molecular systems by the integration into layers. A fixation of the layer including the molecular architecture can even result in a complete restriction of the mobility, e.g. by cross-linking.

A specific molecular recognition is a required condition for the construction of molecular aggregates from subunits. Natural proteins that are integrated into complex aggregates typically exhibit several recognition sites for coupling to adjacent groups. For unrestricted three-dimensional mobility of the molecules, the probability that some molecules meet with their recognition site by chance is rather low. This requires an adjustment of not only the translation, but also the rotation of each molecule involved. Through the restricted surface portion responsible for coupling, many binary combinations of molecular orientations are possible that will not lead to a coupling. If the two molecules are on a membrane or at an interface their rotational mobility is significantly decreased. The pre-orientation of two binding partners at the interface or on a membrane eliminates $2 \times 2 = 4$ rotational degrees of freedom of mobility. The complementary recognition and binding reactions are arranged relative to the surface plane or perpendicular to it. The rotation around the remaining axis will often lead to recognition and binding.

The generation of three-dimensional structures is both goal and method. The constructions should be sufficiently stable to resist changes in the environment or re-

1) cf. Section 5.4.6

moval of the liquid phase. Spatially connected binding topologies restrict the internal flexibility of molecules to avoid collapse, so that intramolecular cavities are preserved even in the case of changes to the surroundings and functional groups preserve both reactivity and positions. A certain density of connections inside molecular constructions is required to ensure that for every linear unit not more than one structure capable of rotation exists, so that no more than two bonds able to rotate exist between them.

The conditions for complex molecular architectures are either a highly precise sorting principle separating all byproducts of the individual reaction steps or a series of synthesis steps with virtually complete efficiency (which means reaction probabilities approaching 1). The latter seems to be more practicable. Because every individual step requires a selectivity, it is advisable to work with pairs of coupling groups that are independent of each other. The greater the number of non-interfering coupling pairs, the larger the possibilities with respect to the design of synthetic architectures. Just one pair yields no structures that are defined in length and structure, only after the introduction of protection group chemistry is an indirect second chemical function represented. Using two coupling pairs facilitates both chain-like and branched structures of exact defined structure and shape.

Cyclization is required to increase the rigidity of molecular architectures. In principle, all two- or multi-fold functionalized units can be used for cyclization and the fabrication of chains. The tools of supramolecular chemistry are applied therefore. Processes of molecular cyclization are assisted by low concentrations and the intermediate formation of coordination compounds. Electron pairs or π-electrons are utilized, so that the chain surrounds the metal ion like a chelate ligand. For rings with greater diameter, one metal ion is not sufficient, because stable coordination bonds require certain distances. Cluster ions or metal ions (bridged by bifunctional ligands or using metal organics) could assist by providing extended core structures for coordinative compounds, so that further extended ring structures are also supported. For the construction of larger rings, the use of multiple core complexes that have an internal central ring formed by a mutual ligand and additionally included outside ligands connected by a group of other metal ions is possible.

Regarding the bond strength, a double strategy is required. Complex architectures require strong and specific bonds in the utilized coupling scheme. In contrast, weaker bonds are used for molecular recognition, using cooperative effects of polyvalent bonds, so that stable connections are only formed if the desired number of coupling groups of two complementary partners is involved. So positions for coupling reactions can be defined that are connected by a local presence of two or more binding groups in close proximity. The occurrence of one of the binding groups has no effect. This technique is of particular importance for the immobilization of pre-synthesized structures on surfaces.

A complex of molecular architectures could be defined that presents not a functional unit on its own but serves only as a framework or positioning tool. The functional elements could be molecular functional units connected to this architecture, so that the original passive construction becomes active. In addition to individual molecules, other structures are also possible for this purpose, e. g., ultrathin layers or nanoparticles. There are two requirements: a wide variety of highly defined geometries is desirable, and the binding locations should be precisely defined.

In the following sections some examples of geometries of molecular-constructive architectures are discussed focusing particularly on objects that are not accessible by conventional lithographic techniques. The initial interest for nanotechnology begins with linear objects of defined length. In addition to compact cylindrical structures, flexible cylinders and hollow tubes of defined dimensions are required. Small areas, dash and spot gratings with periodicities in the medium and lower nanometer range are of interest, but also complex geometric objects such as circles and ellipses, regular and irregular polygons and combinations of these. Besides closed objects, geometries with defined pores and grating structures or with combinations of ring geometries with movable bar-shaped units have been proposed and investigated [104]. Free-standing structures with nanometer dimensions, spacers and presentation structures that bind functional components with high precision with respect to location, and allow for spatial access that is necessary for the mobility or diffusion of reaction partners, are other interesting examples.

A second complex includes architectures that combine constructive and functional aspects: electron-, ion-, phonon-, exciton- and plasmon-guides, photon acceptors and emitters, switches, logic components, memory elements and others. The integration of such molecular functional components is a great challenge for nanotechnology: individual functions have already been realized in smaller molecules. The control and the readout of these elements requires only a limited number of connections to planar technical structures. The size of these connections and the positioning requires a lithographic resolution with the dimension of the molecules used, which means in the lower nanometer range or below. In practice, such small functional elements are addressed as ensembles and not as individual molecules. The ensemble size depends largely on the lithographic precision. The problem of functional connection between single molecules and lithographic periphery is worsened in future developments of a variety of molecular functions integrated in such structures.

An alternative is the vision of a molecular architectures that conducts complex tasks in its interior, and only a few channels are required for communication with the outside world. One could envisage small memory and processor units that communicate via a serial channel with a microtechnical periphery. In such subsystems, the channel could even be in the nanolithographically accessible medium nanometer range. However, the construction of such a complex system through only chemical synthesis using self-organization is not clear. For the further development of nanotechnology, it would be helpful to realize individual molecular functions in a model application and integrate these with a large connecting channel in a planar-technical periphery. With the progress of modular supermolecular synthesis, more and more functions could be integrated into this model application of a molecular self-organized system, until molecular nanosystems are achieved with significant functional advantages compared with conventional systems [105][106].

5.6.3 Biogenic and Bio-analogous Nanoarchitectures

A defined arrangement, shape stability, solubility and position-specific reactivity are important conditions for the construction of molecular architectures. Linear macromolecules that do not form defined secondary structures are less suited to molecular nanostructures. Complex copolymers and systems with fairly interesting nanophases in solids are also not particularly helpful, because the complex geometries are not freely adjustable. Restricting factors that are missing or not defined are positions of binding locations, a high translational symmetry in the linear chain or chain sections, a rather high flexibility in the chain and definitions of branching or connecting points to create position-specific spatial connections. Synthetic polymers often exhibit a distribution with respect to their molecular weight; copolymers show a statistical distribution of monomer units and blocks. A multiplicity of rotating bonds leads to a huge number of thermodynamically similarly preferred conformations in solution or in the melt.

Some biological macromolecules show similar properties compared with the synthetic molecules. Starch and sugar have a broad distribution of molecular mass, and a high degree of translational symmetry in the binding topology. On the other side, nature does provide classes of molecules that fulfill the above-mentioned requirements: proteins and nucleic acids.

Proteins are interesting for nanotechnology both as construction material and as molecular tools. Proteins are already being used by mankind in both fields, such as natural fibers for textiles or enzymes for food and in the washing powder industry. In these fields, proteins are applied in large quantities. However, their molecular properties and functions are only used in large ensembles.

There have been attempts to integrate biomolecules or bio-analogous molecules into technical nanostructures or to apply them directly for nanostructuring. Two-dimensional gratings with periodicities in the medium nanometer range can be prepared by assembling an ordered protein layer on top of a functional layer prior to using the protein layer as an etch mask. Such two-dimensional protein crystals can be found for example in the membranes of microorganisms, e. g., the so-called S-layer in the bacterium *Sulfolobus acidocaldarius*. After transfer of such S-layers onto TiO_2 surfaces, the protein aggregates were used as etch masks in an Ar ion sputtering step (2 keV), resulting in two-dimensional grating structures with about 22 nm periodicity in titanium oxide [107]. In addition to a direct transfer of the layers, a self-assembly approach of isolated and purified proteins was also demonstrated. An S-layer can be used for the preparation of two-dimensional gratings with periodicities in the lower nanometer range, e. g., for quantum dot arrays, by utilizing the protein layer as a mask or as a template for metal deposition [92].

In another example of the application of S-layer protein arrays, a fusion protein based on an S-layer protein containing streptavidin was used. After recrystallizing this protein on substrates, the resulting regular array was used to immobilize biotinylated DNA oligomers, yielding a nanoarrayed DNA chip [108]. The assembly of various nanoarchitectures, including nanoparticle arrays, hetero-nanoparticle architectures, and nanowires, utilizing highly engineered M13 bacteriophages as templates, has been reported [109].

Peptides synthesized in viruses permit the DNA-controlled preparation of various proteinaceous components. Thus, the genome of M13 phage can be rationally engineered to produce viral particles with distinct substrate-specific peptides expressed on the filamentous capsid and the ends, providing a generic template for the programmable assembly of complex nanostructures. Phage clones with gold-binding motifs on the capsid and streptavidin-binding motifs at one end have been created and used to assemble Au and CdSe nanocrystals in ordered one-dimensional arrays and more complex geometries. Initial studies have shown that such nanoparticle arrays can further function as templates to nucleate highly conductive nanowires, which are important for addressing/interconnecting individual nanostructures. This system has also been used for the fabrication of magnetic nanostructures such as Co-Pt crystals [110]. An M13 phage library with an octapeptide library has been used to select binders for cobalt ions in order to prepare fibrous structures with directionally ordered M13 phages through interaction with cobalt ions prior to Co-Pt alloy synthesis. Viruses have also been employed to synthesize and assemble nanowires of cobalt oxide at room temperature. By incorporating gold-binding peptides into the filament coat, hybrid gold-cobalt oxide wires were formed as electrodes for advanced lithium batteries that showed improved capacity [111]. Combining virus-templated synthesis at the peptide level with methods for controlling the two-dimensional assembly of viruses on polyelectrolyte multilayers provided a systematic platform for integrating these nanomaterials to form thin, flexible lithium-ion batteries.

The integration of proteins in artificial nanostructured environments enables the use of functions of individual biogenic functional molecules. However, the application of proteins in nanotechnology is hampered by their natural adaptation into the biological environment. These complex molecules were optimized by natural evolution with respect to their functions in living cells or in their immediate environment. Technical environments, as used for chip or microsensor production, differ significantly from these biological surroundings. Even standard biological requirements, such as the presence of a liquid phase or prevention of elevated temperatures, are even problematic for technical standard processes. Therefore, the integration of molecular-functional proteins in chip arrangements is usually aimed at a connection of protein-compatible settings with solid substrates. Therefore, the biosensor field provides numerous approaches, such as the immobilization in a planar monolayer or the integration into thin polymer or gel layers. Typically, immobilization is connected with a loss in activity. On the other hand, the use of proteins as materials is usually much easier because often no aqueous environment is needed.

Selected classes of macromolecules exhibit dimensions of a few tens up to a few hundreds of nanometers. One has to consider the flexibility of parts of the molecule in the immobilization and orientation of such large molecules. Flexible linear molecules can be extended by stretching in a hydrodynamic flow. For charged or polarizable molecules, the orientation in an electrical field can be used. These approaches do not create stable geometries, because after reduction of the applied forces, the molecules then relax again. So the molecules have to be fixed after positioning, stretching and orientation. This implies that at least two coupling groups are involved in the substrate binding.

Large individual molecules and supermolecular objects are not only of interest as geometrical arrangements, for spatial architecture or mechanical connections. The application of macro- and supermolecules with specific functions is particularly attractive, e. g., chemical, electronic, optoelectronic or chemo-mechanical properties. In analogy with enzymes and natural protein complexes, such molecules can be used as molecular nanotools and nanomachines.

Nanotools have to fulfill the requirements of an individual tool or to serve several complex functions. They have to:

1) be activated by certain assisting substances,
2) recognize their work piece, the substrate molecule,
3) process the substrate molecule,
4) release the product, and
5) emerge fully functional from this process.

Similar to protein receptors in biological systems, supermolecular objects are able to act as general recognition structures or to hold such structures. In analogy with biological structures, oligovalent ligands or macro-ring shapes are particularly well suited to the specific recognition of ions or small molecules. Because of their coordinative properties, rings with ether, thioether or secondary amines between small aliphatic or aromatic groups are appropriate. The geometry and so also the size of the target molecule can be controlled by the number and size of the groups in the ring. The relationships and the distance of the –NH–, –O– and –S– bridges control the coordinative properties of the ligands. The number and arrangement of aliphatic and aromatic rings determines the flexibility of the recognition structure. The introduction of functional and ionic groups into the outer regions of the rings results in the solvation and surface binding properties of the complexes.

The functionalization of side-chains of substrate-recognizing macrocycles by suitable steric arrangements assists significantly in the transfer of atoms (or group of atoms) onto the substrate. In this way some macrocycles act as specific catalysts, with a substrate specificity realized by the binding properties of the recognition ring and the reaction specificity realized by the side chain. Examples are cationic receptor molecules with dihydropyridyl side chains that show increased rates of hydrogen transfer onto smaller molecule [102].

The chemically activated mobility of proteins is utilized by nature in complexes of several protein units to fabricate machine-like supermolecules working on a nanometer scale. Both rotational and translational arrangements are achieved. The systems actin/myosin (muscles) and kinesin/tubulin (cytoplasm) are examples of natural translation systems. ATP synthetase is an example of a rotational system, consisting of rotor and stator similar to technical systems. The propulsion is realized by a pH gradient (proton flux) [112][113][114]. Comparable synthetic systems are not yet available, although supramolecular chemistry shows some promising developments [104]. Also the adaptation of the biogenic supermolecular motors to technical requirements, e. g., by substitution of amino acids or the introduction of more robust molecular modules into these functional protein architectures, remains for future research.

5.6.4 DNA Nanoarchitectures

The modular construction principle for complex molecular architectures has been already developed as supermolecular synthetic chemistry for nucleic acids [115] [116]. This molecule class is particularly well suited due to the existence of a variety of enzymes for construction and manipulation. Most of the reported work was done in homogenous liquid phases, but the underlying principles are also applicable for solid-state surfaces.

The general principle of molecular construction is not limited to nucleic acids. However, these molecules show several properties that make them particularly suitable for the construction of complex molecular architectures. A modular construction principle has already been realized at the lowest level due to only four basic units (four nucleotides with the bases adenine, thymine, cytosine and guanine) connected in a linear manner by identical complementary bonds (sugar–phosphate bonds). The primary modular arrangement leads to a linear and thereby sequential structure, with properties determined by the sequence of the bases in the molecular chain. As a result of the base arrangement, an antiparallel association of two molecules can occur based on attractive interactions of two (thymine/adenine) or three (guanine/cytosine) hydrogen bonds for each base pairing. Because hydrogen bonds are (in contrast to electrostatic interactions) highly oriented, the individual bonds are only formed in the case of fitting geometry, so that in the environment of a complementary base pair, further complementary base pairs are present. Owing to the relatively low energy of hydrogen bonds, the association is only stable at room temperature for about 30–40 simultaneous bonds. This leads to a requirement of about 10–20 subsequent complementary bases without a non-complementary insert to induce such a double-stranded complex. A strand with a given sequence shows a significantly increased binding affinity towards its complementary strand compared with other strands (even when only one or two bases are changed) and a precise assignment of one strand to another with a principal variability V is given by the length of the section n and the number of possible unit types k:

$$V = k^n \tag{5.5}$$

This value amounts to 4^{10} to 4^{20}, depending on the sequence length of n =10...20. Thus DNA is a tool in a system of one kind of molecular material to achieve the connection of subunits in a highly specific manner. For practical nanotechnical applications, this huge number of possibilities will probably not be fully used, due to differences in the stability of base pairs (two or three hydrogen bonds), the occurrence of competing intramolecular base pairing (such as hairpin or loop structures), and maybe to low concentrations of one strand. However, there is a great variety of different usable sequences resulting in numerous possibilities to specifically address self-assembling double-stranded hybrids.

Apart from this specificity, the formation of double strands also results in a significant mechanical stability of the whole system. Instead of a flexible linear chain connected by freely rotating bonds in the sugar–phosphate backbone of the single strand, the double strand exhibits the structure of a double helix with anti-parallel sugar–phosphate chains cross-linked and stabilized by hydrogen bonds. So a relatively rigid complex with restricted degrees of freedom of mobility is achieved as a suitable construction unit for more complex architectures.

Except under strongly acidic conditions, a DNA double strand represents a polyionic complex, because the two-fold substituted phosphoric acid residues donate their third proton so that the DNA represents a polyanion. In strongly polar and in protic surroundings such as water, even larger DNA molecules are strongly charged and solvated, resulting in a high solubility.

Hybridization of single strands yields not only double strands, but also a variety of secondary structures. Single strands with partial self-complementary sequences show back folding, multiple antiparallel sections leading to cloverleaf-like structures as found in natural t-RNA. Larger loops of unpaired sections of such partial self-hybridized DNA can be additionally hybridized with other complementary single strands.

A combination of paired and unpaired sequence parts is also possible in the case of linear molecules, and is the foundation of a simple system to create complex modular architectures. Overhanging unpaired sequences of partial double-stranded DNA creates so-called "sticky ends" that are able to react with another single-strand or the sticky end of another double strand (Fig. 111). The formation of the sugar–phosphate bond between the originally separated units by the enzyme ligase results in a significant stabilization of such aggregates. Such a ligation is also important to suppress the cleavage of this bond in subsequent binding steps. To increase the selectivity of hybridization, and in particular to avoid undesired hybridization of only a few complementary base pairs or under inclusion of non-complementary pairs (mismatches), the hybridization process is started at elevated temperatures, and the hybrids are formed during cooling of the reaction mixture. Non-ligated double strands with a limited number of base pairs (and especially A–T-rich sequences) show a tendency to dissociate at elevated temperatures.

If one or both ends of the double strand exhibit two non-complementary overhanging residues, more molecules can bind so that branched structures are formed. With these structures the possibilities of two- or three-dimensional architectures emerge. If several construction steps are required, a ligation of the backbone is essential, so that constructs already formed cannot dissociate later on. A condition for determined structures is that the coupling sequences are well-defined, and ambiguities (which could lead to a mixture of products) are prevented.

The variety of natural and technical tools for a sequence-specific manipulation and analytical sequencing makes DNA the molecule of choice for practical reasons. Sequence-specific restriction enzymes enable the double strand to be cut at defined positions, usually resulting again in sticky complementary ends that can be utilized in subsequent coupling reactions. The polymerase chain reaction (PCR) allows for a sequence-specific amplification of very small concentrations of DNA for analytical purposes or synthetic applications [117]. Longer sequencing is also a standard technology.

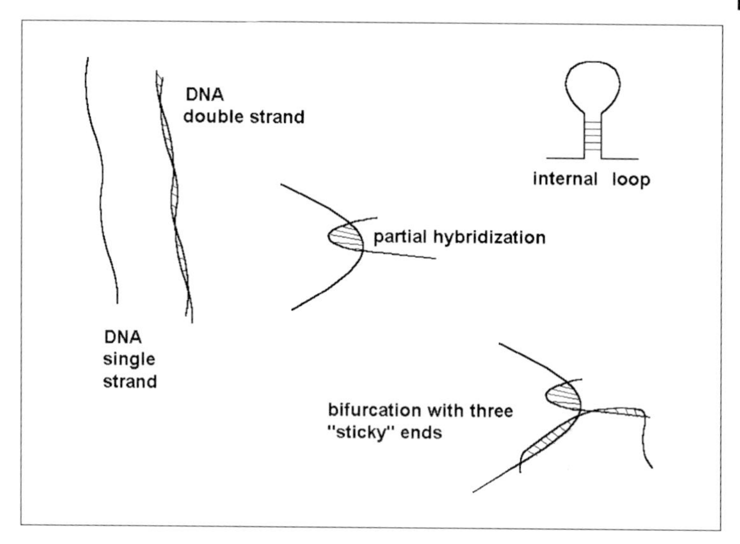

Fig. 111 Elementary structure motifs for molecular construction with DNA

The synthesis of oligonucleotides with a wide range of lengths as well as of oligonucleotides with various substitutes is possible and is commercially available.

To achieve stable two- and three-dimensional DNA architectures on surfaces even after air-drying, a stabilization of the complexes is required. Early experiments demonstrated the successful transfer of DNA architectures assembled in a homogeneous liquid phase onto substrate surfaces. Although these complexes are not yet coupled specifically to the surface, they are an important step towards the combination of DNA nanotechnology with planar technology.

DNA can be utilized to integrate molecular constructs into planar technical structures. Molecules of DNA can be immobilized and (in the case of longer molecules) permanently oriented in order to achieve a correlation between the highly precise positioning inside the molecular construct (e.g., by base-pair-controlled connection of DNA conjugates) and the location of the molecular structure on the chip surface. The first step in such positioning approaches often involves the terminal binding of long DNA molecules through specific interactions (e.g., base-pairing or electrostatic interactions). Techniques based on the directed movement of fluids or interfaces (meniscus) are then applied in order to orient the attached molecules in the desired direction [118]. The last step usually includes some kind of (often less specific) interaction between the oriented, stretched molecule and the surface in order to realize a more permanent immobilization. Based on these kinds of techniques, DNA has been utilized to bridge electrode gaps in multi- [119] and single-molecular [120][121] approaches. The single-molecule-based methods use large molecules (several μm) and small binding areas (lower μm range) in order to limit the binding to no more than one molecule through steric hindrance (e.g., mediated by electrostatic repulsion in the case of negatively charged DNA molecules). These approaches are characterized by the potential for high parallelization, an important feature for broader applications in the future, as demonstrated by the parallel binding of individual DNA structures in microelectrode gap structures prior to nanoparticle binding based on electrostatic interactions (in order to realize nanowires) and electrical measurements without the need for any additional wiring [122].

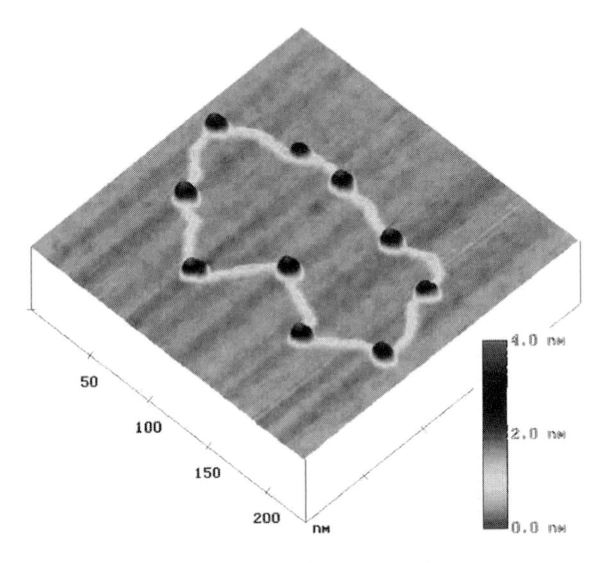

Fig. 112 AFM image of streptavidin–DNA complex (simple ring structure) [124]

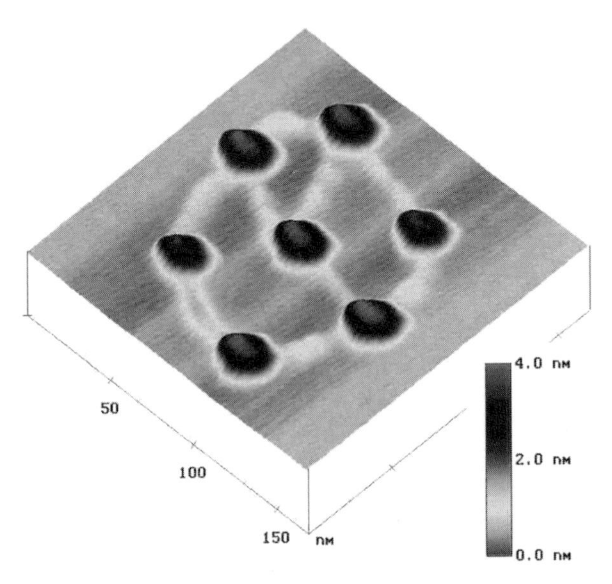

Fig. 113 AFM image of streptavidin–DNA complex (connected ring structure) [123]

Nucleic acids provide excellent properties for the construction of complex hybrid structures using other macromolecules. Thus DNA has been utilized to form adducts of beads-on-a-string structures of DNA-protein complexes based on biotin–streptavidin coupling prior to adsorption and imaging on substrate surfaces [123][124]. Depending on the size of the DNA and the number of streptavidin molecules included, supermolecular complexes with dimensions of 100–200 nm were formed (Figs. 112 and 113).

The principle of base pairing between complementary strands is not limited to DNA. Other classes of compounds, such as RNA, also show this potential. Future developments of compounds with this same potential but which are more suitable for molecular construction than DNA are anticipated. At present synthetic chemistry does not provide the required tools, such as strategies for the combination of the unique enzymatic tools known from DNA biology. So sequence specific cutting (restriction enzymes), template-dependent strand extension (polymerases) or the connection of strand ends (ligases) are addressed for nucleic acids, these being types of molecules found in nature, but have not yet been realized for synthetic systems. Today's approaches for synthetic strand-pairing systems are based on the examples in nature. Thus pyranosyl RNA molecules utilize the connection of phosphate residues with sugar and *N*-hetereocycles, as is known from natural nucleic acids. For this system it was shown, that – in addition to hydrogen bonds – steric arrangements contribute substantially to the pairing of strands [125].

In order to realize complex structures based on DNA units, branches had to be introduced into the conjugates. Seeman established a group of branched complexes called "double-crossover" molecules (DX) with improved stiffness compared to linear DNA [126][127]. Based on a so-called "sticky ends" design, these DX molecules will self-assemble into periodic 2D lattices (Fig. 114) (for reviews, see [128][129]). A variety of rigid, branched DNA tile molecules have been constructed, leading to the assembly of linear arrays, 2D lattices, or tube structures. Concomitantly, additional basic units have been introduced into the molecular "tool box", such as triple-crossover (TX) molecules that can give rise to a variety of final geometries [130]. The novel concept of "sequence symmetry" was introduced for the design of symmetric tile motifs utilizing the fourfold symmetry of a cross-shaped tile and the threefold symmetry of a three-pointed star [131]. This allows a reduction in the number of unique DNA strands required, thereby minimizing both costs and experimental error. By reducing the unique sequence space required, it significantly simplifies the sequence design. At the same time, the symmetry introduced ensures perfect geometrical symmetry of the resulting structures, so that any unpredictable distortion is avoided and much larger 2D DNA arrays are possible. Using only three different strands for the cross-shaped tile (usually nine strands) and three-pointed star (otherwise seven strands), periodic arrays of up to millimeter dimensions were successfully realized [131]. Based on the DX motif, 2D lattices from two strands [132] and long (up to 50 μm) DNA tubes from a one-strand system [133] could be demonstrated.

With regard to future practical applications of this approach, the ability to efficiently self-assemble complex patterns with reduced symmetry and increased addressability is a key issue. There are four general strategies for tackling this problem. The first one

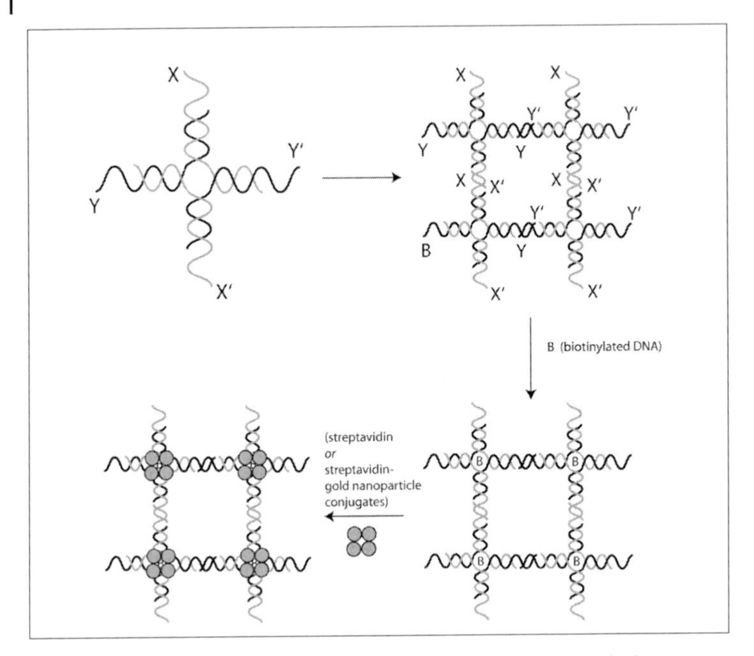

Fig. 114 DNA tile-based assembly relies on the self-assembly of branched DNA junctions with sticky ends (X pairs with X' and Y with Y') in order to realize a 2D arrangement of interlocked tiles (adapted from C. Lin et al. [129]). This arrangement can be modified by using biotinylated DNA as units for the tiles, leading to specific attachment sites, e.g., for streptavidin or a variety of streptavidin conjugates, such as with gold nanoparticles.

relies on a simple "mix-and-go" approach based on a set of unique tiles, each of which bears a unique sticky end, such that each tile is guided into a unique position during self-assembly. The second strategy is one of hierarchical self-assembly: subsets of specific tiles are mixed and assembled separately and then combined sequentially. Thirdly, there is algorithmic self-assembly, which utilizes tiles that are programmed with specific binding domains in order to bind specifically and cooperatively to form complex patterns according to algorithmic rules. The fourth strategy is that of nucleated self-assembly, in which a longer strand is used to initiate nucleation, whereupon other strands associate in order to form complex patterns. Combinations of two or more of these four strategies are often applied.

Although the "mix-and-go" approach is quite successful in yielding complex structures such as striped patterns with variable distances between the stripes based on two to four tiles [134], there is a need to reduce the sequence design effort and the cost of DNA synthesis. A two-step hierarchical assembly allows the reuse of the core part of the tiles since only the sequences of the sticky ends are changed. Here, each individual tile is first formed separately and these are then combined. This approach has recently been demonstrated for two fully addressable fixed-size 2D arrays, a ten-tile array [135] (3×3 square plus an index tile), and a 16-tile array [136] (4×4 square) using cross-shaped DNA tiles. Each tile can be differently accessed (e.g., conjugated), making the system fully addressable. Scaling-up the number of tiles results in a linear increase in the number of different strands with the number of different tiles.

Geometric symmetries allow for the design of DNA tiles with asymmetric sticky ends that self-assemble into symmetric finite-sized arrays, reducing the number of tiles needed according to the degree of symmetry in the final structure [137]. For example, a 5×5 fourfold symmetric square required only seven rather than 25 different tiles. Different sized and shaped arrays can be created using different subsets of tiles. In this way, the number of unique strands required is reduced and the design is simplified, but more complex structures are produced.

As mentioned above, RNA offers a potential for nanoarchitecture construction comparable to that of DNA. Recently, a modular design of artificial RNA motifs referred to as tecto-RNA was introduced, in which each module contains a right-angled motif with a single-stranded 3' tail and two stems each terminated with a kissing loop [138]. These tecto-RNA modules produce a T-shaped three-arm junction, or four of them assemble to produce a tecto-square of variable size, tail sequence, and orientation. These squares can themselves self-assemble into larger assemblies connected by complementary single-stranded tails. A two-step hierarchical self-assembly of these blocks can lead to addressable finite-sized 2D arrays and patterns by preventing association at specific positions.

A reduction in the number of unique DNA strands required is a general objective in these developments in order to scale-up the size of addressable structures. It is possible to reuse the same sequences for sticky ends when these are hidden inside pre-assembled parts prepared in separate steps [136]. A decrease in the number of unique sequences is thus connected with an increase in the number of steps required in the assembly procedure. An example of nucleated self-assembly has been demonstrated using a polymeric nucleating scaffold strand as the starting point and utilizing cooperative binding at multiple weak binding domains of a set of "molecular tiles" [139]. Thereby, a Sierpinski triangle sheet (E. Winfree [139]) and a binary counter lattice [140] were accessed. The first realization of nucleated self-assembly led to an aperiodic patterned DNA lattice (barcode lattice), which was formed by directed nucleation of DNA DX tiles around a long scaffolding DNA strand [141]. The long strand included the binary barcode information 01101 and acted as a nucleation point for the assembly of DX tiles. Combining the concept of nucleating self-assembly and an understanding of branched DNA junction structures led to the recent breakthrough of scaffolded DNA origami [142]. In this approach, a long single-stranded genomic viral DNA (of known sequence) is folded into arbitrary 2D shapes. More than 200 "staple" strands (30–70 nt long) have been obtained by means of computer-aided design and folding of the seven kilobase genome into any desired shape in a single mixing and annealing step. A variety of shapes with dimensions of the order of 100 nm have been demonstrated, such as a square, a rectangle, a five-pointed star, a smiley face, and various triangles. This approach is easy (single step, no purification), high yielding ($> 90\,\%$), and rather inexpensive.

The described approaches for realizing complex nanoarchitectures based on DNA self-assembly can be combined with other components in order to create functional hybrid structures. Usually based on the chemical attachment of coupling groups to DNA, a variety of components are accessible through specific molecular binding. These allow the combination of fascinating 2D or 3D structures with functionalities that can be spatially addressed. In the following sections, examples of the integration of proteins and metal nanostructures are described.

The 2D DNA nanogrids that may be obtained by the self-assembly of cross-shaped DNA tiles represent an ideal scaffold for the attachment of proteins or other molecular complexes. Such a lattice has been used to organize streptavidin proteins into a periodic 2D array utilizing biotin groups attached at the tile center [141]. When a two-tile system was used, the spatial distance and density could be controlled [143]. Although biotin coupling is a robust and established method for molecular construction, it is limited to just one binding pair. In order to address a variety of different proteins at once, other options are required. The use of protein-binding DNA sequences (aptamers) allows us to exploit the full compatibility of the binder molecules incorporated into DNA nanostructures with a large number of potential protein binding partners. Thrombin protein-binding DNA aptamers have been integrated into a self-assembled DNA array so that thrombin could be bound to the DNA grid [137]. A reverse approach utilizes DNA-binding proteins, such as the Holliday junction binding protein RuvA, that may be organized into 2D crystal templates by way of a 2D lattice formed from four-arm junctions [144].

The combination of spatially defined DNA structures with metal nanostructures is an interesting approach for the realization of highly defined functional units for nanoelectronics. Based on the biotin-streptavidin binding pair, DNA grids with integrated metal nanoparticles and control over the tile distances could be demonstrated [134][145][146]. In order to overcome the limitations of the biotin-streptavidin binding pair, DNA-conjugated nanoparticles, that could be addressed by their complementary sequence, have also been used to prepare linear arrays [147] or various 2D lattices [148] of metal nanoparticles.

5.6.5
Synthetic Supramolecules

The majority of synthesized supramolecules are fairly small aggregates with a small number of atoms compared with the examples in nature. Similar to polymer chemistry, supramolecular synthetic chemistry applies the properties of some low molecular weight units to form two or more bonds. Moreover, a strong principle of spatial order is applied. Thus binding topologies are created that extend into two- and three-dimensions in order to realize spatially defined architectures.

There are several strategies for achieving three-dimensional defined geometries. One approach is based on dendrimer synthesis (Fig. 115). As a result of branches, relatively large molecules are formed in a limited number of reaction steps by using a protection chemistry or alternating coupling groups, as known from solid phase synthesis (Figs. 116 and 117). Dendrimers are much more compact than the usually coiled and flexible linear polymers, but their side chains are still flexible.

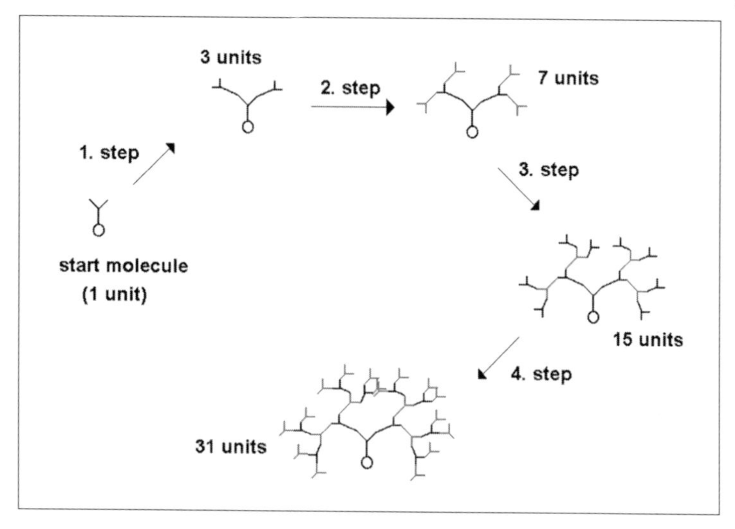

Fig. 115 Scheme for stepwise synthesis of dendritic molecules

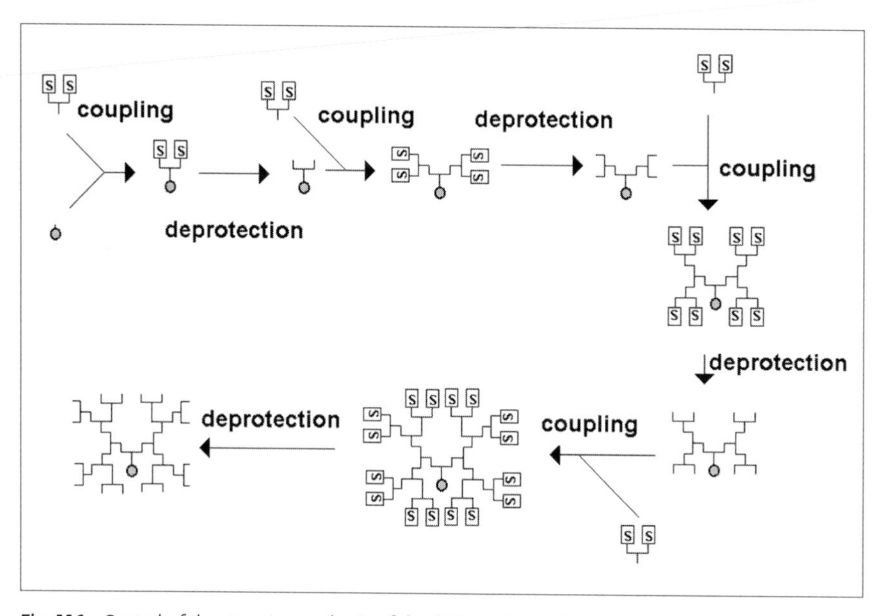

Fig. 116 Control of the stepwise synthesis of dendritic molecules by protection groups

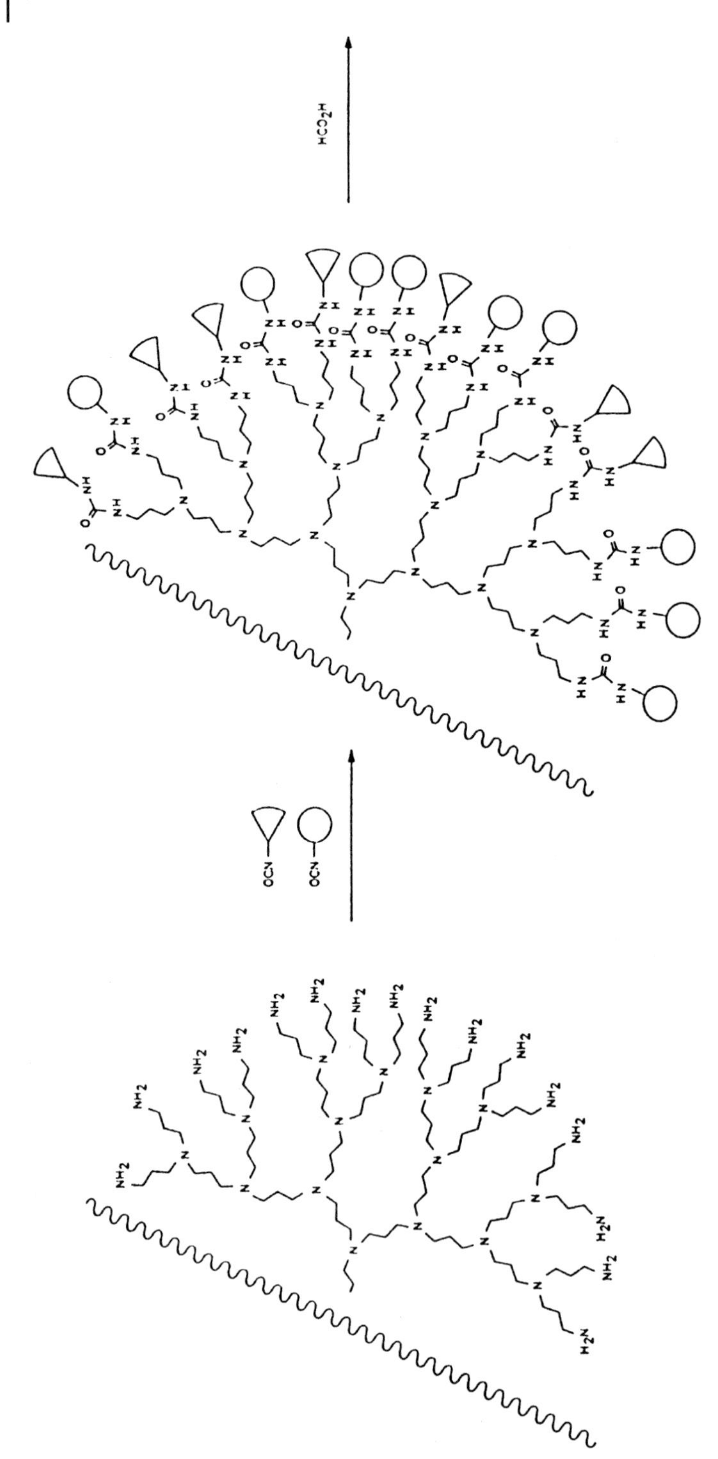
HCO$_2$H
OCN
OCN

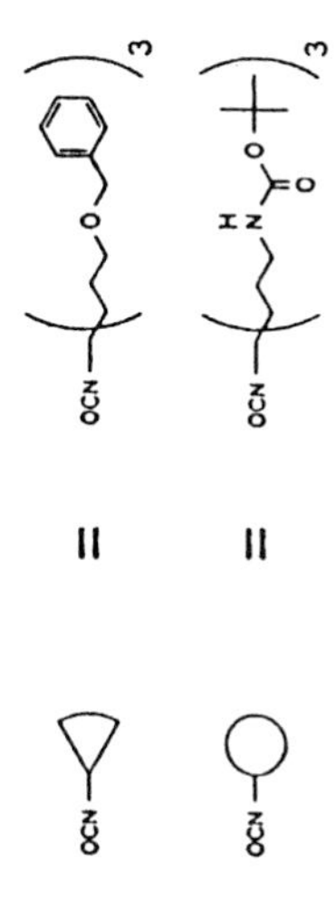

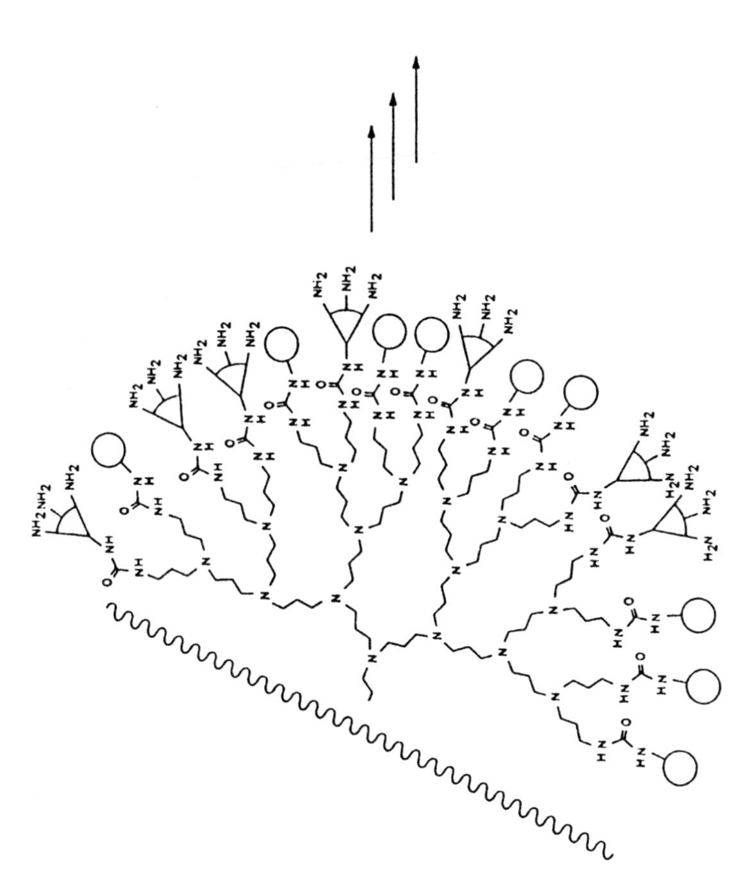

Fig. 117 Example of a dendrite synthesis with different building blocks (M. Fischer and F. Vögtle 1999 [154])

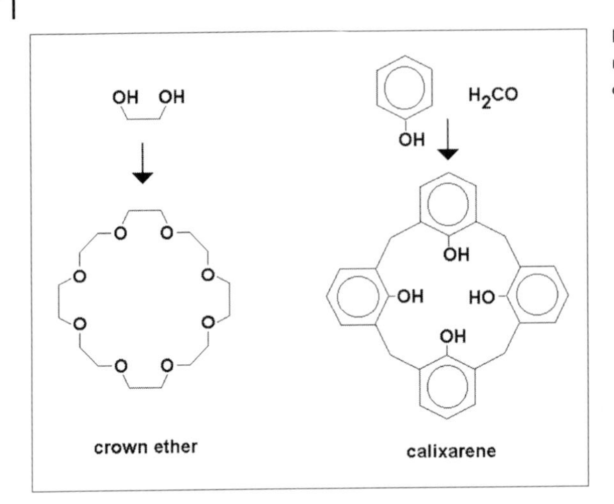

Fig. 118 Examples of simple macrocycles: crown ethers and calixarene

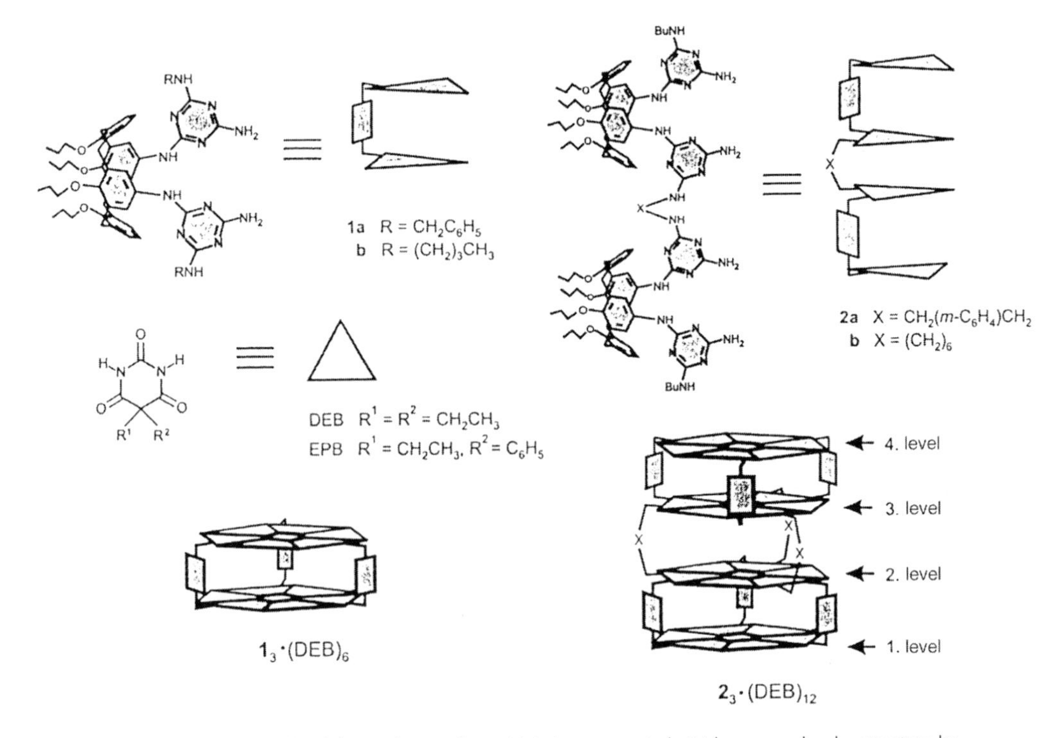

Fig. 119 Example of the synthesis of a multiple-interconnected, rigid, supermolecular structure by modular synthesis [150]

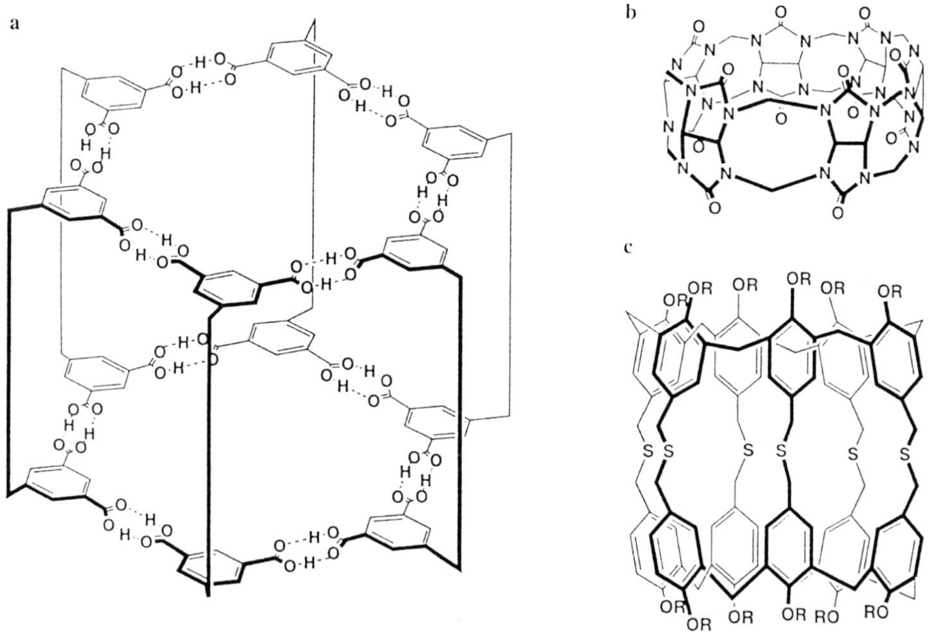

Fig. 120 Examples of synthetic molecular cage architectures [151]

Based on a complexed central ion, two- or three-dimensional ring structures with diameters of two or more binding length are accessible (Fig. 118). Simple rings are found in the crown compounds (coronands), e. g., the crown ethers. Bridged rings lead to spherical structures, the so-called cryptands. The ring structures of crown ethers or cryptands can hold ions or smaller molecules. Metal-catalyzed coupling of halogen-substituted bisalkyl aromatics or gas phase pyrolysis of reactive multiple-ring structures leads to macrocycles without a heteroatom, the so-called spherands. Stable or intermediary metal complexes that induce the formation of two or even three ring structures can also result in covalent not coupled but interlocked rings, the so-called catenanes [119]. Ring-shaped methylene-bridged phenols are described with various substituents and various ring sizes, such as 4, 5 and 6 units. These so-called calixarenes are interesting host molecules for exactly fitting smaller molecules, e. g., for sensory applications. Crown compounds, cryptands and catenanes are all – similar to the natural porphyrins – connected by covalent bonds.

The modular connection of bridged units allows the formation of even large molecules with high rigidity (Fig. 119) [150]. Multiple-bridged three-dimensional structures with larger cavities surrounded by bridges are interesting as hosts for intercalation compounds and are applied for chemosensors [151], but can also be utilized in molecular construction technology (Fig. 120).

While the long-term stable covalent bonds are preferably used for the formation of chemically stable basic molecular modules, the spectrum of bonds aimed at assembling these modules into larger aggregates is much wider. In addition to reaction and activation enthalpies, the binding and the conformational entropy determine the formation and the stability of such aggregates. The following binding types are observed [152]:

- reversible covalent bonds, e. g., disulfide bonds or vanadate and borate ester
- coordinative bonds
- hydrogen bonds
- ionic bonds
- electrostatic bonds between permanent dipoles
- charge-transfer interactions
- π-interactions in stacked aromatic groups
- hydrophobic interactions between amphiphilic molecules
- van der Waals interactions such as in crystalline or partially crystalline n-alkanes

Defined complex structures have already been generated using very simple structural motifs. So poly phenylene compounds represent a class that leads to regular linear, two-dimensional or band structures, with individual molecules being formed by a few up to several hundreds of rings. Typical diameters of rings and the widths of bands correspond to 3–8 phenyl rings (1.5–4 nm). Depending on the type of connection, either flexible or rigid structures are obtained. Polymeric linear arrangements organize, for example, in helical secondary structures, disk-shaped molecules aggregate often into stacked layers forming hexagonal columns [153].

Supramolecular structures can be formed by macrorings due to the restriction of the degrees of freedom of mobility by the ring shape. Chemical structure, shape and diameter of the rings can vary widely. Additional bridges over the rings yield extra stability, as do substituents on the outside of the rings and charges or aromatic states. They contribute to the rigidity as well as to solvation, depending on the solvent. Of particular importance is the relation of cavity size and form to the possibility of binding to a central ion or molecule. Such host–guest interactions are often essential for the synthesis of the macrorings, but also assist in molecular recognition [102].

Macrocycle compounds such as calixarenes are not simply recognition structures for small molecules. Assisted by hydrogen bonds, larger aggregates are formed when sterically defined units in the shape of a U or L, with restricted internal flexibility, are present. When such molecules, which are complementary due to hydrogen bonds, approach, coordinative or π-interactions between them result in assemblage of larger units or chain-like aggregates. Typical diameters of rod-like supermolecules are 3–5 nm, with a tendency to form two-dimensional arrays and three-dimensional stacks [150][155][156]. More complex three-dimensional structures are realized by synthetic aggregation of calixarenes with melamines [155]. Multicenter Ni complexes with chelating ligands form multiple planar rings, which stack spontaneously in aqueous solution into multiple tubes with inner diameters of about 1 nm and outer diameters of 3–6 nm [157]. Formation of non-covalent bonds yields chiral architectures out of small molecular components [158][159].

While compound classes are the primary ordering principle in classical synthetic chemistry, synthesis strategies for supramolecular architectures focus on geometrical and mechanical characteristics and less on the chemical nature. Thus, supramolecular synthesis moves away from a purely chemical approach towards engineering principles. So, for example, a systematic approach based on regular polyhedra can be applied. Such components are accessible through pure organic ring architectures, but also through polyoxoanions, coordinative compounds and Si organic or metal organic molecules [151]. Functionalization by polar or ionic groups leads stable spheroids, such as the Buckminsterfullerenes, into supramolecular aggregation. In addition to nanovesicles, nanotubes with diameters of several tens of nanometers and lengths of up to several micrometers were also observed [160]. Also, classes of molecules different from the biogenic molecules can form synthetic nanostructures through non-covalent interactions. The analogy with biogenic supramolecules is limited to a certain rigidity in the elementary units and the capacity for self-organization of molecular basic units, e. g., by groups of complementary hydrogen bonds. The regular aggregation of smaller molecules into more complex but ordered units is a basic economic principle [152].

Self-assembling structures of dodecyloxybenzylidene-pyrimidinetrione and amino-didodecylamino-triazine are micelle-like but result in a well-defined supramolecular structure (Fig. 121). Three molecules of each component form the supramolecular aggregate of a nucleus containing aromatic rings held together by hydrogen bonds,

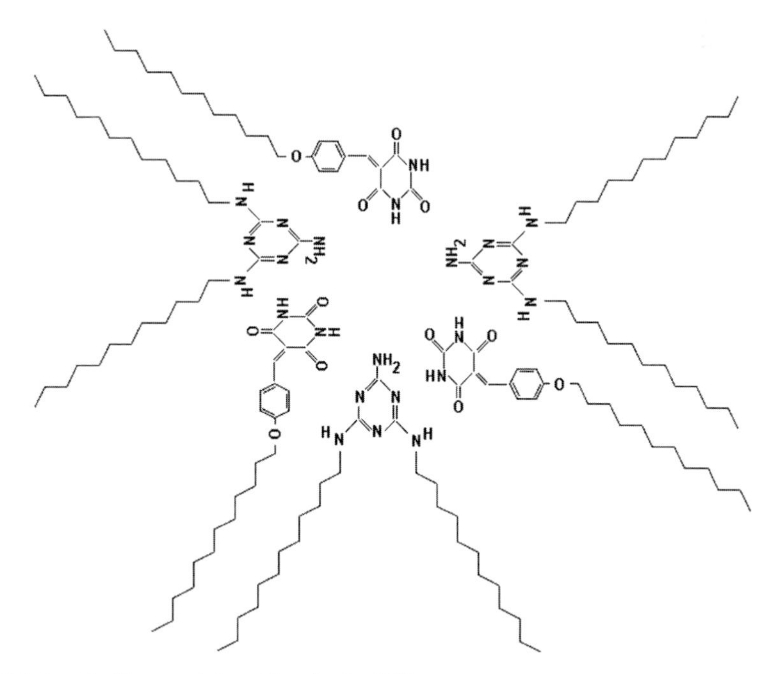

Fig. 121 Disk-shaped adduct consisting of 6 molecules [161]

and alkyl chains screen this nucleus from the outside. These aggregates assemble into long filaments with diameters of a few nanometers and lengths of several micrometers. The aggregation of these filaments yields supercoils with dimensions of 10 μm and 2 μm for length and diameter, respectively. These supercoils rely on the following hierarchical structure:

(4) supercoils bundle of filaments
(3) filaments stacks of low-molecular assemblages
(2) aggregates complexes of several (in this example 2 × 3) molecules stabilized by hydrogen bonds
(1) molecules consisting of a rigid core with H-bond active groups and long-chain aliphatic residues

This system unites aspects of macromolecular synthesis, supramolecular chemistry and the chemistry of nanoparticles.

5.6.6 Nanoparticles and Nanocompartments

From the point of view of chemistry with huge numbers of particles, there are two approaches to the architecture of individual molecules. The first approach restricts the number of particles by decreasing the concentration and the reaction volume. The second relies on subdivision of the reaction volume, so that it is not the overall particle numbers but the quantity of particles in each portion that is decreased. One straightforward technique reduces the number of dimensions in integer steps, from the volume onto the surface and finally ending up in boundaries on the surface, which are utilized for the immobilization of the molecules prior to reactions. Another method is the use of fractal dimensions by subdivision of the three-dimensional volume, yielding micro- and nano-heterogeneous systems. A third possibility is based on domain boundaries, determined by local changes in the ordering state. These boundaries are planes in space where the orientation or ordering of the particle changes.

Such local changes of the ordering state often occurs in solids and can be observed in the liquid-crystalline phase. Analogous states can be found in molecular monolayers and stacks of molecular films. Here both rigid systems (comparable to the solid state and therefore comparable to polycrystalline conditions) and flexible systems (comparable to the liquid state) are observed.

To facilitate highly complex biochemical processes in cells, compartmenting (the creation of subdivided reaction volumes) is an essential principle. It enables highly structured cells for the formation of highly specialized tissues and organisms to exist. Membranes, consisting of a double layer of amphiphilic molecules, subdivide the reaction volumes. Moreover, they ensure the defined transport of charges and particles between the compartments. Proteins inserted in the bilayer typically control this transport. Properties of such systems have already been demonstrated with synthetic structures in artificial membranes. So cation channels could be obtained by aggregates of oligothiophenes integrated into a membrane and bridged as ligands in a coordinative manner by metal ions [162].

In a broader context, supermolecular nanoobjects also include solid-like nanoparticles. Nanoparticles exhibit a three-dimensional binding network. The density of bonds is high, so that moderate changes to the external parameters do not change their shape. Nanoparticles are stable with respect to both internal bond topology and outer shape. The variety of nanoparticles is huge, ranging from species defined by stoichiometry and both number and arrangement of the bonds to non-stoichiometric particles with a certain distribution in parameters, such as size, etc.

Nanoparticles exhibit different types and numbers of reactive surface groups, and they can be connected in specific ways. These are prerequisites for complex architectures. Solid nanoparticles are amenable to secondary functionalization. The functionalization of different groups on defined surface regions of one nanoparticle is more difficult. To facilitate highly defined supermolecular construction based on nanoparticles, techniques for differentiated chemical modifications of surface regions are required.

A wide variety of elementary, inorganic and organic nanoparticles have been developed. Particles with periodic arrangements of the atoms (crystal-analogues) or amorphous (glassy) structure have been described.

Solid nanoparticles are related to the metal clusters. While nanoparticles do not typically exhibit an exact number of atoms or distribution of the particle size, clusters exhibit a defined number of atoms and represent, therefore, molecules. Large inorganic architectures have been created using substituted metal clusters. Based on selenium and triethylphosphine substituents, molecules such as $Cu_{70}Se_{35}(PEt_3)_{22}$ or $Au_{55}(PPh_3)_{12}Cl_6$ have been synthesized [163][164]. The ligand periphery stabilizes the cluster and accomplishes the solubility by surface charges or solvatable organic groups simultaneously.

Supramolecular aggregates of small molecules with the character of nanoparticles are generated when the molecules interact with each other through forces slightly stronger than the solvent interactions. This case occurs typically for molecules with two parts that have quite different solubility properties. It is clearly visible with am-

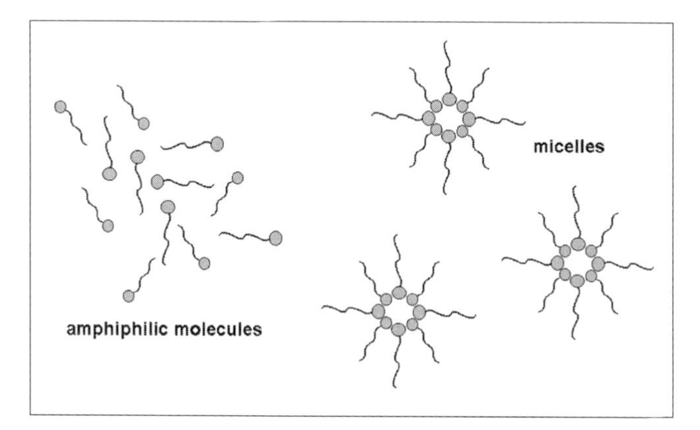

Fig. 122 Formation of simple supermolecular structures by the assembly of amphiphilic molecules into micelles (cross-section of a spherical micelle)

phiphilic compounds: they form lipophilic bonds such as van der Waals interactions using the larger, nonpolar portion, and a smaller polar and often protic portion for ionic or hydrogen bonds. The classical structure of an amphiphilic molecule contains a hydrophilic head and a lipophilic tail (Fig. 122). Typical examples are fatty acids and their anions, but also alcohols of longer chain lengths and amines.

These molecules show a tendency to enrich on interfaces. They occupy the fluid surface due to the minimization of the surface energy, and liquid–liquid interfaces. If amphiphiles are mixed with other solutions, a homogeneous distribution in the volume phase is observed below a critical concentration. At higher concentrations, spontaneous aggregation of the amphiphilic molecules in the liquid phase is observed. Microscopically small internal phases, micelles, are formed. Therefore, the critical concentration is denoted as the micelle concentration. Depending on the mixture, the micelle phase shows various shapes and extensions, such as spheres, rods or layered micelles. Under specific conditions highly complex structures are formed.

5.7 Combination of Molecular Architectures and Nanoparticles With Planar Technical Structures

A basic challenge for molecular nanotechnology consists of the connection of the inner symmetry of molecules with external (accessible through lithographic micro- and nanotechnology) coordination systems. This challenge includes the extension to the six degrees of freedom in space: the three translation directions for positioning, and the three possible rotational axes for orientation. All six parameters have to achieve a position to realize fully functional hybrids between the planar technology and the molecular architecture. Lesser requirements apply when the molecule can be assumed to be a compact symmetrical object. However, the precision of the connection to the planar environment has to be at least the same as the largest extension of the molecules. The case when the connection of planar technology requires a certain location on the molecule is more complicated, so a precision down to atomic dimensions could be needed.

The resolution of the best lithographic techniques is not sufficient in such cases. Although manipulations of atoms are possible for specific systems, it requires huge efforts and is not a standard technique. Electron and ion beam lithography as well as scanning probe techniques reach resolutions in the medium nanometer range, but lower nanometer dimensions are problematic (cf. Section 4.3). So molecular techniques are required not only for the provision of the molecular units, but also in assisting with the positioning and orientation of molecular units in nanotechnical environments, covering the lower and medium nanometer range.

Rigid units are required for a molecular base with a sturdy geometry and geometrically defined connections between the solid substrate and molecular architectures. They have to exhibit a chemical–topological and geometrical stability, also with respect to moderate changes in the environment, such as changes in temperature, pressure, pH, and solvent. In addition to geometry and molecular function, process compatibility and robustness are also important parameters. Even the proteins, which are successful systems in nature, do not fulfill all of these requirements. They are typically sensitive to changes in temperature and the surroundings, and react through denaturing, a process that degrades the three-dimensional structure necessary for biological functions. Fortunately, nature demonstrates how certain proteins also function at elevated temperatures in proteins from cells living under extreme conditions, such as in hot springs. This example also motivates the design of more robust bioanalogous molecules to increase the stability of sensitive molecular systems found in nature.

The fabrication of complex molecular-nanotechnical structures often requires the generation of molecular aggregates on surfaces or in a homogeneous phase prior to immobilization on the substrate surface. Therefore, macromolecular and supramolecular architectures exist in liquids or directly on substrate surfaces (cf. Section 5.4). Another approach produces modules in solution that connect molecular structures with a nanostructured non-molecular solid. Such integrated architectures are similar to large molecules. Metal nanoparticles are generated as a colloidal solution. These solutions are stable and exhibit only a low size dispersion of the particles, when surface groups stabilize the particles. Besides smaller molecules, larger or supramolecules are also possible surface modifications. Gold nanoparticles with a larger organic shell were prepared by modification with alkylthiol-substituted crown ethers [165].

Appropriate functionalization of nanoparticles with complementary surface substituent results in the formation of individual nanoparticles into complex architectures. Based on the sequence-coded hybridization specificity of oligonucleotides, DNA is a convenient way to connect nanoparticles in a defined manner [166].

Nanoparticles of compact molecular architecture are possibly based on dendrimers. In addition to organic groups, metal ions can also serve as cores for the dendrimers. The immobilization of individual dendrimers on a plane surface is achieved by integration into a complete monolayer of a smaller molecule. Often the ideal properties of alkylthiol layers on gold are utilized. The integration of alkyle sulfide-substituted palladium dendrimers in a decanethiol monolayer was detected using AFM [167]. The introduction of chemically active single molecules in an inert monomolecular layer allows the generation of disperse particles on surfaces. Thus, substituted bishydroquinone inserted into a decanethiol layer generated nanolocalized Ag_4 seeds prior to enhancement into an Ag nanoparticle [168].

A serious challenge for the chemical synthesis is the principle of hierarchical bond strengths applied by nature, especially for three-dimensional structures using weaker interactions. A bio-analogous synthetic chemistry has to transform the example of the proteins into technically usable molecular units and tools with robust three-dimensional networks. Again nature provides an example with disulfide bridges that are utilized to connect the cysteine residues of different polypeptide chains or from different domains of a long polypeptide chain in order to stabilize the three-dimensional structure. Similar approaches in synthetic chemistry use for example light-activated coupling groups in the desired positions to achieve covalent cross-links inside folded molecular chains or between molecules that are connected only weakly. Micelles and phases of liquid-crystalline states provide only limited alternatives to macromolecules, because they are geometrically less defined and not sufficiently stable against changes in the surroundings. Such supermolecular complexes that are held together by only weak binding forces require additional internal bonds for the stabilization of nanotechnical applications. Internal nanophase bond formation transforms nanophases in real nanoparticles in a way that preserves their shape and their material composition, even after considerable changes to the surroundings. Photochemically or thermally activated bridge formation or polymerization can be used for an internal stabilization of nanophases as nanoparticles.

Solid-like nanoparticles exhibit a dense network of bonds. The bonds of metal and inorganic nanoparticles are so strong that moderate changes in the environment are usually tolerated and the shapes remain constant. Molecule-like nanoparticles can present a defined number and type of reactive surface groups, which can be used to connect the nanoparticles with each other in a specific way. So molecule-like nanoparticles are suitable for the construction of complex architectures. Solid-like nanoparticles can also be functionalized, but the arrangement and density of reactive surface groups is distributed more broadly, and it is usually problematic to generate different chemical functionalities simultaneously on one surface. Only a statistical approach can be applied: certain densities of surface groups or mixtures are created, and subsequently the resulting derivatized particles are separated due to their differences in functionality. In order to realize defined molecular constructions with nanoparticles, processes for a differentiated chemical functionalization of surface regions of nanoparticles have to be developed.

For the construction of complex architectures, at least three different coupling functions per particle are required. In the simplest case, every nanoparticle exhibits exactly one of each type. Depending on construction strategy and steric requirements, sometimes several groups are also possible, as long as they are confined to certain areas. It should be possible to separate particles prepared in this mixing synthesis, e. g., by three columns that each react reversibly and specifically with one of the three functions A, B and C. So the first column (A′) holds only particles with A, and so on. Thus three-fold functionalized particles can be separated from two-, one- or non-functionalized ones by first using column A′, then rinsing and elution prior to column B′, then rinsing and elution prior to column C′ (Fig. 123). However, this procedure does not guarantee a homogeneous distribution on the particle surfaces or a defined quantity of binding groups of a certain type. It would result in something similar to the selective binding of molecule-like nanoparticles. This approach is promising, because fairly large rigid objects could be used in nanoconstruction technology, without the huge synthetic efforts required for similar objects prepared by supramolecular synthesis.

The assembly of micro- and nanoparticles can be supported by lithographic structures. Here, suspensions of surface-functionalized nanoparticles are deposited on lithographically generated spot arrays that facilitate binding by presenting complementary chemical functions.

It has been demonstrated that functionalized nanoparticles can undergo specific chemical reactions with each other, and also that larger nanoparticles are utilized similarly to molecules [169][170][171]. Particles of both the same and different sizes with complementary functionalization have been used to form supermolecular-like nano-

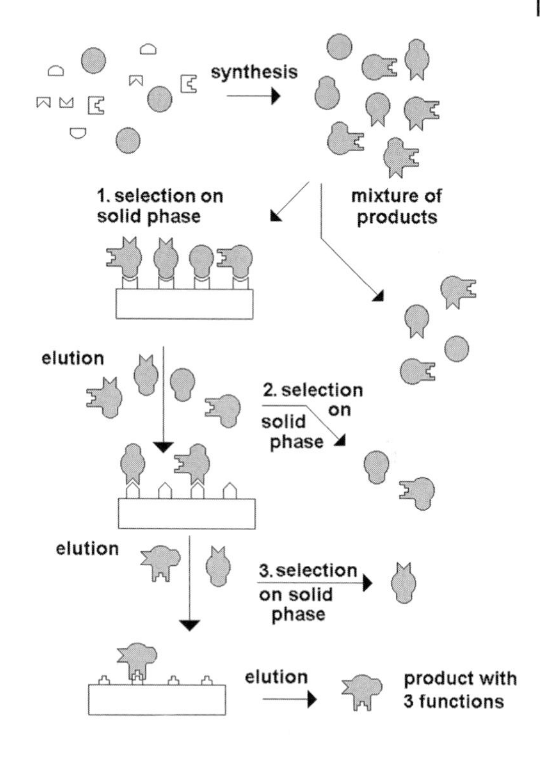

Fig. 123 Random synthesis of multi-functionalized nanoparticles and selection of particle types by specific surface binding and elution

particle aggregates in solution or suspension (Figs. 124 and 125) [172][173]. One has to ensure that the tendency towards unspecific binding, which increases with increasing particle diameter is suppressed by the choice of solvent and the chemistry of the chip surface. Gold nanoparticles of 30 nm diameter still exhibit a high specificity and a high

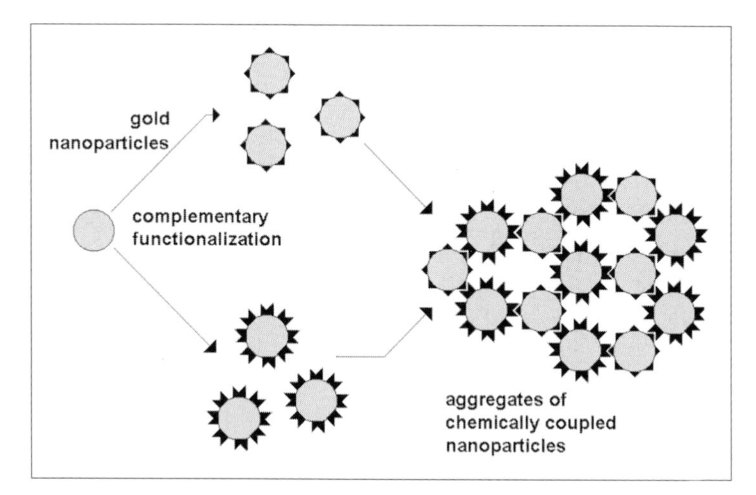

Fig. 124 Formation of larger clusters by the reaction of two classes of complementary functionalized nanoparticles [172]

density in coupling by DNA hybridization. Therefore, complementary DNA single strands are immobilized on the substrate (chip) and the nanoparticle surface (Fig. 126). The area of binding can be restricted using lateral lithographically structured substrates (Fig. 127). Binding areas with the dimensions of a nanoparticle diameter should limit the nanoparticle binding to exactly one particle. So a geometrically defined connection between lithographically (planar technology) prepared substrates on the one side and molecular objects on the other should be achieved.

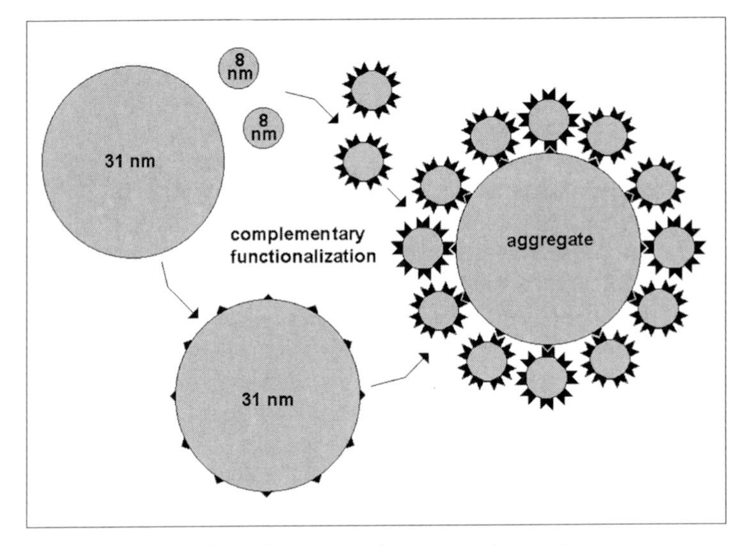

Fig. 125 Formation of complex nanoparticle aggregates by complementary functionalization of smaller and larger nanoparticles [172]

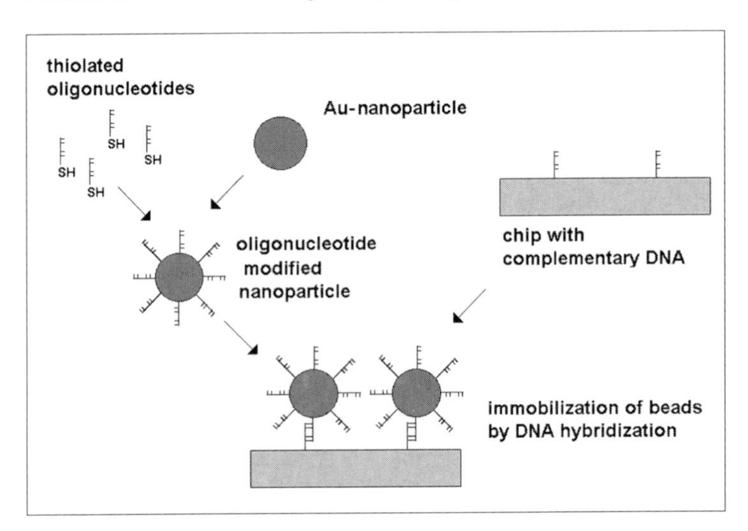

Fig. 126 Binding of surface-functionalized gold nanoparticles on complementary functionalized chip surfaces

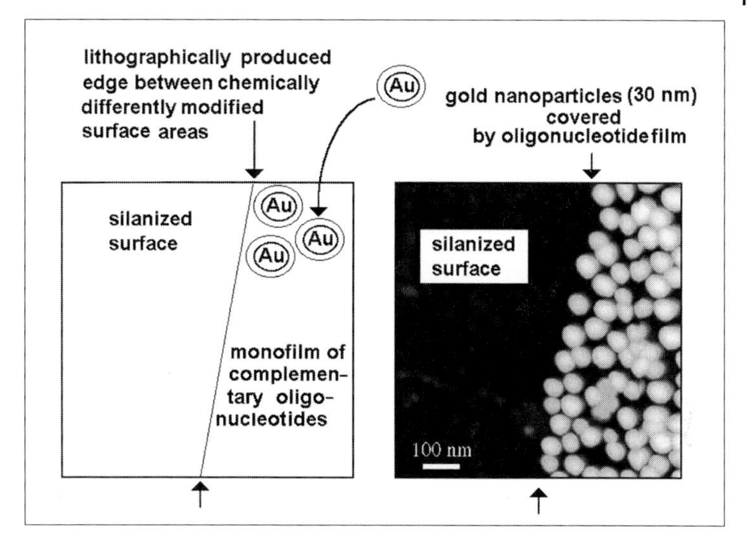

Fig. 127 AFM image of a lithographically prepared edge between a passivated and a functionalized surface region after selective binding of complementary functionalized gold nanoparticles [176]

A further prerequisite for the fabrication of nanoarchitectures from molecular or molecular-like units is their ability to be manipulated in the liquid phase. Particles that have to be chemically manipulated, immobilized before or after surface binding orientation, fixed and modified must be soluble in the liquid phase to be transported by convection and diffusion. In particular, rigid molecules have to provide a sufficient number of surface groups that will interact with the solvent in preference to with each other. In principle, these molecules could be identical with the coupling groups, especially when they are utilized during or after immobilization. If these groups are saturated in solution, uncontrolled aggregates or even precipitation are possible. So it is preferable to separate the solvation process from the coupling chemistry. In aqueous solution, ionic surface groups in particular increase solvation. Ionic surface groups are connected through local dipoles in the molecules, supporting interactions with the small dipolar molecule water. Furthermore, the ionic character supports the formation of hydrogen bonds, and similarly charged surfaces minimize non-specific aggregation. As well as ionic groups, OH groups or HSO_3 rich side chains also enhance solvation by protic solvents.

In apolar, aprotic solvents it is the less polarizable flexible residues that mainly assist in solvation. For manipulations in such media, rigid supermolecular aggregates require assistance, e. g., to form aliphatic side chains. Owing to their high flexibility, oligo- or polyethylene glycols are suitable. Their contribution to the stabilization of the shape is negligible, but the high density of freely rotating bonds guarantees an increased solvation.

Assembly can also be achieved if arrays of hydrophilic spots on a hydrophobic background are exposed to aqueous nanoparticle solutions [174]. Regular lithographic sur-

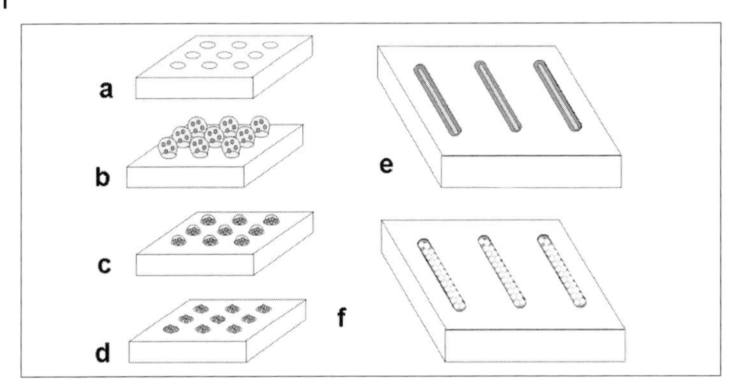

Fig. 128 Assembling of nanoparticles suspended in aqueous solution: a) hydrophilic nanospots generated on a hydrophobic substrate surface, b) self-organization of droplets containing nanoparticles on the hydrophilic surface spots, c) reduction of droplet volume by evaporation of water, d) nanoparticles assembled in groups on the lithographically prepared nanospots, e) lithographically prepared trenches with a hydrophilic surface etched into a substrate with a hydrophobic surface, f) assembling of nanoparticles in the hydrophilic trenches after solvent evaporation [174]

face patterns on chips lead to regular arrays of nanoparticle assemblies if the solvent is evaporated from suspensions on the chip (Fig. 128, a–d). The motion and aggregation of nanoparticles is caused by capillary forces. As a result, rather regular ensembles of up to about a dozen spheroidal nanoparticles on each spot in an array can be obtained. The number of particles per spot depends on the ratio of spot diameter to particle diameter. Instead of circular spots, linear structures and nanogaps can also be applied for the trapping and assembling of the nanoparticles. In the case of larger trenches, the formation of double rows of nanoparticles was observed due to the capillary-force-induced trapping of nanoparticles in the bottom edges of trenches (Fig. 128, e, f).

Silica nanoparticles have been assembled by a spin-on process in trenches generated by RIE of SiO_2 films using a lithographically prepared gold mask [175]. Spheroidal silica particles with diameters between 15 and 78 nm have been deposited in holes and trenches with cross-sections between about 50 and 100 nm. Strips of nanoparticle multilayers could be obtained by direct adsorption of the silica nanospheres on hydrophilic surface elements generated by interference lithography and subsequent etching and hydrophilization steps.

The coupling of molecules that undergo weak but specific bonds with other molecules on rigid nanoparticles has two effects: apart from a mechanical stabilization, there is also a significant increase in chemical stability. For two fairly flexible molecular chains that are interacting (independent of whether they are unspecific or highly specific) the longer the interacting region is and the larger the independent sections of the molecules are the more stable the bond is. The activation energy for a dissociation of both chains is lower for a short range of cooperative movements of rotating groups in a chain than for a wider range. Hence rigid molecules form more stable aggregates for a given type of interactions than flexible ones. When short-chain flexible molecules are coupled to nanoparticles, the flexibility in the chains remains, the connection to the

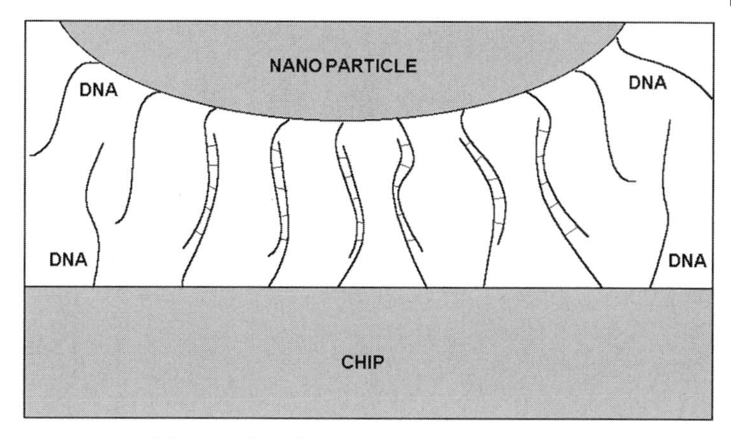

Fig. 129 Immobilization of metal nanoparticles on chip surfaces by DNA hybridization

particle creates a rigid environment. This effect is enhanced considerably when several short-chain molecules able for weak bonds to other chains are immobilized on the nanoparticle, and so are forced into cooperative effects. It becomes even clearer with the case of rigid connections of both sides of the reaction partners (Fig. 129). The specific coupling of nanoparticles on microstructured surfaces (Fig. 130) can be used as an alternative labeling technique for biochips in addition to fluorescence methods [171][176][177][178][179][180][181].

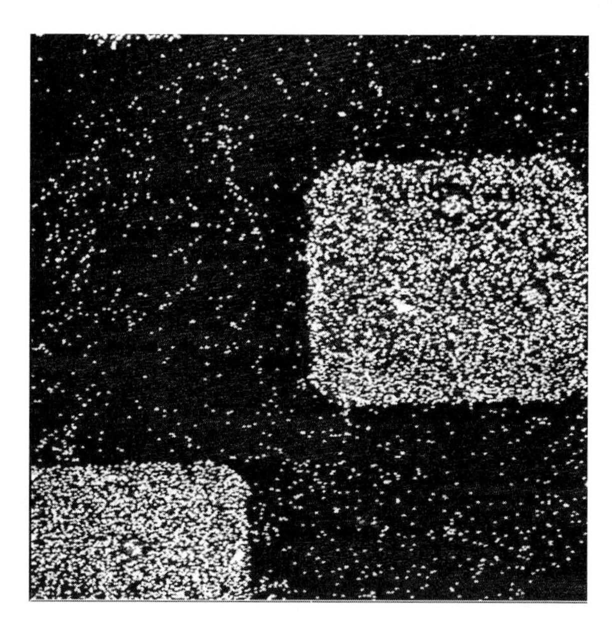

Fig. 130 Labeling of microstructured oligonucleotide spots by nanoparticles [176]

6
Characterization of Nanostructures

6.1
Geometrical Characterization

6.1.1
Layer Thickness and Vertical Structure Dimensions

For transparent layers, the thickness of complete layers can be determined optically. Light can be used to provide measurement possibilities even below its wavelength, if the phase information is used. Every interface between optically transparent media that has a change in refraction index results in a reflection of some fraction of the transmitted light. Two subsequent interfaces induce the reflection of two fractions of the light beam that are phase shifted with respect to each other. Interference occurs for coherent light, so that an amplification (constructive inference) or reduction (destructive interference) is observed. As a result of double reflections, reflected light rejoins the main beam and interference can occur.

The degree of amplification or reduction by an interference is caused by the phase difference of the reflected beams. If the difference is exactly one wavelength or a multiple of it, then 100% amplification occurs for beams of equal intensity. For a half-wavelengths or half-wavelengths plus multiples of whole wavelengths, a total reduction is observed for equally intensive beams. For ultrathin layers and the application of visible light, the differences between the different beams amount to just a few fractions of the wavelength. Hence the intensity of interfering beams is not that different from a non-interfering beam. These small intensity differences change with layer thickness even in the lower nanometer range, and they are measurable. As a result of the high precision of the intensity measurements, differences in layer thickness down to below 1 nm can be determined.

Interferometer thickness measurements require constituent layers with lateral dimensions much larger than the wavelength. The precision increases with larger areas. An absolute determination of geometrical layer thickness requires additional information on the refractive index of the layer material. As an approximation, the value for an analogous bulk material can be used. However, it should be remembered that anisotropic crystal arrangements and the orientation of molecules in ultrathin layers yield optical anisotropies, so that the refractive index differs significantly from that of the bulk material, and moreover depends on the angle of the incoming light.

Nanotechnology. M. Köhler and W. Fritzsche

ISBN: 978-3-527-31871-1

If light at an angle of incidence is partially reflected on an interface, the transmitted and reflected beams experience different polarizations. The degree of polarization depends on the incidence angle of the light. Through the partial reflection on both interfaces of thin and ultrathin layers, this effect is overlaid by the interference of both reflected beams, so that the phase difference of the reflected beams contributes additionally to the intensity, depending on the direction of polarization. This phenomenon is utilized in measurements of the polarized beams from a double reflection at two different angles of incidence. This ellipsometric technique yields both refractive index and layer thickness. It requires a measurement area of a few square millimeters for conventional instruments for a precision of 1 nm or better. The determination of the refractive index also yields information on the composition of the layer. Ellipsometric microscopy provides information regarding the refractive index and layer thickness in the lower nanometer range with a lateral resolution of a few micrometers.

Geometric characterization of nanostructures is mainly carried out using ultramicroscopic methods. Apart from the layer thickness at least one lateral dimension is in the nanometer range, thus light microscopy is of limited use and is usually only applied as a qualitative method. A quantitative determination of nanostructures requires techniques with higher resolution. Interferometers allow the determination of structure heights down to the sub-nanometer range. However, this technique requires two or more reference areas with dimensions above the optical diffraction limit. Hence this highly precise optical method is applicable to the measurement of the thickness of thin layers, but is not suited to the determination of the topography of lateral nanostructured layers.

In addition to general height information, topographic characterization also requires details on the local differences in structure height. In the example of one structured layer on a plane substrate, which is complete and highly selectively structured compared with the substrate, the topography of the resulting structure is determined by layer thickness and the lateral pattern. Hence topography can be reduced to the determination of the lateral structure. This ideal situation is rarely found in reality. In nanotechnology, the substrate relief is usually not negligible, and structure flanks are important. Deposited layers are sometimes not smooth, but exhibit a surface relief due to their internal morphology. Etching processes (especially dry etching) are not ideally selective, so that a certain removal of the underlying material is observed. Often not just one layer, but stacks of structured layers are utilized for structure generation. So the layer thickness alone is not sufficient to describe the geometry of a sample surface, but methods for a three-dimensional determination of surface coordinates are needed.

Because of its high local resolution, the scanning electron microscope (SEM) is a suitable technique for the elucidation of nanotopographies. Secondary electron images can already give an impression of the nanotopography of a sample for samples slightly tilted towards the electron beam. Instead of a complete topographic map, images taken at different angles are for example combined or edge preparations of typical sample regions are prepared and imaged.

SEM allows in principle the generation of complete 3D datasets of sample surfaces for sufficiently textured substrates. Automated methods of image correlation (area correlation) are then applied to calculate the height coordinates (z) for the lateral coordinates (x, y) from two or more images of the same surface area taken at different tilt angles. This procedure is one adapted from aerial photography and yields topographic data sets in the nanometer range [1]. For precise measurements in the medium and particularly in the lower nanometer range, edge preparations are mandatory. SEM characterization of edge and cross-section preparations yield good relative structure heights, but the absolute heights could be altered during the preparation. For nanostructures with high aspect ratios, SEM provides the best possibilities for visualization and measurements.

In the last two decades mechanical nanoprofilometry and especially scanning probe techniques have become very important methods for the determination of surface topography. The measurement of the deflection of a mechanical probe that is raster-scanned over a surface is in principle able to determine lateral height differences in the sub-Angstrom range. The lateral resolution is mainly due to the geometry of the probe tip. In general, the measured signal reflects a convolution of the surface topography with the sample geometry. So topographies with small aspect ratios are described fairly well, but structures with high aspect ratios are changed (Fig. 131).

The instruments for mechanical profilometry differ particularly in the shape of the scanning tip, the mechanical force applied to the tip and the precision of the readout. The best resolution and the lowest forces are achieved with the scanning force microscope. A change in the measurement results due to layer deformation depends not on the force alone but on the pressure, so that probes with larger diameters are more suitable in such instances. For trench measurements, the tip has to penetrate down to the bottom of the trench (that is to the substrate) for accurate results (Fig. 131). Through layer thickness measurements on the edges, the changes to the geometry can be neglected.

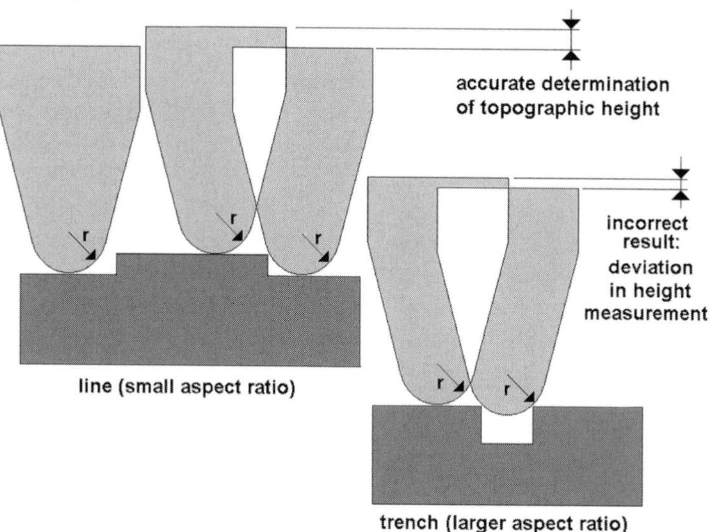

Fig. 131 Limits of height measurements for trench structures with high aspect ratios

Surface topographies are obtained by scanning the surface in parallel, adjacent line scans. The various techniques apply different methods to reading out the z-information. Classical profilometers use a direct readout, e. g., by electric, piezoelectric or electromagnetic transformation. The extremely high precision of scanning probe techniques in the z-direction is achieved with a light pointer in combination with fast and high-precision control of the probe–sample distance by piezo actuators.

Molecular monofilms and ultrathin inorganic layers are essential components of nanosystems, so that their characterization is a central area in nanotechnology. The thickness of the layers ranges from one shorter molecule up to the lengths of several small molecules, which means about 1 to some tens of nanometers. The method of readout of the z-position determines the precision significantly. While classical profilometry has a z-resolution of 10 to several tens of nanometers, the scanning probe techniques achieve a precision of better than 1 nm. For a layer thickness measurement of ultrathin layers, mechanical-profilometric techniques require an exposed structure edge. This edge should include both the exposed substrate as a reference and areas with the complete layer thickness. The preparation of this edge, including a clean reference area, is a general problem for mechanical profilometry. The measurement precision is increased by the presence of several edges, as in the case of a microstructured layer. Using such samples, the effects of tilted samples and inhomogeneities of the substrate and layer surface can be minimized. A tilt in the angle of the light is already problematic, because, for example, with a 100 μm scan a tilt angle of one arc minute induces an error of about 30 nm, which is ten times the thickness of a typical ultrathin layer.

Scanning probe techniques are particularly suitable for the elucidation of topographic structures due to their enhanced precision in the z- compared with the x- and y-directions. With STM, topographic information is accessible with sub-atomic precision. AFM normally achieves vertical precisions in the region of one atom diameter.

Owing to its high resolution, transmission electron microscopy (TEM) is in principle appropriate for providing topographic data in the nanometer range. The thin cross-section preparations required for informative images are so difficult to fabricate that only in certain exceptions are they prepared to provide detailed information on the topography of a layer stack. In addition to the high resolution, TEM provides the advantage that not only the sample surface, but also the interior, e. g., the arrangement of the layers, is visualized due to the contrasts in the material. On the other hand, a digital stereogrammetric processing of image pairs or series comparable to the SEM is complicated, because the local grey value is not only determined by the surface but also by the local interior. So a correlation of the grey value is problematic. Only in the case when just one interface out of a series exhibits textures, is a correlation possible. Then the topography of even an internal interface can be determined.

6.1.2 Lateral Dimensions

Lateral dimensions present the central geometric feature of nanostructures and typically they determine the function. Metrological techniques are essential for the development of structuring techniques and also for quality control of the fabrication of micro- and nanotechnical devices [2][3][4]. Electric control structures are also applied to the development of solid-state circuits to address typical production failures [5].

Electron microscope techniques are more important for lateral measurements than for layer thickness or topography determinations. However, this technique is of limited use as a non-destructive tool in nanotechnology. Although special microscopes with larger sample chambers have been demonstrated to facilitate the use of whole wafers to avoid sample destruction, the sample surfaces are frequently altered by the electron beam or by an additional conductive contrasting layer of metal. To achieve high-resolution electron micrographs, the sample surface has to be electrically conductive. For nanostructures of thin layers, a conductive underlying layer or substrate can be sufficient, and no additional conductive layer is needed. In many cases, such a conductive layer is required. Platinum, carbon or gold coatings are typically applied. They exhibit both an excellent conductivity even for layers of low thicknesses and only a minimal additional texture due to grain structure. The resolution of an SEM for lateral measurements is between about 2 and 10 nm. For very small dimensions, the material contrast determines the visualization, because high-resolution images need a certain beam voltage and the penetration depth of the electron is in the range of several tens of nanometers. Fine structures of heavier elements give rise to a high contrast on an appropriate background. Hence for elements with high atomic numbers, a resolution of 1 nm is possible, compared with lighter elements that exhibit a greater penetration depth and low material contrast, so that typical resolutions are about 10 nm. Organic material and biomolecular nanostructures in particular are not resolvable in the lower nanometer range. Often – in particular when using oil pumps – a contamination of the sample surface by carbonaceous layers is observed on the substrate surface and under the influence of the electron beam can obstruct the images of small (especially organic) structures. Owing to the high surface sensitivity of the signal at low beam voltages – especially in the BSE mode – organic structures are visible down to the molecular monolayer with lateral resolutions in the sub-μm range.

TEM achieves a better resolution than SEM. Samples prepared in an ideal manner allow resolutions down to the sub-nm range. Such measurements require a minimum thickness of the samples for penetration of the beam. So for the determination of lateral dimensions in one direction, cross-section preparations with thicknesses between 0.1 and 1 μm are needed. For measurements in both lateral dimensions, the thickness of the reverse side of the substrate has to be reduced appropriately. To obtain high-resolution images of organic nanostructures, a chemical labeling with particular substituents (elements of higher atomic numbers) is introduced. As a result of the complicated sample preparation method, TEM is not used as standard method.

Mechanical scanning probe techniques have also become an essential tool in the characterization of lateral dimensions in nanotechnology [6]. Instruments with large sample platforms are now available that allow the non-destructive characterization of extended substrates (e. g., in a wafer format). While in principle STM provides resolutions down to the atomic range, it is restricted to use with conductive substrates. AFM, which can be applied on any substrate, has therefore become the preferred scanning probe technique for the characterization of lateral dimensions of nanostructures. A key prerequisite for AFM studies is a level substrate below the nanostructure of interest, to avoid topographical "noise" that could mask the signal leading to the vertical dimension of the nanostructure. Scanning probe techniques are applied for measurements of lithographic structures and for the calibration of SEM measurements [7].

Investigations of the precision of the measurement of lithographic structures show that absolute precision is not guaranteed by a single technique. Calibration structures are required. A good combination is AFM with SEM measurements. For structures with dimensions of 150–350 nm (measured at the edges of broken samples), both methods differed by less than 10 nm; for measurements on complete samples, SEM exhibited a significant deviation of 50–100 nm [8].

6.1.3 Structures that Assist Measurement

Nanolithographic structures themselves are applied to solve measurement problems in the nanometer range that are connected with the fabrication of the smallest structures. In particular, when several structures have to be aligned with each other in numerous steps, the precision of the geometric relationship is essential. The alignment problem of multiple lithographic levels is a key microtechnical problem, but it is even more pronounced in nanotechnology where small absolute positional or angular errors result in the failure of devices.

For positioning with a resolution of 2.5 nm, a grating structure with a periodicity of about 200 nm was developed, which works in a manner comparable to a nonius, through small deviations in the grating periods. It relies on the phenomenon that it is easier to distinguish small relative deviations in the positions of lines quantitatively than to establish absolute length measurements. When both grating structures are fabricated in close proximity, the number of registered grating lines in both structures yields the positional deviation of both structures. Using this tool, gate electrodes for single electron tunneling elements can be positioned with a resolution of 10 nm [9].

The feedback of structure measurement and positioning is essential for planar technical fabrication of multi-level structures, as applied in practically all chip production for IT. A precise construction of lithographic mark structures with symmetrical edges allows a reduction of the scattering in the alignment of X-ray lithography (the so-called 3-σ value) below 10 nm [10][11].

6.2 Characterization of Composition of Layers and Surfaces

6.2.1 Atomic Composition

In nanotechnology, the characterization of the composition of thin layers and surfaces is of similar importance to the determination of geometrical parameters of small structures. A wide spectrum of methods can be applied to determine the essential parameters of layers, addressing, for example, geometrical as well as chemical or physical-functional parameters [12]. Probe techniques that address only a small surface area are particularly suitable for the characterization of small structures. The same probes that are used at higher intensities for nanofabrication are also utilized for characterization: focused particle beams and short-wavelength electromagnetic radiation. In contrast to lithography, for characterization purposes the focused beam should not induce changes in the layer. This requirement can only partially be met for focused probes of energetic particles or short-wavelength photons.

Electron-induced X-ray Spectroscopy

Electron-induced X-ray spectroscopy is a method for the determination of the atomic composition of a thin layer. Through the impact of energetic electron beams on a target, the electrons are slowed down by interactions with the target atoms. The deceleration of the charged particles induces the emission of electromagnetic radiation. With increasing kinetic energy of the electrons, the emission edge of this radiation shifts to a shorter wavelength range. As a result of the typical electron energies of around 10 keV, the spectrum starts in the medium X-ray and extends through the UV to the visible range. Individual sharp emission bands are overlaid around this broad continuum of Bremsstrahlung radiation. They originate through the direct interactions of the beam electrons with electrons of the inner shells of the target atom. With sufficient energy transfer, the shell electrons leave the solid and a relaxation process occurs that includes the movement of electrons of the outer shells into the vacant inner shell positions. The generated energy is released as photons. Owing to the fixed energy levels because of the shell arrangement of the atoms, only fixed values are observed and the released photons exhibit discrete frequencies that are characteristic of the originating atoms. Thus the type of atoms in the target can be determined, so that the local composition of a material is accessible.

The area that emits characteristic X-ray radiation is significantly larger than the diameter of the focused electron beam probe due to the multi-step energy transformation from the kinetic energy of the electrons through the electronic energy of the target atoms to the photon energy. The emission area typically grows with increasing energy of the electron beam and decreasing atomic number of the target atoms. While the electron beam can be focused down to 1–5 nm, the effective diameter for electron-induced X-ray spectroscopy is usually 0.2 to 1 μm.

The depth of generation of characteristic X-ray quanta also depends on the material. Typical values are 0.1 μm for heavier atoms such as Au, Pt or Pd, and 1 μm for lighter elements such as H, C, O, and N in organic polymers and biomolecules. Thus nanometer resolution is not achievable for the lateral dimensions, but micrometer and sub-micrometer resolution are, and this method can therefore also be designated as an electron probe micro analysis (EPMA) technique. Depending on the type of X-ray spectrometer, two different methods can be distinguished: wavelength dispersive X-ray spectroscopy (WDX) and energy dispersive X-ray spectroscopy (EDX) [13].

Auger Electron Spectroscopy

Auger electron spectroscopy (AES) is a method for the determination of the composition of the upper layer of a material [14][15][16]. This analytical technique is related to X-ray spectroscopy. Instead of the characteristic X-ray radiation, electrons with a specific energy are used for material identification. To induce electron emission, excitation is required. When a layer is radiated with energetic electrons, electrons are emitted from the inner shells. However, in this case the energy of electrons from outer shells refilling the vacancies is not emitted as electromagnetic quanta but as accelerated electrons. The kinetic energy of these emitted electrons corresponds to the difference between the energies of the electron energy levels involved and the work function. Because these three parameters are specific to each element, the resulting energy distribution reflects the elementary composition of the material.

If Auger electrons are generated at great depths in the material, they are usually involved in secondary collision processes, so they will not reach the surface. Thus the measured Auger electrons originate from only a thin layer on the material surface. Auger electron spectroscopy is therefore a method suitable for the characterization of very thin layers or of the changing compositions in the upper layers of a material. The depths at which information can be obtained using this technique decreases with an increase in the average atomic number of the studied material, e. g., 0.5–2 nm for metals, 1.5–4 nm for oxides and 4–10 nm for polymers.

Because the excitation is with an electron beam, AES can also be applied for measurements with high lateral resolution. Owing to the direct conversion of the energy of the beam electrons into emitted secondary electrons, the X-ray radiation emitted by the excitation volume can be neglected. Therefore, in principle, the lateral resolution of Auger spectroscopy exceeds that of EPMA. However, in practical applications, the beam is usually not particularly well focused.

AES is often utilized to characterize gradients in the composition of layers, as for example in inter-diffusion regions. Here some atomic layers of the substrate surface are removed by energetic but inert ions (e. g., Ar^+) in short intervals, before the layer composition is probed by AES. This procedure of sputtering and probing is applied repeatedly.

Secondary Ion Mass Spectrometry

Secondary ion mass spectrometry (SIMS) is a method that yields information about the composition of regions near to the surface of a material with high lateral resolution. It relies on the impact of energetic ions or neutral particles on surfaces, which induces the release of particles that are ionized in the gas phase and subsequently characterized in a mass spectrometer.

SIMS provides three-dimensional surface information. Using a particle beam, the surface is raster-scanned in the *x*- and *y*-directions, comparable to SEM. Every point in this array yields a mass spectrum, so that the mass spectra of the secondary generated ions represents the third coordinate. Thus the elementary composition of a material can be resolved with high lateral resolutions that lies, typically, in the sub-µm range [17][18].

Most SIMS experiments do not result in complete three-dimensional images. Either the composition in a small surface region is of interest, so that only the spectrum is used, or a surface image for just one peak in the mass spectrum is generated to visualize the lateral distribution of the emission of one ion, and thereby the distribution of an element.

The depth information of the primary generated signal results from a region of a few atomic layers of about 1–3 nm. In contrast to AES and ESCA, which both show hardly any removal of material during measurements, SIMS is principally a destructive method. Because of the material removal during scanning of the surface, it provides the possibility of achieving a depth profile of the surface without an additional sputtering source. However, the interaction of the energetic particles and the different sputtering efficiency of various atoms in the material lead to a SIMS-induced chemical change in the regions near to the surface.

In addition to the chemical composition of thin layers, it is specifically the arrangement of the atoms in the solid, the presence, density and shape of internal phase boundaries, grain boundaries and the structural regularity of the solid that determine the properties of a layer. These morphological properties depend not only on layer material and the conditions during layer generation, but are also influenced by the impact of nanotechnical manipulation and the use of the final devices (e. g., by elevated temperatures due to particle radiation or the electrical current).

A fundamental parameter is the average inter-atomic distance, the statistical distribution and the anisotropy of the distribution of these distances. Because the atomic distances are in the Ångstrom range (0.1 nm), a direct measurement requires extreme ultramicroscopic techniques such as STM, TEM and field ion microscopy (FIM). Characteristic lengths of morphological parameters related to local changes in the distribution of inter-atomic distances are often in the nanometer scale. Domains in polycrystalline or in textured amorphous material exhibit diameters in the medium and upper nanometer range. SIMS yields global information on the orientations (morphologically anisotropic) and the wide-scale ordering in thin layers.

SEM is also applicable as a qualitative method to yield information about the composition on the nanometer scale. The local intensity of backscattered electrons (BSE signal) and the secondary electron (SE) signal depends (in addition to the instrumentation and nanotopography) on the material. Hence the SEM image represents an overlay of geometrical and material parameters. Scanning electron microscopy detects ultrathin layers and even molecular monolayers consisting of the light elements C, H and O for reduced beam voltages. However, the rigidity of the beam and therefore the lateral resolution is decreased. Because the efficiency of secondary electrons and backscattered electron is thus greatly influenced by the surface, even very thin layers result in a material contrast (Fig. 132).

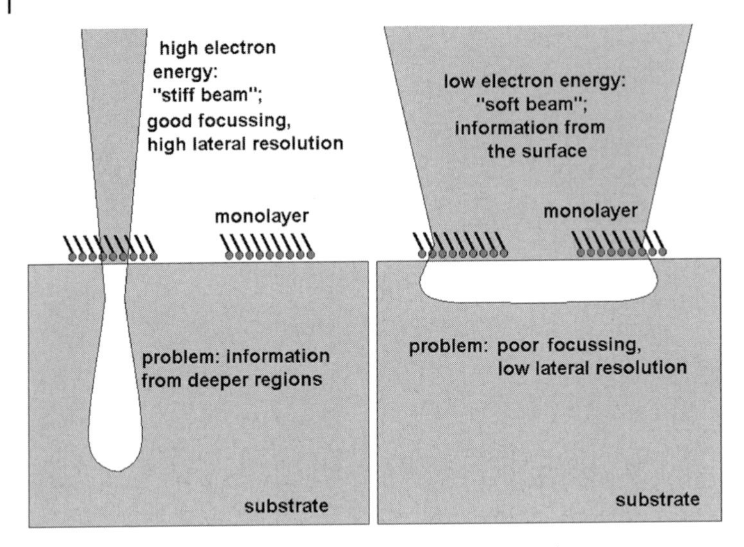

Fig. 132 Limits of SEM imaging of lateral nanostructured monolayers

The ability of transmission electron microscopy (TEM) to yield high-resolution images gives this technique an advantage over SEM for the characterization of the local material composition of samples. TEM is able to visualize individual planes in crystals; this information can be used to deduce the type of material. A main disadvantage of TEM is the complicated sample preparation, because the thickness of the region of interest has to be reduced to several dozens to several hundreds of nanometers, depending on the applied beam voltage, in order to be sufficiently transparent to the electron beam.

6.2.2
Characterization of the Chemical Surface State

Because of the importance of the interface between two materials in nanotechnology, but even more so due to the importance of individual binding places on the surface or in molecular monolayers, knowledge of the chemical composition of the surface and the chemical reactivity of the surface is essential for nanotechnology.

REM, ESMA, SIMS and AES just provide information concerning the chemical composition of a surface region, and only to a certain extent on the specific chemical properties of the surface, but not on the reactivity and binding state of the elements on the surface.

Indirect methods are often applied in nanotechnology to determine the chemical properties of a surface. Thus, lateral resolution is sometimes sacrificed for chemical information.

X-ray Photoelectron Spectroscopy

In contrast to ESMA and AES, X-ray photoelectron spectroscopy (ESCA or XPS) provides information on the atomic composition of a material but also information on the binding state of surface atoms [19]. They are able to differentiate between materials with the same or similar atomic compositions, as in the case of different organic compounds. Because their information results only on a limited depth, they are used for the determination of chemical changes on surfaces of materials.

XPS relies on the effect that X-ray radiation of a material induces electron emission with a kinetic energy that is the difference between the energy of the photons and the binding energy of the electrons (including the work function). If electrons of an outer shell are removed as a result of this interaction, the kinetic energy of these electrons will thus provide information about the binding state of the atom. So not only elements, but also the state of the binding orbitals or the proportion of atoms of the same type but binding in different ways can be determined with a sensitivity of about 1 atom-%. The depth resolution of 1 to 10 nm is comparable to AES [20]. XPS is therefore suitable for the chemical characterization of ultrathin layers and in particular of molecular monolayers. Unfortunately, it does not show the high lateral resolution of for example ESMA. Typical measurement areas are in the square millimeter range.

Contact Angle Measurements

A straightforward technique to characterize changes in surfaces, e. g., in connection with chemical modifications, is the measurement of contact angle. It is optimal for liquids with molecules demonstrating high cohesiveness, therefore with a high surface tension and a comparatively high interface energy with many solid surfaces. Water meets this requirement and represents a suitable measurement liquid.

Surfaces with a high density of hydroxyl groups or even stronger acidic groups such as carboxyl or sulfonyl groups show a strong interaction with water, and wetting of the sample surface results in a small contact angle. Aprotic non-ionic surfaces show only weak interactions with water, resulting in a low wetting and a large contact angle.

The contact angle itself provides no information about the composition and the binding state of a surface. However, the nanotechnician usually has previous knowledge regarding the surface state. Thus the contact angle could yield information as to the probability of the success of a process. A replacement of hydrogen in Si–O–H surface groups by alkyl silyl groups changes a previously hydrophilic into a hydrophobic surface. The low contact angles present after the reaction suggest problems, high angles indicate success of the surface-chemical process.

Contact angle measurements rely on free surface areas and display an optical readout. Therefore, droplets larger than a few tens of micrometers are required. Hence the lateral resolution of this method is limited, and it is strictly not a nanotechnical method.

Surface Labeling for Radiometric and Fluorimetric Characterization

The best results with respect to a chemical characterization of surfaces with a resolution in the micrometer range were achieved by molecular markers. The use of radioactive labeled molecules is well established. They undergo specific surface reactions, and their distribution reflects the local distribution of binding properties of the surface towards a particular molecule.

Fluorescence labels have the advantage of being easier to manipulate. They are also highly sensitive, so that they can be detected in the sub-monolayer coverage of surfaces. The specificity of recognition is provided by the binding of the molecule connected to the fluorophore; so any interference of this connection with the binding properties should be minimal.

The use of fluorescence labeling is widespread due to its sensitivity and wide chemical applicability. However, it exhibits several shortcomings. The fluorescence intensity depends not only on the density of dye molecules, but also on the quantum yield, which in turn is influenced by the chemical environment of the molecule that can differ for different surfaces. For a low dye concentration or a low quantum yield, either a long collection time for the optical signal or a high intensity of excitation is required to achieve an adequate signal-to-noise ratio. However, the dyes themselves show only limited photostability, which means that they are destroyed photochemically by repeated absorption of photons. This photo-bleaching effect results in a decrease in dye density and thereby in a decrease in the optical signal, so that quantitative measurements are seriously hampered.

The lateral resolution for all techniques with optical visualization as the readout method is diffraction limited, so that in principle fluorescence labeling does not represent a nanotechnical technique. Only the application of optical near-field probes allows for a definite nanometer resolution in the UV/VIS range.

Surface Labeling for SEM and Scanning Probe Microscopy

As well as fluorophores, large molecules, supermolecular aggregates and nanoparticles are also applicable as surface markers. Such markers are particularly well suited to visualization with high-resolution topographic techniques, or techniques with high-resolution material contrast when the marker particles within their composition or their physical parameters differ significantly from the sample surface. Scanning force microscopy is ideal for pure topographic visualization, therefore synthetic macromolecules and biomolecules as well as synthetic organic and metallic nanoparticles are applicable as markers.

Surface labeling for electron microscopic visualization usually uses heavy metal molecules or metal nanoparticles directly due to the high material contrast. Because they are small, they show a small lateral error with respect to their positioning. Additionally, these small particles are accessible, as molecules are, and also show the same highly specific binding, in contrast to larger particles that often bind non-specifically. Thus a high chemical selectivity is combined with high contrast in an ultramicroscopic technique. The combination of a selective nanoparticle marker with ultramicroscopy represents a real nanotechnical characterization technique.

Chemical Force Spectroscopy

The combination of high chemical selectivity with high lateral resolution is realized in chemical force spectroscopy (CFS). It uses the specific chemical interaction of a probe with the sample surface. A prerequisite is the sufficiently close proximity of the probe to the surface, because the outer electrons of the surface groups have to interact with the recognizing part of the tip. Scanning probes allow the mechanical measurement of individual molecules [21]. Such measurements point not only to the identity of the studied molecules, but also provide the response of the molecule to applied external forces. Chemical force spectroscopy has been applied to recognize individual polysaccharide molecules and differentiate between native and denatured states [22]. One approach has been developed in which the unbinding forces required to break intermolecular bonds are measured in a differential format by comparison with a known reference bond (here a short DNA duplex) [23].

6.3 Functional Characterization of Nanostructures

The functional characterization is seriously hampered by scaling down to the nanometer range, even more than the geometrical details are. The presentation of the required interfaces to a measurement system is even problematic. Moreover, the measurement signal is often too low or the nanoscopic object is significantly influenced by the interactions of the measurement process. So a physical-functional characterization is often limited to a classification of the functional properties of a nanostructure in a nanotechnical device, which means within the framework of the geometrical and functional integration.

Some physical parameters of nanostructures can be deduced indirectly from the dependence of signals of ultramicroscopic probes on the geometry or material. Thus electron microscopy and STM yield information on the conductivity of nanostructures. Structural alterations due to the imaging with the scanning force microscope point to a low mechanical resistance, which correlates with the macroscopic hardness. Regular structures in STM and TEM images indicate regular arrangements of the particles in the nanostructured material and the nanomorphology is therefore elucidated. Hence a differentiation between amorphous and crystalline nanoregions is possible.

A chemical-functional characterization in the classical sense is very problematic in the case of nanostructures, because the amount of material in a nanostructure is very small and is not sufficient for a variety of analytical techniques. The amount of sample material required for standard analytical techniques often exceeds the quantity of the whole nanotechnical device. In principle, only non-destructive and high-resolution probe techniques are applicable, such as ESMA, SIMS and AES (cf. Section 6.2). While for μm volumes these techniques exhibit a resolution with respect to concentration in the percent range, decreasing the thickness and width of nanostructures causes serious problems regarding the detection limit. Within the dimensions of individual particles, functional and geometrical characteristics are inseparable.

In the lower nanometer range, the characterization of the chemical composition of structures is usually obtained indirectly through the physical detection (e. g., AFM) of the binding or the release of marker molecules. Labeling by molecules or nanoparticles that can be identified based on topography or material contrast is not only applicable for an integral characterization of surface properties, but also for the visualization of features in the nanometer range. In addition, labeling with fluorophores, absorbing or fluorescing nanoparticles, are utilized to detect chemical functions or resolve even single binding groups.

7
Nanotransducers

7.1
Design of Nanotransducers

Nanotransducers include all transducers that achieve a signal transformation from an environment through structures that fulfill the definition of nanostructures, which means in the lateral direction at least one dimension is smaller than 100 nm and in another dimension smaller than 1 μm. The term sensory transduction is therefore not just limited to the transformation of a signal into an electrical one, but includes all other types of signal transformation, if they comply with the reception and processing in an information-processing system.

Nanotransducers differ from conventional and also microtransducers in the fact that the primary signal is usually only received at low power. Nanosensors are therefore interesting for all processes and objects that only transmit low power or provide only sufficient low power for signal generation, so that the object or the process is not disturbed by the measurement. In contrast to the low total power, the power density (power per area) should be high, so that for small areas a processable and resolvable signal is also produced.

Sensory applications with nanostructured transducers are not really applicable to processes and values with large active areas and low power densities. Therefore, nanotransducers include only such sensors that belong to a nanotechnology based on geometrical dimensions. Sensors that measure in the nanorange, such as power measurements in the nW range, distance measurements in the nm range, temperature measurements in the nK range, concentration measurements in the nM range, but which possess dimensions above the nanometer scale, are excluded.

The starting point for the design of a system or an individual transducer is always the required function. In nanotechnology the principle that the choice of shape and material are the most important decisions to be taken with respect to the desired function also applies. The design has to consider the geometrical, functional and technical environment of the nanodevice. In contrast to other technical fields, material and shape, and also the specific physical and chemical properties are not independent of each other in nanotechnology. Therefore, the atomic or molecular arrangement has to be considered in the design of nanodevices and units. For traditional design tools, which do not include the internal structure of the material, the designer has to provide

Nanotechnology. M. Köhler and W. Fritzsche

ISBN: 978-3-527-31871-1

information on the relationship between geometrical structure and the binding structure of the devices.

For a technical layer arrangement and the lithographic shaping, nanotechnology can apply the same tools as are used in microtechnology. Several types of CAD software are utilized those support the work with levels, the implementation and control of the lithographic alignment by marks and multi-level design, and transform the internal data into output formats that can be used for the generation software of structure-generating machines.

A quite different world of structure design developed for the design of molecular structures: molecular modeling. It is based on the topological framework of primary chemical structures, and not on construction levels and rigid geometries as are known from traditional CAD approaches. For the generation of applicable, discrete nanostructures, the secondary and tertiary structures are more important than the chemical primary structure. In particular, in the field of protein modeling, recent years have witnessed tremendous progress in the correlation of tertiary with primary structures. Today's chemical modeling software also provides possibilities for the calculation of the arrangements of molecules and their visualization.

In addition to the calculation of geometries, chemical software is able to model processes of molecular mobility and molecular transducer function. There is a need for software that allows for the combination of microtechnical and chemical software. Furthermore, software aimed at the stability problem of nanotechnical devices for higher integration and extended application times while taking into account the chemical stability criteria, such as reaction probability, is missing.

The design of nanodevices needs an even stronger connection with the technical requirements than the design of microdevices. As yet no general rules have been established. The methods for molecular design and synthesis planning are quite different from the methods for microsystem design. Adequate standardization of shapes, materials and processes has not been established in microsystem technology. The initial design requirements were formulated for selected types of elements.

Two classes of nanoelements can be distinguished with respect to the relationship between structure dimensions and function: nanoelements in the sense of nanoscaled elements are elements developed through scaling from the macro or micro range whilst preserving their function. Nanoelements in the sense of nanofunctional elements are units that achieve their function only because of their size, and that have no analogy in the micro- or even in the macroworld.

7.2 Nanomechanical Elements

7.2.1 Nanomechanical Sensors

In many conventional and microtechnical applications, a single increment of a signal transduction is not sufficient to detect parameters or to induce actions. In addition to the amplification of signals, often two- or multiple-step signal transformations are applied to adapt the internal signal processing to the outer conditions or to the characteristics of the incoming signals. This adaptation problem increases on moving from micro- to nanoscale elements. For the same external conditions, the disproportion between environment and element increase with respect to the transfer of forces, energies and signals.

Absolute powers and forces are reflected in the size-dependence of very different power densities and pressures in macroscopic, microscopic and nanoscopic elements. Furthermore, nanoelements utilize single sources, action targets of nanomechanical actuators and transducer principles, which do not arise in conventional techniques or that are not readily accessible. So nanotechnology is challenged much more than classical system techniques by the problems of processing very different signals and of establishing the connection between an information-processing measurement and control system and its environment through secondary transduction. Nanomechanical transducers usually provide important connections in a sensory transduction chain.

Scanning force probes are especially suited to the measurement of extremely small forces and the mechanical activation of small surface regions. The smallest differences in position or changes in surface forces are detected by the deflection of small tips or micro cantilevers.

If such cantilevers are coupled with supramolecular units, this nanomechanical readout can be enhanced to provide a more sensitive or chemically selective technique. Chemically active scanning probes with a high aspect ratio were constructed by connecting a carbon nanotube onto an AFM tip prior to functionalization of the nanotube. Functionalization was achieved by modification of the carboxyl groups of the tube with, for example, aryl amines. This probe was able to image microstructured SAM of alkyl thiol carbonic acids and of streptavidin (after biotin-modification of the tip) [1].

Nanomechanical transduction has also been applied to the measurement of very low thermal effects. For the detection of the smallest quantities of heat down to the pW range, micro cantilevers consisting of layers of different materials with different thermal extension coefficients were proposed. In analogy with the bimetallic effect, changes in temperature induce a slight bending of the cantilever, which can be detected with high sensitivity using a light pointer. Arrays of such cantilevers have been used for chemical calorimetric nanosensors, detecting temperature differences of about 10 μK and energies down to about 1 nJ [2].

Magnetic scanning probes are able to visualize magnetic domains of less than 100 nm diameter based on a magnetic–mechanical transduction [3][4]. These probes can also be used for the magnetization of very small regions.

7.2.2 Nanometer-precision Position Measurements with Conventional Techniques

For measurements of micro- and nanostructures, and also for the precise control of motion, measurement techniques with nanometer precision are needed. Sensors are often micro- and nanostructured. A precision of better than 100 nm can be achieved with a variety of measurement principles (Table 7).

Tab. 7 Examples of measurement techniques to determine positions in the nanometer range [5]

Principle	*Technique*	*Resolution/precision*
Electrical	strain gauge with resistive detection	about 1 nm
	capacitive sensor	about 0.1 nm
Optical	with optical scale/microlithographic markers	about 10 nm
	interferometry	about 1 nm
	light pointer	about 0.1 nm
Magnetic	eddy current sensor	about 0.1 nm
	Hall sensor	about 0.1 µm

The electrical and magnetic measurement procedures require a direct mechanical contact or at least a very small distance between the object and the transducer. So thin film resistors for a strain gauge have to be integrated into the object, or an extendable helper object with strips has to be mechanically coupled to the object. For capacitive measurements and the readout of magnetic increments, parts of the transducer system – a capacitive electrode and a magnetic strip with a magnetic pattern or a micromagnet – have to be in close proximity. On the other hand, optical techniques allow for greater distances in the measurement system, but the object must still be accessible by the probe. For interferometry, a reflective surface is sufficient. However, with marker systems, structures for optical readout – such as microstructured strips or edges – are integrated into the object.

7.2.3 Electrically Controlled Nanoactuators

Even more so than in microtechnology, the electrostatic principle becomes increasingly important with decreasing dimensions. Because of the short distances, moderate voltages also result in high field strengths and therefore a significant deflection of electrostatically controlled objects. Tweezers have been prepared based on carbon nanotubes with a thickness of 50 nm and a length of about 1 μm. The nanotubes were then attached to the end of a glass needle, which had two thin film electrodes so that a field could be applied between both nanotubes. Voltages below 10 V resulted in a closure of the gap between the nanotubes (Fig. 133). It was possible to manipulate for example polymer nanoparticles [6].

An electrically controlled mechanical resonance structure with high resonance frequency was nanostructured as a freestanding crystalline Si structure. The structure of 2 μm × 2 μm was connected laterally by bars, and generated a grating with a periodicity of 315 nm and a minimal structure width of only 50 nm. The resonator is capacitive addressable and exhibits a resonance frequency above 40 MHz. Such resonators could be potentially interesting as mechanical transducers for molecular sensors or for the material characterization of the smallest sample volumes [7].

Nanometer gaps can create traps for very small particles. Such a trap for particles with diameters below 10 nm was generated by a gap in two nanostructured freestanding structures of an Si_3N_4/AuPd double layer. These structures have a thickness of 30 nm and a width of 100 nm, and initially generate a gap of about 100 nm. This opening is subsequently reduced by electron beam induced deposition (EBD) on both structures, resulting in a gap width of between 5 and 16 nm (Fig. 134). Using these gaps, Pd nanoparticles of 20 nm diameter were trapped electrostatically [8].

Beside nanostructured actuators, also larger actuators play a role in translation tasks on the nanometer scale. They should work continuously, with small step sizes, and exhibit extremely high precision. Piezo actuators are of particular importance. They facilitate the movement by an extension or a compaction of the crystal planes due to an applied electrical potential, and achieve precisions down to the sub-nm range. This extreme resolution is applied in scanning probe techniques. An individual piezo ac-

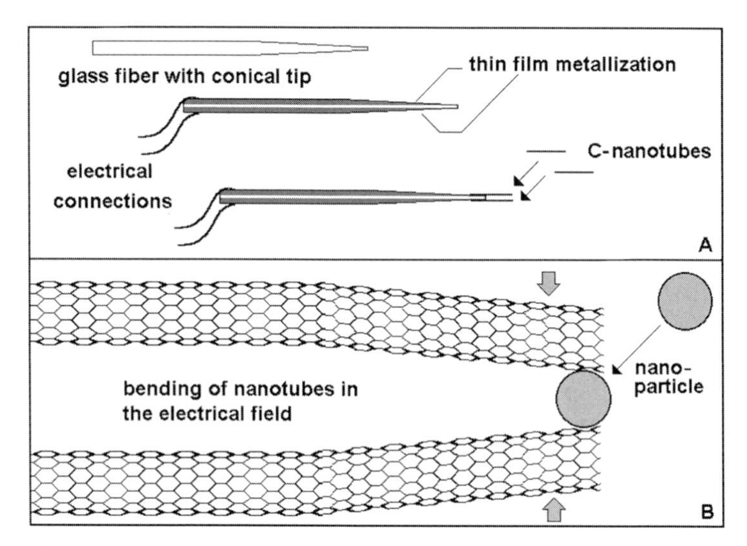

Fig. 133 Arrangement (A) and application (B) of nanotweezers consisting of carbon nanotubes [6]

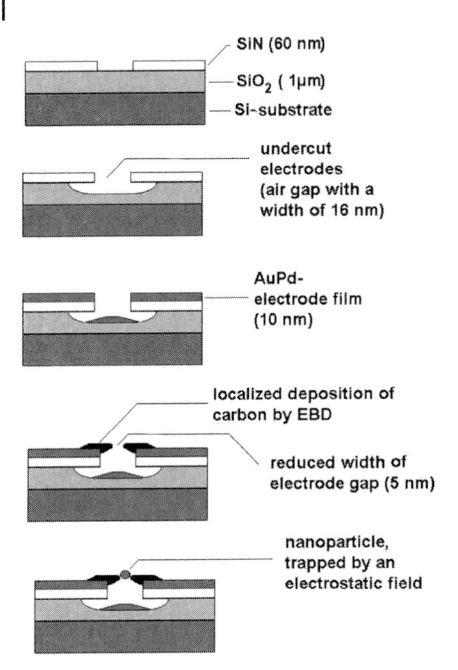

Fig. 134 Preparation of free-standing nanometer gaps ("nanotrap") for electrostatic trapping of nanoparticles [8]

tuator has a limited distance of extension, so piezo actuator systems with stacked individual actuators are applied. Apart from column systems for movement in one direction, combinations with lever elements in order to increase the working distance, or combinations of several actuators acting in different directions for x–y-tables are also known. Small changes in the angle of the positioning table can be realized by piezo actuators, so that multiple axis tables are possible [5]. Piezo-driven actuators and tables with nanometer and even sub-nanometer precision play a role in many technical fields, such as in microscopy, lithography, optics and communication industries.

7.2.4 Rigidity of Nanoactuators

Within the nanoactuators, cantilevers are of particular importance. They can be prepared by planar and thin film technology. These techniques are suitable for the preparation of various combinations of layers. The rigidity of the cantilevers of thin film materials (a thickness of between around 100 nm and 1 µm) is essentially determined by macroscopic material parameters, which means the E-module, of the bulk material. Decreasing thicknesses will lead to differences, and as a result of the increased influence of the interface or because of orientation effects, an anisotropy in the mechanical behavior is expected.

Nanorods and nanotubes lead from the thin film and nanocantilever to the supermolecular architectures. Inorganic nanorods and nanotubes are solid objects, consisting of a number of atoms of a few elements. Because of the small dimensions, the shape of these small solids and therefore also the location and arrangement of the individual atoms with respect to the surface is essential, thus quasi-single molecule properties become important. This effect is also apparent in the mechanical behavior. Hence tubes of SiC are more stable than the bulk material, and multi-walled carbon nanotubes (MWNTs) are more stable than SiC rods [9].

7.2.5 Rigidity and Stiffness of Nanomechanical Elements

Nanosprings of diamond-like carbon nanotubes have been prepared by FIB-CVD. The diameters of the springs ranged from about 0.3 to 1 μm. The diameters of the rods used to form the springs were in the sub-100 nm range. Springs of lengths amounting to several microns were prepared. The stiffness of these nanosprings was measured by means of silicon microcantilevers. Spring constants between 70 and 470 $mN\ m^{-1}$ were found [10].

In comparison with polymeric linear or slightly branched molecules, many biological macromolecules exhibit a much higher rigidity, which makes them interesting for nanoactuator applications. This applies to the proteins, especially to the filamentous supermolecules that provide the molecular basis for biological transport and actuator systems, such as tubulin and actin filaments. Double-stranded DNA also exhibits a considerable rigidity that is of interest to nanoactuator applications.

7.3 Nanoelectronic Devices

7.3.1 Electrical Contacts and Nanowires

The reason that electronics have become the fundamental technique of the information age does not only lie in the discovery of the transistor effect and the miniaturization of these devices. It is also based on the special properties of the electron, as the lightest elementary particle accessible for technical applications, which are key factor in all chemical interactions. An electron contains a charge, and it is distinguishable from all ions and also the proton in its extremely large charge/mass ratio. The transfer rates are controllable by electric fields, and they are extremely sensitive to the medium, so that electrical conductivity can be varied by many orders of magnitude by changing the material.

The first technical devices for controlled electron transport were vacuum devices (tubes). Their replacement by semiconductor devices began in the 1950s. These devices became key components for controlled electron transport in the micron and submicron range, especially in the area of transistors. Semiconductor technology is also interesting to the nanometer range, and the technology of integrated circuits can now reach below the 100 nm mark for memory and processor applications.

Semiconductor devices rely on the formation of spatial charge domains. Miniaturization of the devices requires a minimization of the dimensions of these domains. So combinations of high and extremely low conductive materials gains importance. Beside classical semiconductors, the ongoing miniaturization has lead to highly doted semiconductor materials as well as nanostructured metal–isolator systems. The latter are of particular interest for the controlled transport of individual electrons by tunneling barriers and via conductive islands. In comparison with conventional semiconductor devices, such single electron devices would have the advantage of much less heat production at the same switch rates because of their extremely low power dissipation.

Circuit paths with nanometer dimensions are elementary components of nanoelectronics in general, and especially for switches and transducers that control the transport of single electrons.

Contacts with widths in the medium nanometer range can be fabricated by electron beam lithography (EBL) wiring using a molecular monolayer as the positive resist. If a conductive substrate is utilized, the EBL-structured lines in the monolayer can be enhanced galvanically [11]. Copper contacts of 50 nm width were fabricated by a combination of EBL, dry etching and chemical-mechanical polishing. Therefore, trench structures were EBL-written into a thin PMMA layer prior to transfer by plasma etching into an SiO_2 layer. After removal of the resist layer, these trenches were completely filled with a TaSiN or TaN adhesion layer prior to a thicker copper layer. Chemical–mechanical polishing removed the copper on the planar SiO_2 surface, resulting in metal contact structures in the pattern of the trench structure [12].

A simple technique for the fabrication of narrow metal contacts relies on the nano-epitactical deposition of metal strips along monoatomic steps of single crystalline dielectric materials, which are cut at a defined angle to the crystal planes (Fig. 135). These

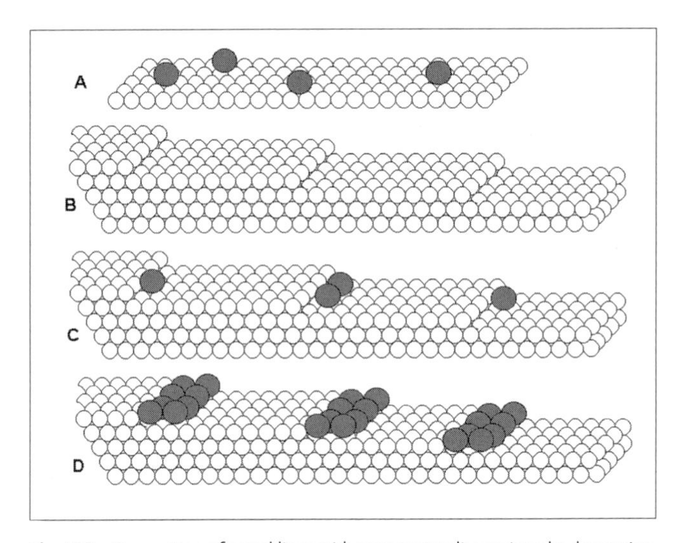

Fig. 135 Preparation of metal lines with nanometer dimensions by decoration of elementary steps on single-crystalline surfaces [13]

strips can be miniaturized down to a width of 2–3 nm [13][14]. Unfortunately, the geometries of such contact lines are predetermined by the arrangement of the crystal planes, so that they cannot be chosen arbitrarily.

The electrical conductivity of molecules with mobile electrons is a required condition for the preparation of nanowires using single molecules. The macromolecules typically prepared by polymerization or polycondensation often exhibit a linear geometry with a greater number of rotating bonds, so that they can form a variety of unpredictable conformations. For nanoconstruction and the fabrication of nanowires with predetermined geometries, rigid molecules are required. Therefore, compact rod-like molecules have been developed with conjugated ring-shaped structures, which combine increased rigidity with high intermolecular electron mobility.

The question of DNA conductivity is still a matter of dispute [15]. The experimental evidence points to electron transport over short (lower nanometer scale) distances. A single poly(G)–poly(C) double strand of 10.4 nm length positioned in a nanogap of 8 nm yields a non-linear current–voltage plot. The steps clearly observed at room temperature point to the transport of individual electrons, as would be expected for such a molecular wire. The current can be increased up to 100 nA, indicating a high conductivity. The transport rate of the charge was about 100 electrons per μm and picosecond [16]. A theoretical and experimental study indicated that sufficient conductivity is only achieved after a certain exposure to energetic electrons [17].

Instead of the direct use of individual macromolecules, in nanolithographic processes molecules can also be applied as masks in order to form single molecule-based nanowires. One example is the fabrication of Au nanowires by dry etching of a thin Au layer masked by an adsorbed microtubule molecule [18]. Such a microtubule molecule can also act as the template for a nanowire for an electroless Ni or Pd metallization [19][20]. This principle is also applicable to DNA [21][22][23][24].

Of special interest is the combination of chip surfaces or lithographically fabricated microelectrodes with molecules and metal nanoparticles. An assembly of surface-functionalized metal nanoparticles could yield electrical contacts with nanometer dimensions. With the application of an electric field, rod-like metal micro- and nanoparticles can be oriented and positioned on substrate surfaces. Hence gold particles of 70–350 nm diameter and 8 μm length were positioned between gold electrodes out of a dielectric solvent [25]. Layers of chemically functionalized gold nanoparticles were generated in a stepwise process [26]. The deposition of ensembles of metal nanoparticles on immobilized molecules can be applied for the detection of molecular interactions [27][28][29]. DNA molecules individually positioned between microelectrodes could be an interesting objective for such an approach [30].

Nanowires can be based on molecules by using an extended chain-like molecule as a template for the binding of metal nanoparticles. One example utilizes supermolecular aggregates of DNA streptavidin conjugates that were decorated with biotinylated gold nanoparticles, resulting in chains of gold nanoparticles. Through the integration of branched DNA, more complex supermolecules are accessible [31]. The applied supermolecular streptavidin-DNA aggregates are also interesting for the highly sensitive detection of biomolecules by immuno-PCR [32].

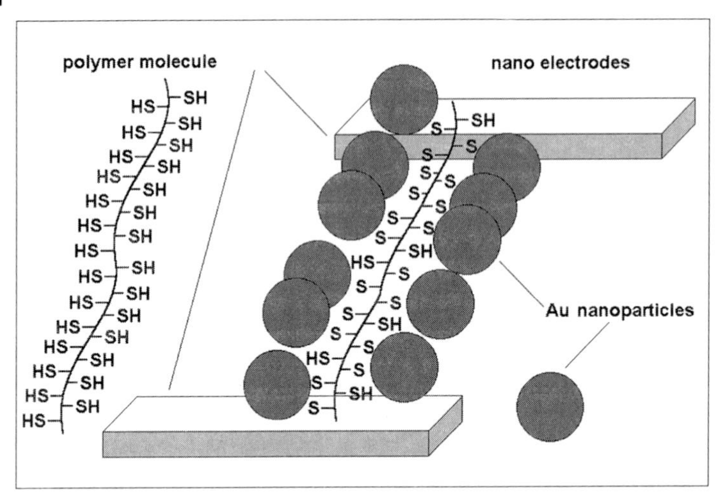

Fig. 136 Single-molecule based preparation of conducting paths of a chain of nanoparticles by immobilization of a chain-like molecule between lithographically structured electrodes and coupling of particles by thiol groups [34]

Free-standing nanowires with a width of about 20 nm were fabricated by the electron beam induced deposition of carbon needles at the edge of a prestructured gold layer, resulting in free-standing nanodiodes with a gap width of about 5 nm [8].

Electrical contacts consisting of chains of conductive nanoparticles are of both technological and functional interest. Particles below 100 nm diameter are comparable to molecules with respect to their binding behavior, which means that their non-specific binding to surfaces can be suppressed and specific bonds can be generated through complementary coupling groups on the surface of the substrate or other nanoparticles. Hence contacts can be formed by chemical self-organization, e. g., by binding via thiol groups (Fig. 136). The conductive particles present barriers to the electron transport at their contact points. Dielectric molecular surface layers on gold nanoparticles can further increase the barrier effect, and the barrier height can be controlled by the choice of thickness and composition of the molecular monolayer. Therefore, such particle chains are not only interesting as simple wiring elements, but also for controlled electron transport. Chains of assembled nanoparticles of 10 nm diameter span gaps of 30 nm and 150 nm width, respectively [33][34]. On the other hand, fairly large electrode gaps of several millimeters can also be connected through the dielectrophoretic assembly of metal nanoparticles [35].

Semiconductor nanowires represent a versatile field for nanoelectronics applications [36]. A general strategy for the controlled growth of these nanowires (NWs) has been developed in recent years [37]. An important step was the control of the critical nucleation and subsequent elongation steps of NW growth using metal nanoparticles as catalysts. In this way, a broad range of NWs with homogeneous composition and single-crystal structure could be prepared. The successful incorporation of specific dopants represented a major step towards achieving an ability to control the fundamental electronic properties of NWs in order to develop active electronic and optoelectronic nanodevices. Another critical breakthrough was the recent demonstration of the controlled growth of axial [38] and radial heterostructures [39], where the composition and/or doping was modulated down to the atomic level either along or perpendicular to the axes of the NWs. Heterojunctions in NW can be created by alternating the flow of different reactants and/or dopants. In order to prepare radial NW nanostructures, after growth and elongation of a crystalline NW core, conformal radial shell growth is achieved by altering the conditions to favor homogeneous deposition on the NW surface as opposed to reactant addition at the nanoparticle catalyst. Subsequent introduction of different reactants and/or dopants produces multiple-shell structures of controlled composition. Homogeneously doped NWs represent a key building block for electronic devices, such as the NW field-effect transistor (NWFET) [40]. In the case of this device, high-temperature growth processes are used for the preparation of high-quality single-crystalline material, and then low-temperature assembly and contact deposition enables the rapid design and fabrication of a host of single- and multi-NW device structures on virtually any substrate [37]. A simple fluid-based assembly combined with patterning has been developed in order to realize more complex structures, e.g. ring oscillator circuits [41]. Biosensors represent a promising application for individual as well as arrays of single NWFETs, because binding to the surface of an NWFET is analogous to applying a gate voltage, which leads to the depletion or accumulation of carriers and subsequent changes in the NW conductance [42][43]. Crossed NW architectures are also accessible [44]. These can be configured from one NW as the active channel and the second crossed NW as the gate, separated by a thin SiO_2 dielectric shell on the SiNW surface, with the gate on the surface of one or both of the crossed NWs. The crossed NW concept allows the creation of p-n diodes by crossing p- and n-type NWs, leading to rectifying behavior and even the creation of multicolored light-emitting diodes (LEDs) [45][46].

Nanotubes

Specific interest has focused on carbon nanotubes as possible wiring elements in the nanometer range [47]. These carbon filaments can be generated by various methods, such as laser ablation, arc discharge or chemical vapor phase deposition (CVD). Their electrical properties depend greatly on the geometry, particularly the diameter of the tubes, which is typically between about one and a few tens of nanometers. Applying four-point measurements, electrical resistances of individual nanotubes were found to be between $0.2\ k\Omega\ m^{-1}$ and more than $500\ M\Omega\ m^{-1}$. A great increase in resistance with decreasing temperature was observed for selected tubes, others exhibited a linear decrease of the resistance between 4 and 300 K of only 10%. In general, a semiconductor-like temperature dependence of the electrical behavior is observed [48]. Logic circuits and a room-temperature single electron transistor were fabricated based on carbon nanotubes [49][50].

Carbon nanotubes exhibit several fairly different conformations: tubes can have a single wall, but they can also be double- or multi-walled, or narrower and sometimes helical tubes integrated into tubes of greater diameter [51]. In addition to pure carbon tubes, synthetic nanofilaments with conjugated bond backbones and with aromatic-

and aromatic-heterocyclic units, such as polyphenylenes, polythiophenes or bisalkylthiophene-bridged ureas, are also candidates for organic conducting wires with nanometer dimensions [52][53][54].

Carbon nanotube preparation processes generally produce mixtures of semiconducting and metallic nanotubes, which has hindered the development of these systems for use in large-scale electronics. Zhang et al. [55] have described how a methane plasma and annealing treatment can be used to selectively remove the metallic nanotubes. By also controlling the diameter of the nanotubes during growth, pure semiconducting devices can be reliably obtained, as demonstrated by the fabrication of high-current transistors.

Two Fe/Ti contacts separated by a 1 μm gap can be bridged by individual carbon nanotubes prepared by gas phase deposition. The thin Fe layer thereby acted as a catalyst for the nanotube generation. Most of the nanowires of about 1 μm length and a diameter of 40–50 nm yielded electrical resistances of between 50 and 80Ω [56]. Directed growth of carbon nanotubes results in freestanding nanowires between microstructured silicon columns [57].

Biomolecular assemblies can assist the positioning and integration of CNTs. A DNA scaffold molecule has been utilized to provide the point of address for the precise localization of a semiconducting single-wall carbon nanotube as well as the template for the extended metallic wires contacting it, resulting in a DNA-templated carbon nanotube field-effect transistor [58]. A complementary metal oxide semiconductor (CMOS)-type five-stage ring oscillator (including 12 FETs in total) built entirely on one 18 μm long single-walled carbon nanotube has been demonstrated [59].

7.3.2
Nanostructured Tunneling Barriers

Well-defined tunneling barriers are necessary for all devices with electron tunneling processes. This requires superconductive devices that rely on the mutual tunneling of electron pairs over barriers by the Josephson effect, but also single electron tunneling (SET) devices. Such barriers should be thinner than 2 nm, because the probability of electron transfer decreases steeply with the barrier thickness due to the exponential decay of the tunneling conductivity. So thicker barriers provide no applicable switches or transducers. The adjustment of the barrier thickness is usually realized by the adjustment of a thin layer. The direct deposition of dielectric layers with a highly defined thickness in the range of 1–2 nm inside a stack of layers is challenging. Therefore, instead of directly deposited layers, layers of a less noble metal are often oxidized in a spontaneous reaction with the oxygen or water contained in the air into a dielectric layer that can be used as a tunneling barrier. Such a chemically controlled formation of the tunneling barrier layer can be regulated by the oxygen or partial pressure of water, the temperature and the duration of the incubation in the reactive atmosphere.

In addition to the thickness, the lateral extension of tunneling barriers is also important for SET devices. It should be low to achieve high integration densities. The individual switching process depends on the capacity of the contact, which should be low. The capacity depends on the barrier thickness, but this thickness has upper limits for which tunneling is still possible. Another parameter influencing the capacity is the area. Thus to realize low capacities – and this applies especially to room temperature devices – nanostructured barriers are required.

Small lateral dimensions of tunneling barriers are achieved with metal nanoparticles or clusters that are separated by ultrathin dielectric layers from metal or semiconducting contacts. The isolating layer can cover a larger area, but it acts as a tunneling barrier only in the contact region with the nanoparticles. Barriers with an efficient diameter only 4 nm have been prepared, for example by the coupling of 4 nm Au clusters via a monomolecular *o*-dithiolxylol layer onto a GaAs layer [60]. SEM imaging presents the contacts as small flat structures.

Another approach to the fabrication of nanostructured tunneling barriers utilizes local oxidation of a metal contact by STM or conductive AFM tips (Fig. 140). Therefore, the metal layer is anodically polarized. After the approach of the scanning tip, the metal is locally oxidized and generates a localized barrier to the electron transportation. This approach requires easily oxidizable metals, such as aluminum. To achieve electron confinements for the controlled single electron transport, a metal layer is prestructured as a narrow contact prior to writing an island through local oxidation.

Nanostructured semiconductors can also be utilized as barriers for the electrostatic single electron transport. So a freestanding nanolever of highly doted silicon can be fabricated with a length of 800 nm and a cross-section of 24 nm $\times$ 80 nm. A gated

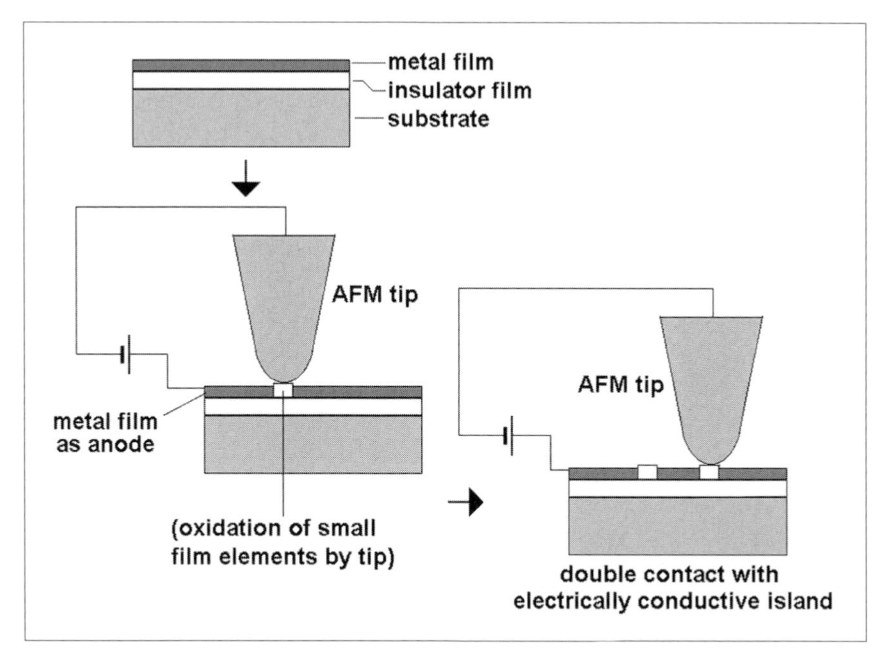

Fig. 137 Preparation of a pair of tunneling contacts by local anodic oxidation of a metal layer using a conductive AFM (E. S. Snow et al. 1996)

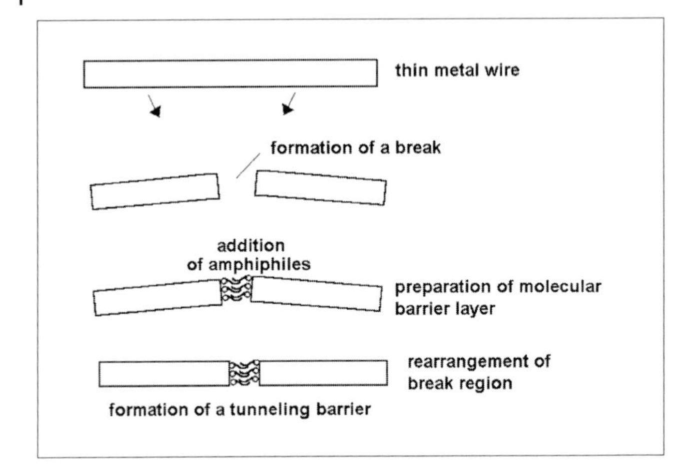

Fig. 138 Preparation of molecular tunneling barriers by the formation of SAMs in a break junction of a wire [64]

electrode in close proximity allows for the possibility of a controlled single electron transport up to temperatures of about 100 K [63].

Molecular tunneling barriers can be prepared by the generation of double gaps and the subsequent filling of the gaps with molecular mono- or double layers (Fig. 138). The preparation of vertical molecular tunneling barriers was carried out with the formation of an SAM monolayer of *p*-dithiohydroquinone on a thin gold wire. After monolayer formation, the wire was mechanically broken in one location. A reconstruction of the monolayer on both sides of the gap occurred, prior to reconnection of the two parts of the wire by evaporation of the solvent. In this manner an SAM double layer is formed [64]. This breaking technique probably facilitates the fabrication of contacts with charge transfer via individual atoms [65].

A precise control of nanogaps is possible by in situ gap impedance tuning [66]. In this way, electrode gaps of some tens of nanometers can be reduced to a gap width of less than 10 nm.

Theoretical considerations demonstrate the possibility of single photon detection with the help of nanostructures. A transduction chain is required that transforms photons into an electronic signal. Such a photon counter may be read with a superconducting quantum interferometric device (SQUID) [67].

7.3.3
Quantum Dots and Localization of Elementary Particles

The transformation from classical electronic behavior to quantum behavior in association with the miniaturization of devices is clearly visible for a reduction of two dimensions (quantum wire), but is significant for structures confined into all three directions, the quantum dots (QD). This so-called quantum size effect (QSE) applies when the dimensions of a solid drop below the de Broglie wavelength of valence electrons.

Quantum dots are generated by extremely high-resolution lithography or through the control of the segregation in solids or growth processes of solids in the early stages of seed formation, e. g., with vacuum deposition processes. An elegant alternative is the generation of clusters. It has the advantage of highly defined particles of a fixed size and sometimes even a fixed number of atoms in the clusters. Thus the generation of chains of superclusters of 6 nm diameter from a liquid phase of clusters has been observed, and each supercluster consists of 13 subunits of the cluster $[Au_{55}(PPh_3)_{12}Cl_6]$ [68].

Nanostructures are applied to visualize quantum effects on nanostructures. It is possible to create metal islands that contain individual electrons and allow the visualization of these particles with their dual wave and particle characteristics. A ring-shaped quantum coral of 48 Fe atoms was fabricated on a crystalline Cu surface and represents one of the most stunning results of early nanotechnology. STM can be used to visualize the structure of a standing wave with concentric potential extrema inside a ring of 14.3 nm diameter, at low temperatures, and this was interpreted as being the wave structure of a surface-bound electron [69]. Nanostructured islands, which trap individual charges, are key structural elements for single electron devices.

7.3.4 Nanodiodes

Nanodiodes consist of a solid with two nanostructured and electrically conductive electrodes that are separated by a less conductive zone. Instead of a reduced classical conductivity, a tunneling barrier is also possible.

Nanodiodes can be prepared by local writing of barriers using local oxidation with conductive AFM. Thus a metal–isolator–metal (MIM) contact was fabricated by oxidizing 3–4 nm thick niobium structures in about 70 nm wide strips. Local oxidation of 3 nm thick titanium layers by STM yielded tunneling diodes that are of potential interest for switching processes in the sub-picosecond range [70][71].

Instead of individual islands or nanoparticles, small arrays of nanoparticles are also applicable as electron islands in an electrode gap for SET devices. These devices utilize the Coulomb blockade, that is the change in transfer probability for electrons through two serial barriers due to the change in the charge of a small conductive island by just one elementary charge (cf. Section 7.3.5). A Coulomb blockade has been demonstrated in a nanostructured 30 nm gap with gold clusters of 1–3 nm diameter [72].

A nanodiode with a molecular monolayer as a tunneling barrier and a gold nanocluster for electron confinement was prepared by deposition of a biphenylenedithiol layer on an Au substrate prior to incubation with Au nanoparticles on this SAM. By utilizing an STM tip positioned on an Au cluster, controlled transport of a single electron could be demonstrated [73]. A strong diode effect was found on an Au and Ti/Au wired monolayer of *p*-thioacetylbiphenyl. This structure was prepared through a 30 nm aperture in a freestanding Si_3N_4 membrane. The diode exhibited a current of more than 400 nA at 0.4 V in contrast to –1.3 nA at –0.4 V. The asymmetry in the molecular structure is probably the main reason for the extremely asymmetric

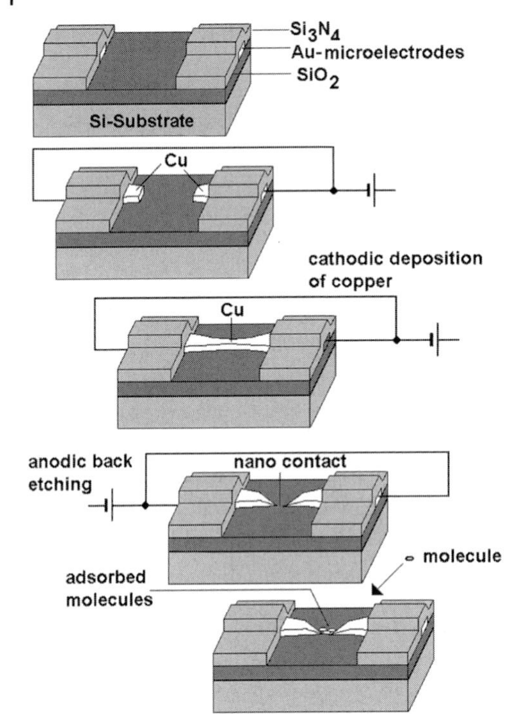

Fig. 139 Electrochemically controlled preparation of nanogaps for trapping of molecules [75]

current–voltage plot [74]. Investigations into the electron transport in monolayers of ordered organic molecules with a high proportion of π-electrons led to the suggestion that the individual molecules could be considered as electronic nanoconfinements separated by tunneling barriers. Therefore, a molecular monolayer of such π-electron rich molecules shows a similar behavior to that of a two-dimensional array of tunneling contacts [9].

Nanoelectrodes with extremely small spacings were prepared by cathodic deposition of Cu on covered planar gold microelectrodes and anodic back etching. These electrodes were used for the adsorption of small ensembles of molecules (Fig. 139). The adsorption could be detected by changes in the conductivity [75].

Variations of the substituents on benzene adjust the barrier height of tunneling diodes with molecular tunneling barriers. An Au/GaAs diode with such an organic barrier exhibits a low tunneling barrier with electron-withdrawing substituents (–CN) and a high barrier with electron-donating substituents [76].

7.3.5
Electron Islands and Nanotransistors

Nowadays, nanotransistors are present as integrated elements in electronic chips encountered in everyday life. Beside the fabrication of conventional field-effect transistors using down-scaled optical lithography, new techniques have also been developed for the nanofabrication of electronic devices. Not only established technologies such as EBL can be used for the generation of electronic nanodevices, but also more recently developed techniques such as nanoimprinting (see, for example, W. Zhang and S. Y. Chou [77]).

The fabrication of transistors with critical dimensions in the medium nanometer range depends on sufficiently thin layers of isolating material for the separation of the gated electrode. An Si field effect transistor with a gate length of 25 nm and a distance of 70 nm between the source and the drain electrode was realized using a molecular monolayer of alkyltrichlorosilane as the gate barrier [78].

The important principle for single-electron transistors is control of the electron transport over a small conductive island (electron confinement) that is wired by two contacts (source and drain) [79]. The loading of such an island by an individual charge increases the barrier for the transfer of a second charge onto the island significantly, if the island is sufficiently small. This Coulomb blockade effect can be modulated by external fields, so that further electrodes can be used as gates (Figs. 140 and 141). A requirement for such switching devices is a layer thickness below 5 nm for the tunneling barriers [80]. Contact pairs that separate an electrically conductive confinement from the wiring electrodes can be realized by in situ oxidation of prestructured electrodes and subsequent deposition and structuring of the confinement material (Fig. 142) [81].

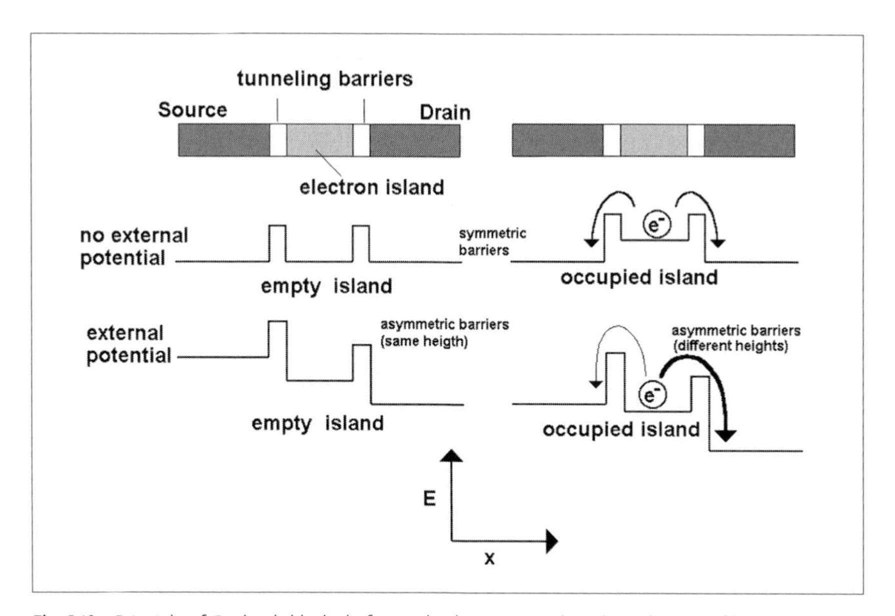

Fig. 140 Principle of Coulomb blockade for single electron tunneling through pairs of barriers

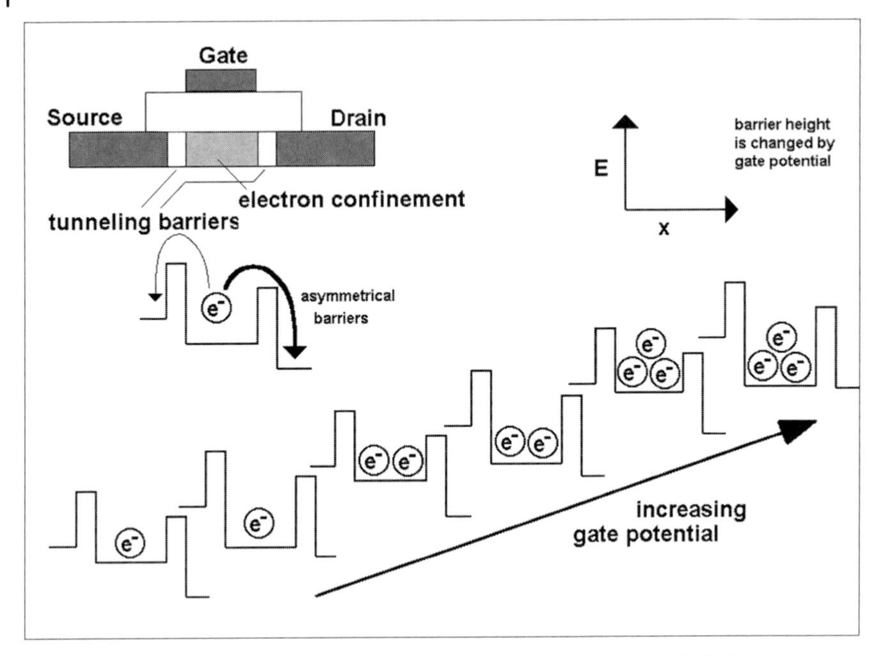

Fig. 141 Dependency of the charge of electron confinement on the gate potential of SET transistors

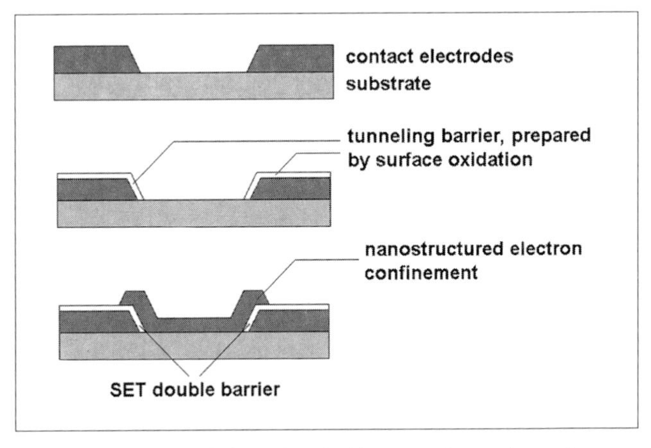

Fig. 142 Preparation of a front plane double contact for SET devices (courtesy of K. Blüthner)

An SET transistor working at 6 K has been prepared in highly doted Si. The primary structure was written by electron beam lithography into a PMMA mask layer and transferred by RIE with a Ti mask into the 15–20 thick functional layers. Oscillating voltage–current characteristics were observed that confirm an SET process [82]. Nb/ Nb oxide structures written by AFM oxidation into 2–3 nm thick Nb layers yielded an

SET transistor working at 100 K. The width of the lines oxidized by AFM was 10–25 nm, but a barrier width that is efficient for the tunneling process is significantly smaller [83]. A room temperature SET transistor was produced in granular palladium using an Al gated electrode dielectrically coupled by an Al_2O_3 layer. Source and drain electrodes were prepared by a shadow mask, which was fabricated by EBL in a PMMA double layer about 0.1 μm above the electrode level. The tunneling barrier was geometrically adjusted by stepwise changes in the deposition angle [84].

A decrease in the width of the electron confinements is achieved by controlled oxidation starting from the edges. A transistor has been fabricated with P^+ doted Si (SIMOX process) and an island width below 10 nm that exhibited SET-like voltage–current characteristics up to 129 K [85]. A room temperature SET effect has also been observed with an Si memory structure of a transistor with a floating gated structure of 30 nm width [86]. Such devices have great potential for future electronic memory with 1 bit coded by a single elementary charge [87].

Most SET devices use lithographically well-defined coulomb islands. Ensembles of randomly distributed metal particles between electrodes provide an alternative to these lithographic structures.

So-called satellite structures of several tens of nanometers width that are deposited in a gap between the source and the drain can perform as electron confinements and lead to SET transistor arrangements that show strong oscillations of the conductivity dependent on the gate voltage at 28 K [88]. Particularly small islands were prepared by the deposition of extremely small metal droplets from a metal ion source (Fig. 143). These islands with diameters of about 1 nm show Coulomb blockade effects up to 200 K in a 10 nm gap of a quadrupole electrode arrangement [89]. Evaporation also yields small islands with fairly regular size distribution, which can be used as electron confinements in SET devices [90]. This technique is based on the fact that for lower energies of the deposited atoms and low mobility of the deposited atoms on the surface, a multiplicity of seeds is formed. In contrast to standard thin film preparation, the process does not proceed to cover the surface completely, but is interrupted before a conductive connection between the electrodes is formed. With this technique, tunneling contacts were prepared with small ensembles of grains with about 2 nm diameter and

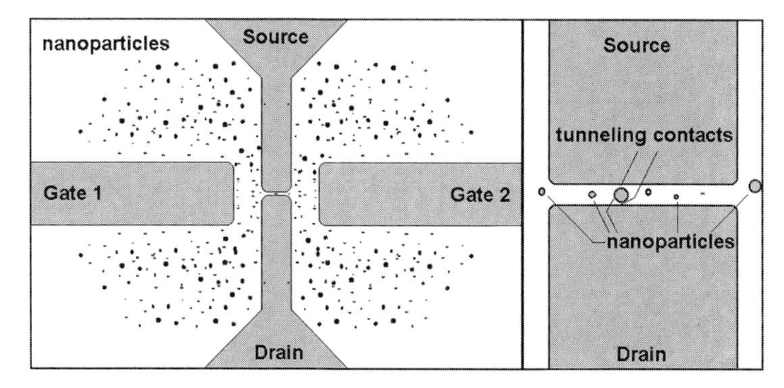

Fig. 143 Preparation of an SET transistor by deposition of small metal droplets in a 4-electrode arrangement with nanometer gap [89]

grain distances below 2 nm in electrode gaps of 5–50 nm. The resulting networks of tunneling barriers and islands exhibited voltage–current characteristics typical for SET processe [91].

A problem regarding quantum dot devices, such as quantum dot cellular automata (QCA), is the possibility of different tunneling paths and co-tunneling in the case of multi-particle SET structures. Investigations by X. Luo et al. [92] have shown that co-tunneling can be suppressed in the case of Al/AlO_x junctions, as evidenced by coulomb blockade oscillations (CBO) in the tunneling current depending on the gate voltage.

J. A. Liddle et al. [93] used capillary forces to trap single gold nanoparticles of diameter 30 nm, or pairs of these, inside a nanogap between two lithographically prepared gold electrodes (Fig. 144). The obtained device showed nonlinear current-voltage characteristics.

A gold/chromium layer architecture on SiO_2 has been applied for the construction of a SET transistor with a gold island [94]. For this, CrO_x lines with a width of 70 nm were fabricated using an EBL and lift-off technique. The modulation curves obtained (changing gate voltage) proved the principal function of the SET device (at 0.3 K). In some cases, the devices showed a multitude of frequencies, indicating additional islands with a coulomb blockade effect due to the roughness of the CrO_x resistors.

EBL in conjunction with a bilayer resist process has allowed the fabrication of 15 nm structures in silicon. In this way, an SET device was prepared, which showed the typical CBO structures in the current-voltage characteristics on the gate voltage at 4.2 K [95].

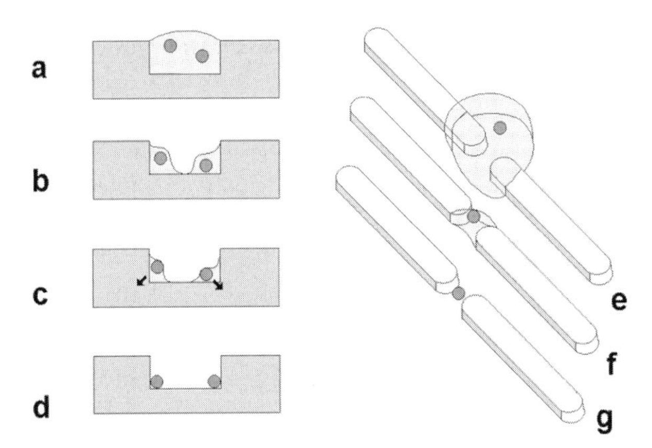

Fig. 144 Trapping of hydrophilic nanoparticles in trench structures and nanoelectrode gaps: a) suspension containing nanoparticles in a micro- or nano-trench, b) wetting of bottom edges due to volume reduction of the liquid during solvent evaporation, c) nanoparticles are apparently attracted by the bottom edges of the trench structure due to the wetting behavior of the residual liquid, d) nanoparticles fixed in the bottom edges, e) self-positioning of a nanoparticle-containing droplet in a gap region of a nanoelectrode arrangement, f) wetting of the gap by the residual liquid during evaporation, g) resulting fixation of a nanoparticle from solution in the nanogap after complete evaporation of the solvent [93]

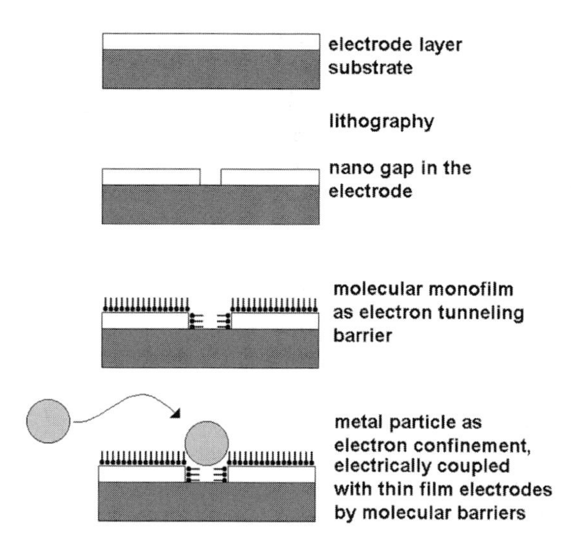

Fig. 145 Preparation of a double contact for SET devices by binding of a metal nanoparticle in a gap between electrodes covered by molecular monolayers

Particularly small SET devices can be realized by using nanoparticles or nanotubes. An SET device capable of working at room temperature was obtained by the deformation of a carbon nanotube, which had been previously deposited between two electrodes using an AFM tip [50]).

In addition to noble metal particles, nanocrystals of compound semiconductors such as CdS can also act as nanocontacts [96]. Tunneling barriers with better geometric definition are achieved when electrodes and metal particles or islands are separated by a molecular monolayer instead of random seed formation or deposition and separation by air (Fig. 145). Therefore, the electrodes of the contact are covered by a monolayer prior to deposition of the island metal [97]. Control of single electron transport also succeeds in carbon nanotubes. In bundles of single walled nanotubes deposited on an isolating gated electrode, a sharp modulation of the single electron transport has been detected [98]. Single electron tunneling transistors do not require a complete gap between the source and drain. Often a narrowing of the source–gate connection down to nm dimensions is sufficient to reduce the transfer probabilities of electrons significantly. Such tapered structures apparently generate Coulomb islands, so that the restricted wire acts as a system of two external electrodes, two (or more) tunneling barriers and the respective number of Coulomb islands. If such zones are capacitively coupled with gated electrodes in close proximity, the transport of individual electrons can be modulated.

The formation of an electromechanical modulated single molecule structure was achieved through the deflection of a C_{60} molecule using the mechanical pressure of an STM tip. A modulation of the conductivity was measured that corresponded to an amplification factor of 5 [99].

A particularly high density of single functional elements in integrated devices can be achieved when linear structures are oriented perpendicular to the substrate plane. For this, techniques for the preparation of patterns with high aspect ratios are necessary. T. Bryllert et al. [110] developed a strategy based on a combination of lithography (EBL) with catalytic chemical material deposition and vacuum deposition for the preparation of regular arrays of coaxial pillar structures. These consisted of a core cylinder of InAs, grown by chemical epitaxy with the aid of lithographically patterned Au films, which were initially deposited on the flat substrate. The InAs piles had diameters between 40 and 200 nm and heights of up to 5 μm. The cylindrical core electrodes were further coated with an Si_3N_4 barrier layer and an Au/Ti layer. The transistor structures could be contacted after back-etching of the dielectric layer from the upper part by the deposition of a metal film for the drain contact.

SET Transistors with Chemically Modified Nanoparticles as Electron Islands

If an individual or an ensemble of nanoparticles surrounded by a tunneling barrier is situated in a nanoelectrode arrangement, the particles can form electron islands for the transport of single charges (Fig. 146). If the particles create a conductive chain between two of the electrodes in a three-electrode arrangement, with the position of the third electrode being a distance away that is greater than the tunneling distance, then this electrode can be used as a gate, and a nanoparticle SET is produced [101]. An SET-transistor based on an individual metal nanoparticle positioned along an extended DNA molecule located in a microelectrode arrangement has been proposed [102]. If n particles modified by a dielectric surface layer are arranged in a chain, they generate $n + 1$ tunneling barriers. Charges are transported by a hopping process, and the gated electrode controls all electron islands simultaneously (Fig. 147). A Coulomb blockade effect has been demonstrated at 4.2 K with individual gold particles of 2–5 nm diameter arranged in a gap of less than 10 nm between two planar gold electrodes [103].

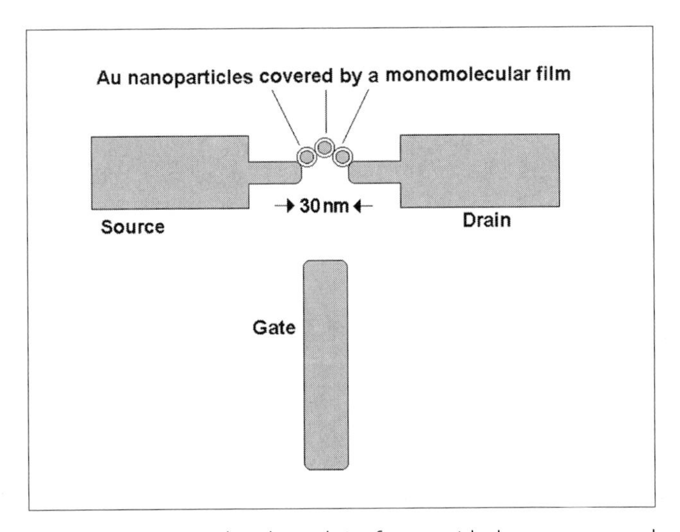

Fig. 146 SET transistor based on a chain of nanoparticles between source and drain electrode

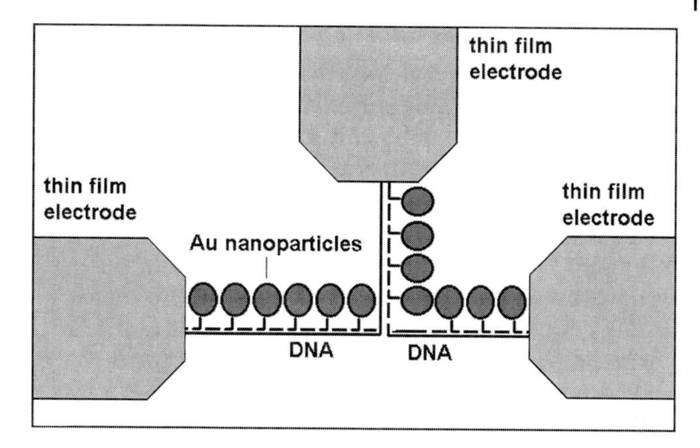

Fig. 147 Proposal for an SET transistor utilizing DNA as the positioning tool for oligonucleotide-modified nanoparticles

In principle, SET devices controlled by mechanical means should also be feasible through movement of the electron island between the source and drain electrode. Such a device requires that the turning points of these oscillations – which means on the points of highest electron tunneling probabilities – the dwell time is longer than the electric time constant of the RC-circuit [104].

7.3.6
Nanoswitches, Molecular Switches and Logic Elements

Switching processes can be supported by electronic, mechanical or material changes. When approaching the dimensions of single molecules, the borders between these three principles disappear. Molecular movements are linked to electronic processes, and on the other hand, electronic transfers lead to, at least temporarily, changes in the chemical structure.

Nanomechanical and nanoelectronic systems with free-standing functional structures can be realized by a combination of FIB and CVD technologies. The local deposition of different materials in the focus of an ion beam allows the formation of complex geometries. Different materials can be deposited by the application of different process gases. Thus, systems can be assembled from different nanometer-sized functional elements [105].

Nanoswitching processes are not limited to tunneling effects or single molecular processes. In addition, the nanomorphology of complex materials provides possibilities for nanoelectronic switching. One example is the use of the spontaneous spatial organization of domains in block-polymers with a sequence of electron-conductive sections to control the electrical conductivity. Mixtures of pentadecylphenol with polystyrene-poly-*p*-vinylpyridine block-copolymer that has been protonated with methylsulfonic acid exhibits thermally controlled electrical conductivity. This effect is caused by molecular reorientation processes from a lamellar domain structure with character-

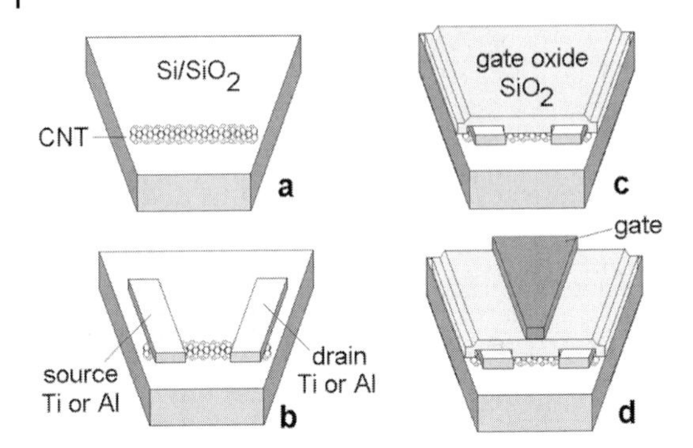

Fig. 148 Steps in the preparation of a CNT-FET: a) single-wall CNT deposited on an SiO_2 layer, b) lithographic patterning of source and drain electrodes, c) deposition of gate oxide, d) deposition of metal film and patterning of gate electrode (S. J. Wind et al. [107])

istic dimensions of 35 nm and 5 nm at 100 °C over a non-lamellared block structure (with increased conductivity) into a matrix structure with integrated columns with distances of about 28 nm at 150 °C [106].

A field-effect transistor (FET) with an integrated single-wall carbon nanotube (CNT) has been reported by S. J. Wind et al. [107]. The CNT was deposited on a silicon substrate covered with SiO_2. The respective ends of the nanotube were connected with source and drain electrodes, and the gate electrode was applied after the deposition of a thin SiO_2 gate oxide layer over the source, the drain, and the CNT (Fig. 148).

Single CNTs and arrangements thereof are particularly well suited for the construction of information-processing devices at the single-electron level. Logic elements (XOR) have been realized by the integration of MWCNTs into thin-film multi-electrode arrangements [108].

Circuits with two or more electron islands, which are coupled via tunneling barriers and capacitively controlled by another electrode, are discussed as basic elements for SET logics [109]. Coulomb blockade effects that lead to the reversal of the current paths were demonstrated at 30 K for a structure of two Si nanoconfinements with lateral dimensions below 50 nm integrated in a T-shaped electrode arrangement. The islands probably interact with each other by capacitive means and influence the single charge transport on the other island, respectivel [110]. A room temperature single electron transistor was constructed by electron beam lithography of alumina [111]. The limited efficiency of the potential on a single gated electrode is compensated for by a second gated electrode resulting in a logical "AND". Array arrangements of electrodes have been produced that transfer single electrons between adjacent islands [112]. Beside logic elements, other electronic functions such as amplification and memory are possible applications for SET devices [113].

Metal clusters of 55 gold atoms surrounded by a tunneling barrier provide the charge transfer inside clusters by two electrons, but the tunneling between clusters requires several electrons. If several clusters are wired to the outside, tunneling resonance resistors (TRR) are generated that cut their electrical resistance on application of the resonance frequency, even at room temperature, in half. So cellular automatons of tunneling elements can be fabricated based on such clusters [68].

The addressing of functional nanoelements in larger arrays is a central problem in the context of integrated nanodevices. Z. Zhong et al. [114] proposed an address decoder based on arrays of crossed nanobars with field-effect transistor functions (cNW-FET). Chemical modifications at the cross-points determined the switching behavior. The feasibility of the concept was proven by electrical measurements on multi-electrode arrangements.

Beside the arrangement of CNTs in an environment of thin-film structures with dielectric and metallic layers, the local modification of CNTs themselves can be used for the realization of specific electronic functions. Energy gaps in the tubes can be introduced by the opening of the π-structure and the formation of dangling bonds by STM (J.-Y. Lee and J.-H. Cho [115]). The quality of nanotransistors obtained using CNTs is strongly dependent on the diameter of the tubes and on the contact between them and the metal electrodes (Y.-C. Tseng et al. [116]).

A memory cell based on SET has been presented by E. G. Emiroglu et al. [117]. It was realized by a serial arrangement of coulomb islands by means of SOI technology. The operation of the device was demonstrated by CBOs and current-gate voltage characteristics showing significant hysteresis.

Meanwhile, a whole spectrum of molecular machines and supermolecular functional systems is included in concepts and theoretical works on nanoswitches, actuators, and information-processing systems. Several such systems have been obtained by way of chemical synthesis and have been tested experimentally. Some of them allow the construction of logic elements and simple circuits at the molecular structure level (see, for example, V. Balzani et al. [177], G. Bottari et al. [118], S. Muramatsu et al. [119]). The most exciting property of these systems is their small size. However, notable disadvantages are the low process speeds due to the time needed for molecular motions, a lack of interfaces for signal and energy transport, and a lack of methods for connecting single nanomachines as well as integrating them into lithographically prepared peripheral nano- and micro-circuits. Molecular logic elements based on chemical reactions and/or molecular motions will thus not be able to fully substitute semiconductor devices or future single electronic devices based on metal/insulator technologies. Nevertheless, they could become of increasing interest for the development of nanosized transducers applied in interface devices for various signal transfers between solid-state electronic circuits and molecular as well as cellular systems.

7.3.7. Particle-Emitting Nanotransducers

The fabrication of small metal tips has led to the successful preparation of devices capable of emitting electron beams by field emission. The extraction of electrons from certain sharp tips by an electrical field is particularly efficient when the tips are as small as possible. Therefore, tips with nanometer- or subnanometer-sized radii are optimal for field-emission devices.

Extremely small tips can be prepared by electrochemical methods using noble metal covering layers on tungsten tips. Tips possessing only one atom at the very end have been obtained by a combination of electrochemical reduction of surface oxide and electroplating [120]. The structure of the tip was scrutinized by field-emission microscopy (FIM). The atomic tips showed excellent field-emission behavior for electrons. Emission currents of about 0.2 nA were observed at a field of 1.2 kV. The brightness of the electron beam emitted by these single-atom tip emitters was about one order of magnitude higher than the beam of a three-atom tip and about two orders of magnitude higher than that of an eight-atom tip operated at the same voltage.

CNTs are of particular interest for the construction of field-emission devices. A very high emission density has been achieved by arranging perpendicularly oriented CNTs in regular arrays on a plane substrate using nanoimprinting technology [121].

7.4 Nanooptical Devices

7.4.1 Nanostructures as Optical Sensors

Chloroplasts of green plants have two complex supermolecular architectures which have two main objectives: the efficient absorption of photonic energy and the transformation and efficient use of this energy in chemical processes. Chromophore systems are used for the first objective: they absorb the photons in antenna arrangements, and transfer the photon energy as electronic excitation energy into the photosynthesis center. Synthetic molecules have been developed for this antenna function. They include dendrimers with a central conversion element that is linked by branched connecting elements with a periphery of chromophorous absorption elements. The outer chromophores and the inner chormophorous element are thereby completely connected by covalent bonds. The feasibility of this concept was demonstrated by the fluorescence of a central dye unit [122]. Coumarin and coumarin 343 were used as donor and acceptor dyes, respectively. The energy transfer between an antenna and a conversion unit is probably based on a dipole–dipole mechanism.

Molecular nanoarchitecture and complexes of molecules and nanoparticles are utilized as key elements for novel optical sensors. The chemical specificity of the binding of functionalized nanoparticles has been combined with the high optical absorption of nanoparticles in optical biosensors [123]. As a result of their light absorption and the cooperative effect of the reflection and diffraction of the light, they are utilized in labeling of molecular binding reactions [124][125]. The nanoparticle thereby acts as a marker for a coupling reaction on a solid substrate surface. Optical detection is enhanced by an arrangement of layers including a mirror layer below a binding layer for the coupling of the nanoparticles. A thin optically transparent distance layer between these two layers amplifies the signal even further [123]. A significant improvement in sensitivity of the optical readout is achieved when the metal nanoparticles bind in the proximity of optical apertures that have diameters with the same dimensions as the nanoparticles. The probability of photon transmission through holes or apertures in the sub-micrometer range is significantly decreased by the presence of metal nanoparticles, so that an individual binding event could lead to a significant effect.

Interesting possibilities for optical nanosensory applications are provided by the combination of optically active inorganic solid structures and molecular chromophores. Energy transfer between inorganic and organic elements can be highly efficient. Theoretical calculations indicate that the energy of semiconductor quantum wells can be transferred over short distances (5 nm) onto organic chromophores in a short period of time (10–100 ps) [126].

Metal solids with dimensions in the nanometer range as well as thin layers exhibit characteristic plasmon resonance. Nanostructured layer elements or metal clusters can be utilized for the control of optical processes, because their dimensions, shape and orientation influence both the wavelength of maximal light absorption as well as the interaction cross-section depending on the polarization direction [127].

Fluorescent nanoparticles are an interesting alternative as markers for the detection of molecular binding events. For diameters in the lower nanometer range (2–5 nm), molecules labeled with these particles are still able to react highly specifically, because of the low strength of non-specific interactions. The particle diameter controls the fluorescence wavelength. Choice of particle size and material allows a broad optical range between the UV and the NIR to be addressed. Applicable materials are, e. g., the compound semiconductors CdSe, InP, ZnS, CdS and InAs [128][129]. In addition to these inorganic particles, organic nanoparticles are also applicable. Therefore, fluorescence dyes are incorporated into polymer particles that are surface-functionalized. Thus a well-defined and relatively rigid chemical environment is provided for the organic fluorophores, in order to enhance the relaxation behavior and the stability of the quantum yield. By the use of different dyes, a multichannel analysis is possible [130].

7.4.2 Nanostructured Optical Actuators

The term optical actuators includes devices that generate and emit photons. The layer thickness, optical semiconductor luminescence and laser diodes are all within the nanometer range. Low vertical dimensions are essential for charge recombination and the emission of the photons generated by this process. The layers extend laterally usually over several hundreds of micrometers, so that these devices are not included in nanotechnology in the narrowest sense.

Very small dimensions in all three directions are required for quantum point light sources. Emitters with diameters comparable to the diameter of exciton radii are distinguishable through the high stability with respect to the wavelength, extremely high robustness towards changes in excitation current and temperature as well as by a relatively low threshold of the excitation current density [131]. A high light yield can also be realized by compound semiconductors. Three-dimensional arrays of small tetragonal pyramids of InAs with dimensions of 8 to 14 nm embedded in a GaAs matrix show laser characteristics at working temperatures of between 100 and 130 K [132]. In addition to InAs, InGaAs and InP were also used as quantum dot laser materials (cf. Table 8).

Tab. 8 Quantum dot lasers. [131]

Material	*Structure size*	*Emission wavelength*	*Current threshold density*	*Working temperature*	*Year*	*Reference*
InAs	7 nm	0.9 μm	1000 A cm^{-2}	300 K	1994	[133]
InGaAs	30 nm	1.26 μm	7600 A cm^{-2}	77 K	1994	[134]
InGaAs	20 nm	0.92 μm	800 A cm^{-2}	85 K	1995	[135]
InP	25 nm	0.7 μm (optical excitation)	25 kW cm^{-2}	300 K	1996	[136]

It is likely that systems with delayed fluorescence or electroluminescence of individual photons are possible through single nanoparticles [137]. Instead of inorganic solids, organic polymers are also appropriate materials for luminescence diodes where there is a sufficiently high density of conjugated double bonds. For example, polyparaphenylenevinylene exhibits a high electrical conductivity and is used as a material for organic light emitting diodes („OLEDs“) [138]. By integration of electrically conductive polymers such as substituted polythiophenes in a nanoporous membrane and wiring by a metal base electrode and at least a partially transparent membrane electrode, organic light diodes have been constructed [139].

Small dimensions in two directions are typical for field emitter devices. They generate photons via the field-based emission of electrons from a cathode, which are then accelerated over a short distance onto a luminescent material. To achieve large electrical field strength (about 0.3 V nm^{-1}) with moderate voltages, the gaps between the cathode and counter electrodes should be as small as possible. Therefore, nanolithographically prepared cathode tips were constructed opposite plane anodes, separated by a distance in the sub-μm range. For the diameter a hole in a gated electrode of 1 μm, a distance to the emitter tip of 500 nm, a gate voltage of 130 V and an anode voltage of 600 V had to be used. Concentric positioning of the cathode in micro hole apertures of the counter electrode requires tip diameters in the medium to lower nanometer range. In addition to the concentric arrangement, microtechnical edge emitter arrangements have also been described with gap sizes in the sub-μm down to the nanometer range. These techniques are also applicable for the preparation of whole arrays of similar emitters [140]. The fabrication of field emitters in small gaps in metal electrodes can be achieved, for example, by oxidation on tapered structures. Such p Si field emitters have been prepared with gap widths of 10 and 160 nm, resulting in electron emission currents of up to 12 μA per individual element [141].

Nanostructured electrical conductors of carbon exhibit field emitting properties, in a similar manner to small metal structures. Therefore, carbon nanotubes can be utilized as field emitters. High radiation yields of up to 100 μA cm^{-2} can be achieved with a screen electrode positioned 20 μm above a layer of carbon nanotubes [142]. Superior field emitting properties were found for carbon nanotubes with tips opened up by laser treatment [143].

Light-emitting diodes based on organic materials (OLEDs) are of increasing interest for use in micro- and nano-sized optical transducers. Efficient OLEDs can be produced by combining nanopatterning of inorganic electrode structures with spun films of optically active organic semiconductors such as substituted poly(phenylene-vinylene). This material can be used to form light-emitting areas with a spot size far smaller than half the wavelength of the emitted light. For example, OLEDs with an aperture of 60 nm have been prepared for emission at 590 nm [144].

7.4.3
Nanooptical Switching and Conversion Elements

The exact mechanical control by a micro actuator is used in waveguide-type optical devices that rely on the so-called nanomechanical effect of integrated optics. A wave passing through an exposed optical waveguide is sensitive to objects that enter the region of the evanescent field even without touching the waveguide. The nearer the approach, the larger the interaction due to the diffraction indices or a resonant coupling. Typical distances are in the range of a few nm up to about 100 nm. Such nanotechnical light modulators are, e. g., prepared through Si micromechanics by bonding a movable component with a planar surface onto a lithographically prepared gap in the waveguide substrate [145].

Transducers based on this principle can also be applied for all types of influences that induce small changes in the location of micro- and nanocantilever tips. In addition to direct mechanical force, these are, for example, small changes in temperature, variations in the surface charges, magnetic or electrostatic effects.

A two-dimensional polymer photonic crystal based on azo-dye-doped PMMA has been prepared by EBL [146]. EBL of the polymer layers allowed the generation of a very regular pattern and, therefore, the formation of photonic crystals of high quality. Arrays of holes of diameter 330 nm with material walls between them of thickness ca. 50 nm were obtained in dyed PMMA films of thicknesses between 0.8 and 1.2 μm.

The coupling of photons with single-electron tunneling is not only of interest in relation to the UV and the visible regions of the electromagnetic spectrum. Josephson devices use, for example, the connection between microwave frequencies and superconductivity for ultra-sensitive magnetic measurements or highly precise voltage definition in magnetic tunnel junctions (MTJ). Beside these effects of long-wavelength radiation on the transport of Cooper pairs of electrons through tunneling barriers, single-electron transport processes can also interfere with low-energy photons. So-called Kondo peaks have been observed in the case of SET transistors exposed to microwave radiation [147].

In general, optical tweezers can be regarded as optically driven nanoactuators, if particles inside the optical trap are moved. The principle of optical tweezers can also be applied to realize an optically driven nanorotor. It could be shown that nanorods were rotated when the plane of polarization of a polarized laser trap was rotated [148]).

Magnetic resonators of small size are of interest for the modulation of high-frequency electromagnetic radiation. Such structures lead to effective material properties corresponding to negative permittivity and negative permeability (so-called "double-negative materials", DNM). These effects can be realized by the preparation of free-standing nanoloops or nanobridges. For this, a multi-step deposition and etching strategy, including a sacrificial layer technique, is necessary. S. Zhang et al. [149] have reported on the fabrication of such Au/ZnS nanostructures by using auxiliary structures composed of silicon nitride and a photoresist. The patterning was achieved by means of an IL technique. The width of the loops was of the order of about 130 nm, the height was between 180 and 280 nm, and the structure width was in the range 15–30 nm.

7.5 Magnetic Nanotransducers

The discovery of the giant magnetoresistive (GMR) effect in extremely thin magnetic layers separated by a non-magnetic layer in a stack made nanolayers the focus for lateral microstructured magnetic devices [150]. At 4 K, a magnetoresistive effect of 100% was observed on an Fe/Cr multilayer system [150]. The thicknesses of individual layers were about 1 nm, which means just a few atomic layers. While such multilayer systems of ultrathin inorganic layers are interesting for sensors and highly-integrated magnetic memory, soft-magnetic structures provide the possibility of producing electron spin valves as fast switching devices, which excel because of the extremely low power dissipation and the low voltage requirements [151].

Magnetic materials nanostructured in all three dimensions were constructed by cathodic deposition of alternating metal layers in nanoporous polymeric matrices. Therefore, nuclear trace membranes with 40 nm pore widths were applied onto a conductive substrate. Alternating deposition of Co and Cu layers resulted in the fabrication of nanowires that exhibit the GMR effect (Fig. 149). The magnetoresistive effect for 0.8 nm thick Cu layers was 15% measured along the nanowires [152].

High aspect ratio structures of CoFeNi and FeNi for magnetic recording devices have been generated by EBL and electrodeposition on an AlTiC substrate [153]. Aspect ratios of 5–8 at critical dimensions between 22 and 30 nm were achieved.

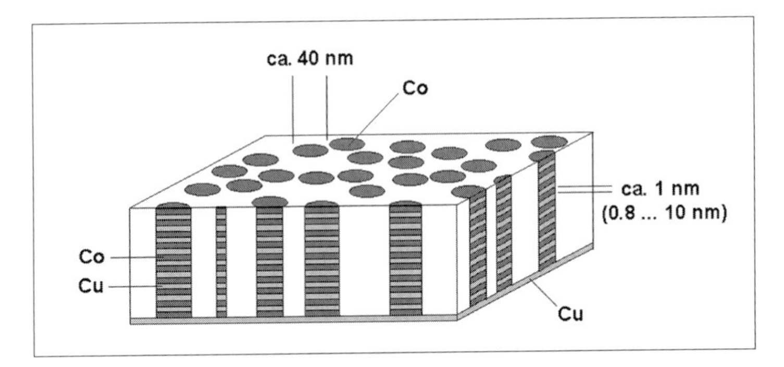

Fig. 149 Scheme of a lateral nanostructured column-like GMR-layer stack for highly sensitive magnetoresistive sensors [152]

Magnetic storage and readout with pixel densities of around 10 μm^{-2} can be realized by magnetic structures prepared by nanoimprint technologies. Therefore, a resist layer is deposited on a magnetic layer and shaped by a stamp. Subsequently, the resist is removed from the thin regions by a dry-etching step. A second etch step transfers the resist pattern into the magnetic layer. This procedure results in a periodic pattern of magnetic elements with their magnetization orientation changed and the readout is by magnetic scanning probes [154].

Magnetic scanning probes represent nanostructured elements. They can be applied either as a sensor or as an actuator [3][4]. They are able to magnetize soft-magnetic materials down to the sub-µm range [155].

The detection of small local magnetic fields by microstructured GMR sensors can be applied for the generation of highly sensitive transduction chains for molecular sensors based on a magnetoresistive readout (Fig. 150). Therefore, instead of the fluorescence or radioactive nanoparticles normally used, magnetic nanoparticles are applied for molecular labeling. After binding of magnetic nanoparticle-labeled biomolecules onto the surface of the GMR sensor, the change in the local magnetic field induces a change in the conductivity of the multilayer resistor system, which can be easily recorded by electrical means [156][157][158]. To achieve an optimal signal for small nanoparticles, GMR sensors with minimal lateral dimensions are needed. Such sensors would probably provide the means for the detection of a small number or even individual small nanoparticles, with their advantage of reduced non-specific binding compared with larger particles with higher magnetic signals.

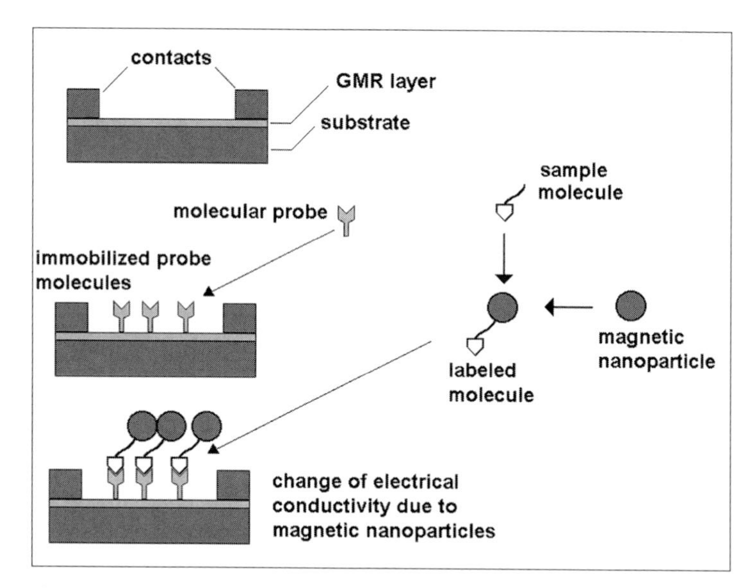

Fig. 150 Principle of a magnetoresistive detection for biosensory applications [156]

7.6 Chemical Nanoscale Sensors and Actuators

The geometrical matching of molecules and their functional groups plays a central role in the specificity of biologically functional molecules, such as antibody/antigen or enzyme/substrate systems. This key–lock principle also can be applied for the recognition of molecules with synthetic sensors. One approach towards molecular recognition structures is using three-dimensional matrices with embedded and subsequently (after stabilization of matrix) dissolved template molecules. The remaining cavities in the matrix represent the geometry of the template molecules. This molecular imprinting technique applies for example to spatial cross-linked polymers (Fig. 151). A drawback of these three-dimensional networks is the difficulty in accessing these cavities from the outside. A significantly enhanced accessibility is achieved by two-dimensional imprint structures that are prepared by SAM and which are compatible with planar technology. Through this approach, a high specificity for the recognition of barbituric acid was achieved with the help of thiobarbituric acid integrated into a matrix of dodecanethiol on gold (Fig. 152). Such structures are candidates for the construction of artificial sensors [159].

Very low concentrations can be detected by the adsorption of molecules onto the surface of micro- and nanostructured cantilevers [160][161]. The adsorption of the molecules results in a change to the mechanical strain of thin cantilevers. This effect yields a slight deflection, which can be detected by a light pointer, as in the case of

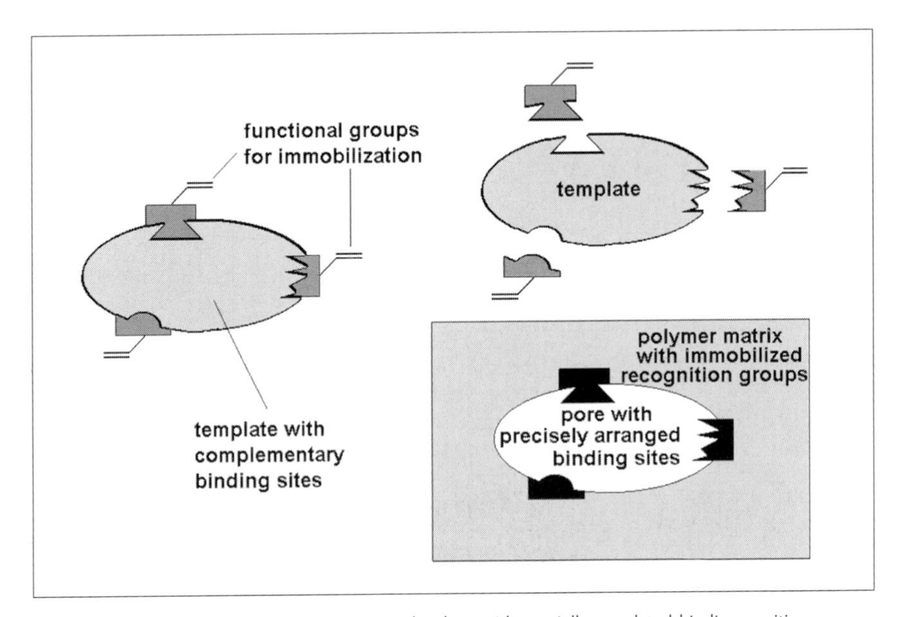

Fig. 151 Principle of molecular imprinting technology with spatially correlated binding positions. A, Template molecule with binding positions. B, Complex of template molecule with complementary-bound molecules. C, Cavity structure after imprinting of B into a polymeric matrix and dissolving the template A. [181]

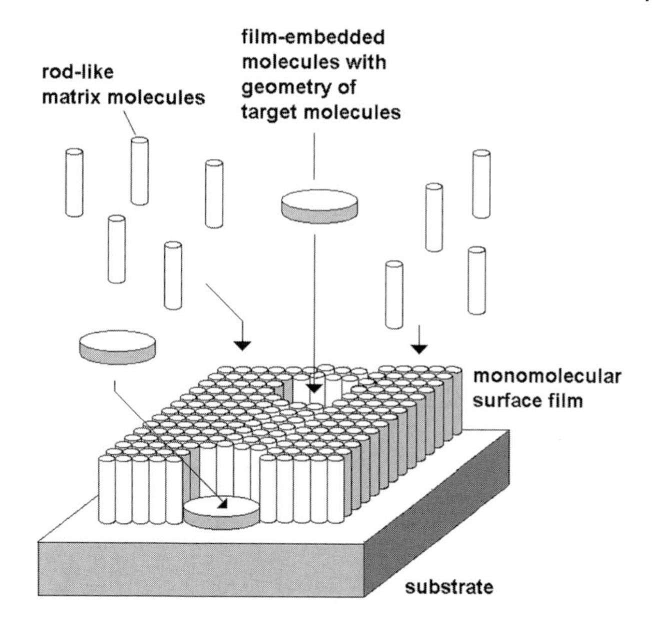

Fig. 152 Molecular imprinting in molecular monolayer [159]

scanning force microscopes. Standard sizes for cantilevers are thicknesses of 1 μm and widths of about 100 μm, which means that the mechanical transducers are not on the nanoscale. However, their detection principle relies on the adsorption of molecules in a sub-monolayer and a detection of deflection in the lower to medium nanometer range.

Biosensors already represent classical devices as examples of micro system technology. They include typically ultrathin, often mono- or sub-monomolecular layers as active surfaces of nanostructured components. Owing to the universal importance of electrical parameters in signal processing, the integration of individual molecules in nanoelectrode arrangements is of particular interest. The combination of nanolithography with selective immobilization techniques enables the integration of molecular ensembles or single supermolecules into planar electrode arrangements [21][75]. These methods are interesting for both the physical characterization of individual molecules and nanowiring approaches (Fig. 153) [19]. Nanobiosensors include devices that achieve specific molecular recognition through the nanogeometry of larger

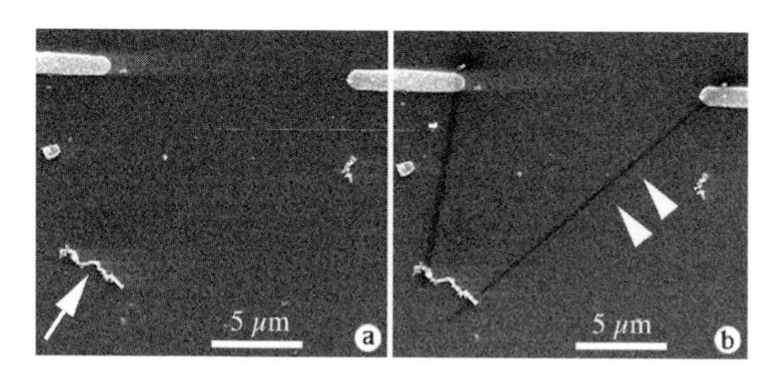

Fig. 153 Wiring of an electroless metallized tubulin assemblage (arrow) to microelectrodes by EBD nanoelectrodes (arrowheads) (W. Fritzsche et al. 1999b [38])

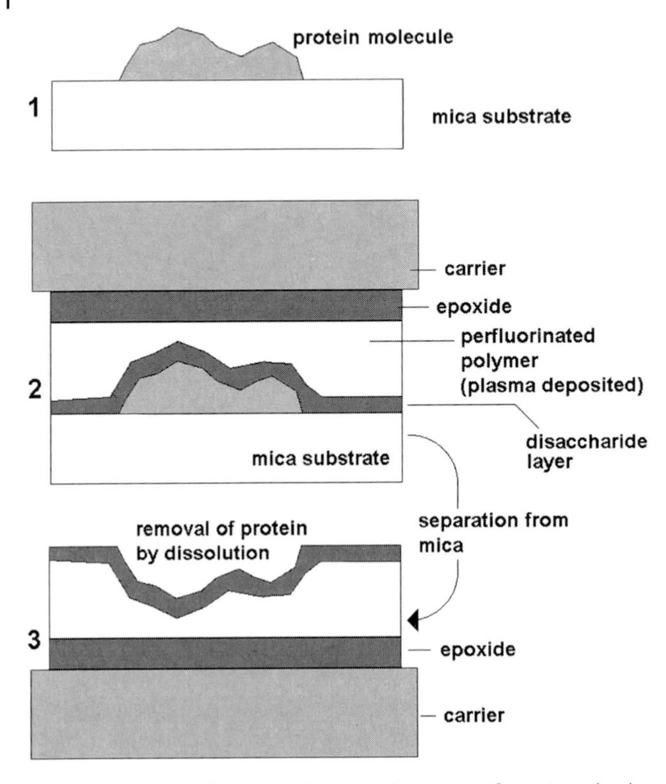

Fig. 154 Imprinting of the three-dimensional geometry of protein molecules for molecule-recognizing surface topographies [163])

molecules, based on a topological match but also on the geometrically defined arrangement of several functional or specific binding groups. The geometry of biomolecules can be transferred by imprinting processes in a negative mold. This technique lies between that of molecular imprinting techniques and the embossing of solid substrates. The imprinting of proteins has to be very gentle to preserve the three-dimensional structure of these sensitive molecules. Molds in a Teflon-like matrix have been fabricated by first adsorbing the proteins on mica prior to covering with a thin disaccharide layer. The sugar layer protects the protein in the subsequent generation of the matrix of a perfluorohydrocarbon by plasma deposition of perfluoropropene. After stabilization of the matrix by a fixed glass substrate, the mica was removed and the protein dissolved. A cavity remains that matches the shape of the protein and which can be used for the recognition of this protein (Fig. 154). Instead of the plasma embossing, also the use of microcontact printing procedures for the generation of such surfaces should be possible [163].

The efficiency of electrical detection of nanoparticles or biomolecules can be enhanced if two-electrode arrangements with a slit structure are applied (Fig. 155). The extended width of a long gap improves the probability of analyte molecules or nanoparticles being trapped in the gap [164]. The immobilization of recognition mo-

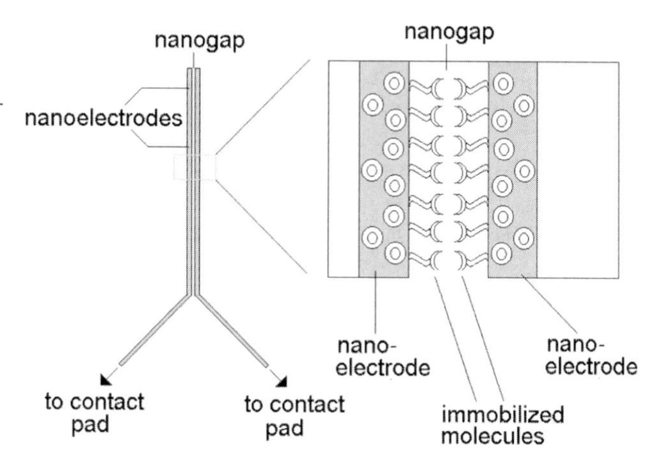

Fig. 155 Detection of nanoparticles and biomolecular interactions using an electrode nanogap arrangement [164]

lecules at the electrode surface and in the gap leads to an efficient change in the electrical conductivity of the gap structure upon the binding of particles or analyte molecules. For example, the binding of recognition proteins such as streptavidin led to a significant enhancement of the gap conductivity in comparison with that of the bare electrodes. The current-voltage characteristics are further modified if analyte molecules are bound at the recognition sites.

Microcantilevers, as utilized in atomic force microscopy, can be used for chemical sensing. The displacement of cantilevers is influenced by mechanical stress, but can also be influenced by liquid flow, thermomechanical stress, or surface adsorption. These effects can be used for molecular recognition. Highly specific sensing is made possible by the immobilization of DNA receptor molecules on a cantilever surface. In this case, read-out has been achieved by means of the widely used laser deflection principle, enabling interaction-induced bending of the cantilever in analyte solutions of between 3 and 32 nm [165]. Antibody-covered cantilevers permit sensitive stereoselective detection [166]. Phase-dependent measurements on piezo-actuator-driven cantilevers can be used for the determination of binding constants with very small amounts of substances [167].

A very high detection sensitivity in small volumes can be combined with a degree of selectivity if surface-enhanced Raman scattering (SERS) on metal nanoparticles is applied to small analytical samples. This technique has been used for pH measurements with optical read-out [168], but may also be adapted for more specific molecular recognition.

The term nanoreactor has already been applied to devices working in the nanoliter range. The volume unit nanoliter correlates with a length of 100 μm and is therefore outside of the limit of 100 nm for nanostructured technologies. Therefore, nanoreactors are not considered here. However, some devices for the processing and separation of cellular portions or large molecules that are used in nanoreactors exhibit functionally important dimensions in the nanometer range. So arrays of nanostructured columns have been used for the separation of mixtures of DNA molecules in electrical

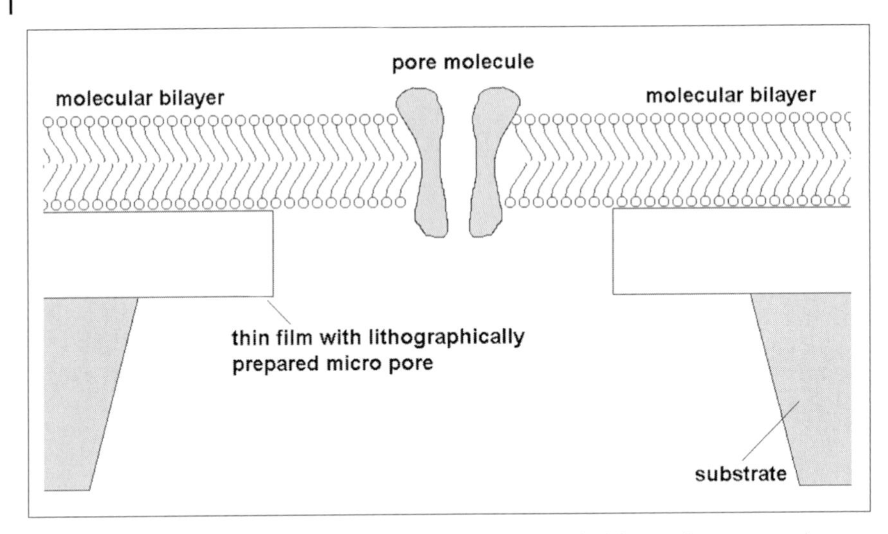

Fig. 156 Lithographically fabricated nanopore covered by a molecular bilayer with an integrated pore molecule

fields [169][170]. The retention or the passage of single molecules is determined by the pore size of the sieve-like nanostructure, and by the size and folding state of the molecules. The passage of DNA through nanopores allowed the characterization of voltage-driven transport at the single molecule level [171]. In addition to lithographically prepared nanostructures, spontaneous processes are also utilized for the fabrication of nanopores in semiconductor layers, such as nanoporous silicon by photoelectrochemical etching of silicon single crystals.

One can expect an increasing number of developments involving the combinations of subcellular and biogenic molecular functional elements with artificial compartments and nanotechnological material management. A particular role in this process is played by architectures of geometrically defined lithographically prepared micro- and nanopores and molecular mono- and double-layer structures (Fig. 156). Membrane proteins with enzymatic or porous functions and also molecular machines could be incorporated into such structures. The integration of cell components in microstructured compartments with controlled exchange of materials will lead to new types of biosensors and bioreactors as well as to new models of cells with minimal biological function. Such systems are interesting for the investigation of cellular mechanisms, of cellular communication and the development and testing of new bioactive components.

The differences between nanoreactor applications and single cell handling techniques diminish for decreasing volumes and numbers of cells. Biogenic macromolecules can be manipulated as single particles using physical and microtechnical tools. Hence DNA immobilization is well established. DNA can be oriented in shear forces induced by fluid flow or by an electrical field due to the negative charges of the phosphate residues along the molecule. It can be manipulated by the tip of a scanning probe microscope. Using fluorescence labeling, one can watch how an individual DNA molecule is divided enzymatically and how the fragments diffuse away [172]. The translation of the products can be controlled by microtechnical channels or wells [173].

7.7 Chemically Driven Nanoactuators

The input of energy into a nanosystem and the organization of entropy exports are two main demands for the further development of nanotechnology. It is evident from the natural world that power supply in spatial connection with information processing can be very efficient. So, there exists an urgent need for molecular processes that can supply power in direct connection with the actions of a technical nanosystem. An important step in the direction of chemically based power supply is the transformation of chemical energy into other forms of energy. Nanometer-sized systems for the transfer of chemically stored energy into mechanical energy are of particular interest.

Nanoactuators can be driven by chemical reactions or absorption, adsorption and desorption processes. The principle of deflection change of micro- or nanocantilevers due to swelling or shrinkage of thin surface layers induced by chemical reaction can be applied for actuating purposes. A prerequisite is a cantilever with an asymmetric profile. The lever has to exhibit at least two different layers or different surfaces to enable asymmetric adsorption or desorption processes to take place.

Electrochemical processes are optimal for the readout of an electrically driven movement, because they can transform the electrical signal into changes in the material in order to achieve the mechanical actuation. Typically, flexible materials are applied in electrochemical nanoactuators, which can be deformed by the action of an electrochemical transformation, or by the insertion or cleavage of small molecules formed electrochemically. Such materials are also described as artificial muscles. Often conductive polymers are applied, and carbon nanotubes can also be utilized after direct or indirect substitution with electrochemically active groups. Movement can be achieved by an electrochemically-controlled intercalation of Li^+ ions into the graphite walls of carbon nanotubes [174]. Cantilevers provide easy access for transduction processes where there are deflections induced by electrochemical reactions [175]. In this case, the actuator represents one electrode. Depending on the potential, current flow induces either deposition, modification or removal of material at the cantilever surface. For multiple applications, the electrochemical process used has to be highly reversible. A second limitation lies in the compatibility of the medium that surrounds the actuator. It has to be an electrolyte that contains all components required for the surface processes on the actuator.

It is rather difficult to construct chemomechanical systems wholly by chemical synthesis. However, recent work has offerred insights into the mechanisms of molecular energy conversion with the potential for introduction into nanotechnology. For example, the rotational direction of intermolecular motion can be controlled if the rotatability is reduced by suitable substituents or the activation of rotation depends on specific molecular interactions. For example, S. P. Fletcher et al. [176] were able to show that the rotational direction of an *ortho*-substituted phenyl group as opposed to that of an ortho-substituted naphthol group could be controlled by the choice of reaction partner in the chemical opening of a lactone bond.

A focus of investigations in supramolecular chemistry in recent years has concerned molecular machines [177]. These can be driven by thermochemical, photochemical, or electrochemical forces, or by combinations thereof. The actuation principles can be proven by spectrochemical investigations and by modeling. It is difficult to arrange these molecular systems about a micro- or nanolithographic periphery. So, there is a lack of chemically driven nanoactuators available for use in nanomechanical applications or the development of chip-integrated nanomechanical devices. The most impressive nanomachines driven by chemical energy are hybrid systems that rely on biomolecules supplied by the natural world. Thus, recent nanomechanical achievements have relied on, for example, the rotary motor principle of ATPase or the linear motor principle of the tubulin/kinesin system.

Nature provides us with several examples of biomolecular motor systems that are chemically driven. They are all based on protein complexes, as is the intercellular transport along the supermolecular microtubule network that occurs as a result of the movable compact protein structures, the kinesin. The process is powered by ATP (adenosine triphosphate). Tubulin acts as the "rails", the kinesin as a "car". This system can be isolated from the cell and operated in a technical environment [178]. Therefore the relationship between the stationary and mobile parts is reversed. It has been demonstrated, that kinesin immobilized on polymer surfaces with a striped relief induces the movement of microtubules along this pattern [179]. This movement can be controlled externally by changing the temperature. For example, a composite surface, at which functional kinesin motor molecules were adsorbed onto a silicon substrate between surface-grafted polymer chains of thermo-responsive poly(*N*-isopropylacrylamide), enabled dynamic control of gliding microtubule mobility [180]. By external temperature control between 27°C and 35°C, the reversible landing, gliding, and release of motor-driven microtubules in response to conformational changes of the polymer chains was demonstrated. Imobilization of the kinesin onto glass substrates enables this structure to be transported by the biomolecular motor system (Fig. 157) [181][182].

Another example is provided by microfluidic systems into which kinesin was seamlessly integrated, which allowed the transport of microtubules through the extraction of chemical energy from an aqueous working environment [183]).

Active control over the movement is possible. By integration of kinesin motor proteins into closed submicron channels and active electrical control of the direction of individual kinesin-propelled microtubule filaments at Y-junctions, molecular sorting of differently labeled microtubules could be accomplished. This was attributed to electric-field-induced bending of the leading tip [184].

Chemomechanical nanomachines represent the central functional parts of a cell. They are not only responsible for the innercellular transport, but also for the generation of macroscopic movements. In analogy with the kinesin/tubulin system, myosin/actin represents a linear motor system. It forms the basis of our muscles, and eventually drives the macroscopic movement of organisms.

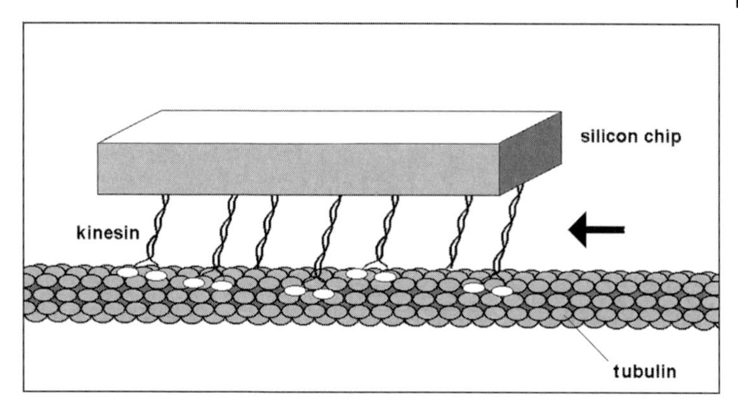

Fig. 157 Transport of technical micro- and nanoobjects along immobilized tubulin aggregates by functionalization with kinesin molecules (L. Limberies and R.J. Stewart 2000 [181])

In addition to these translational working protein structures, a cell also exhibits rotating systems. In analogy with technical engines, these systems consist of a rotor and a stator; the latter is integrated into a membrane and therefore immobilized. Molecular motors in cell membranes power for example the movement of flagella in order to move the organism.

The driving force in molecular machines are concentration gradients, e. g., of protons, which provide the chemical energy by translocation from one side to the other of a membrane. This energy is transformed by the protein complex into mechanical energy. The molecular rotation process can be also used for chemical synthesis. The molecular motor of the ATP-synthetase complex in the membrane of mitochondria produces the energy-rich ATP as the universal energy-source in cellular processes [185][186][187][188].

Synthesis of molecular machinery analogous to the motor proteins described is not yet possible. However, the example given of the use of the molecular movement of tubulin shows that it is possible to isolate a native system, to modify it and to integrate it into technical environments. Also the rotation of the ATP-synthetase and related rotating supermolecules could be used by nanotechnology. A first step in this direction is the coupling of an actin filament onto the rotor of the protein complex (Figs. 158 and 159). Several groups have demonstrated that this system represents a partially synthetic nanomotor, which can be monitored and measured with an optical microscope [189][190][191].

There have been attempts to integrate complex functional protein architectures into synthetic systems. An artificial photosynthesis-active membrane has been prepared that has integrated cytochromes (responsible for the absorption of light and the primary conversion of the electrochemical energy) and ATP-synthetase. The membrane-integrated cytochrome works as a proton pump and creates a light-driven pH-gradient, which in turn drives the synthetase that is responsible for the synthesis of ATP [192].

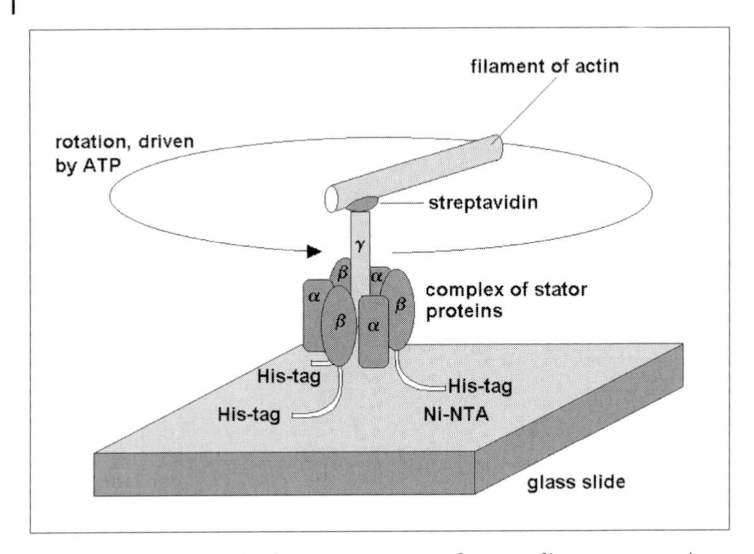

Fig. 158 Modified biomolecular motor consisting of an actin filament connected to the rotor protein of ATPase [189])

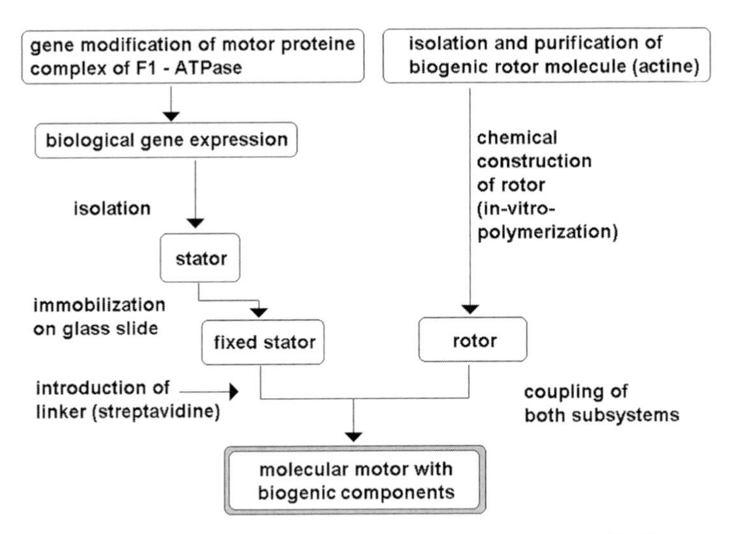

Fig. 159 Main steps in the construction of a molecular motor using modified biogenic macromolecules

7.8 Nanochannels and Nanofluidic Devices

The term "nanofluidics" is used for two different types of devices and applications. On the one hand, it means the miniaturization of fluidic, chemical, and biological operations to the nanoliter volume range. This relates to the significant development of fluidic microsystems, micro reaction technology, miniaturized analytical systems, and, in particular, to micro total analysis systems (μ-TAS) and lab-on-a-chip devices. A volume of one nanoliter corresponds to a cube with a side length of 100 μm. Thus, the geometries of devices for the handling of liquids and gases in the nanoliter range are typically in the sub-mm and micron ranges, but not on the nanometer scale.

On the other hand, there have recently been investigations on fluidic devices having linear dimensions in the nanometer range. The generation of these small structures has been based on various tools of micro- and nanolithography and on molecular self-organization. As a result, cavities, channels, and other elements for fluids can be obtained, which correspond to volumes far smaller than the typical volumes of lab-on-a-chip devices. The picoliter (pL) level is achieved by a moderate miniaturization down to the 10 micron level. The one-micron level corresponds to a cube with a volume of a femtoliter (fL), the 100 nm level to a volume of one attoliter (atL). The ratio of 1 nL to 1 atL amounts to 1 billion. This enormous factor highlights the qualitative difference between the above mentioned field of "conventional" nanofluidics and nanometer-sized systems for ultra-miniaturized fluidic operations.

The planar technology principle allows the formation of nanochannels either parallel to the substrate plane or parallel to the normal direction. The formation of nanochannels in the substrate plane corresponds in principle with other standard lithographic processes for patterning thin films or thin parts of the uppermost region of planar substrates. In most cases, channels are formed by etching extended grooves and linear trenches. Channels are obtained after sealing with another planar substrate or by bonding of two planar substrates with complementary mirror-image half-channel structures (half-shell principle).

Beside these options for nanochannel generation, a third class of channel preparation technologies uses special processes to close channels prepared at a substrate surface. One way relies on the local reflow of patterned polymer material. The deformation of polymer walls along narrow gaps can be supported by capillary forces of liquids inside the gap. In the case of small dimensions, i.e. the submicron or nanometer region, surface-tension-induced pressures can exceed one atmosphere. Movable or deformable polymer walls can be brought together by surface forces during the evaporation of liquids deposited in preformed channel gaps (Fig. 160). Closed nanochannels are obtained after successful contact formation over channel gaps (J. L. Pearson and D. R. S. Cumming [193]).

Another principle for nanochannel formation is based on the deposition of nanoporous material on a substrate bearing material with a negative channel structure as a template. After complete deposition of the covering material, the template material is removed in a dissolution process similar to a lift-off procedure. This strategy has been

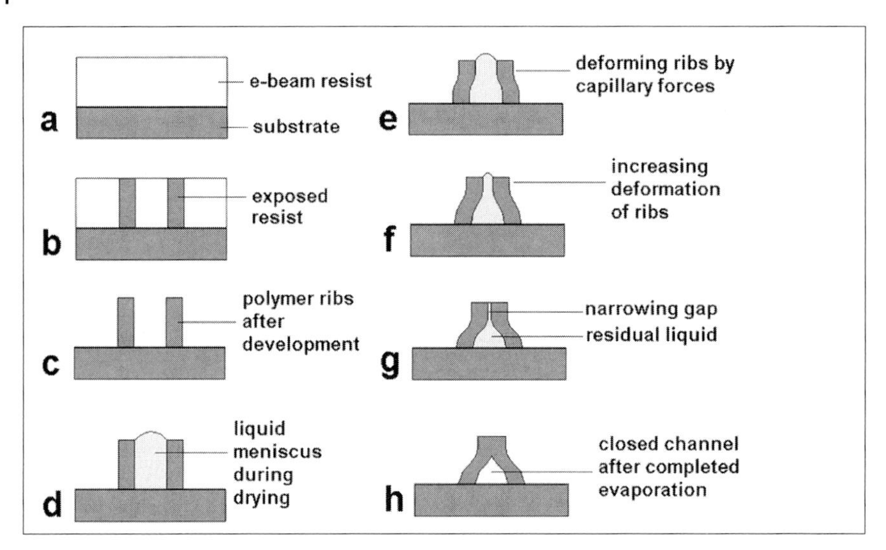

Fig. 160 Formation of closed nanochannels without substrate bonding or sealing using polymer structures deformed by capillary forces: a) polymer film (e-beam resist) on substrate, b) nanopattern generation by EBL, c) small polymer walls obtained after resist development in an organic solvent, d) deposition of liquid inside the gap between two polymer ribs, e) displacement of top of ribs by capillary forces, f, g) reduction of liquid volume by evaporation and narrowing of opening between the polymer ribs due to the rib deformation, h) closed channel after complete removal of liquid (J. L. Pearson and D. R. S. Cumming [193])

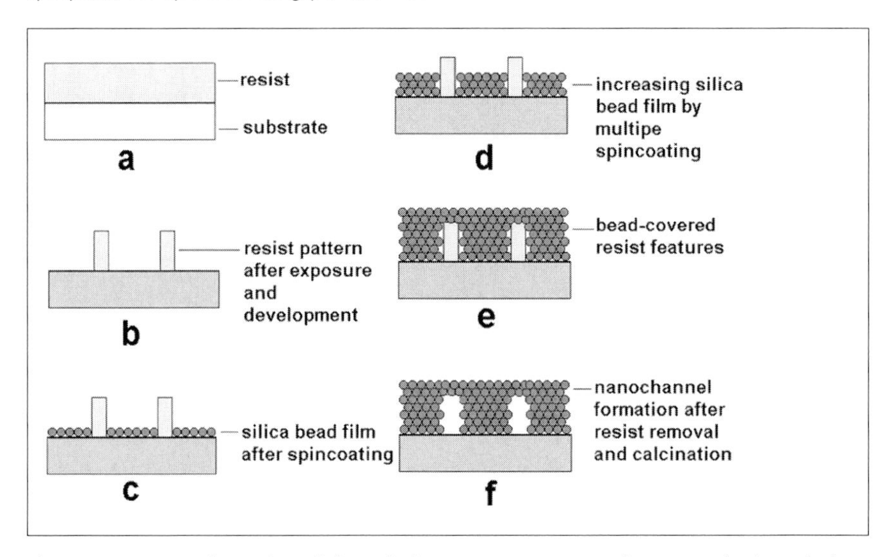

Fig. 161 Formation of nanochannels by multiple spin-coating of nanoparticle suspensions over resist patterns and subsequent calcination and resist removal: a) resist-coated substrate, b) resist pattern after exposure and development (negative of channel structure), c, d) filling of space between resist structures with nanoparticles by multiple spinning steps, e) continuation of suspension spinning after complete coverage of resist structures, f) finally formed channels after resist removal and calcination (D. Xia and S. R. J. Brueck [194])

successfully applied, for example, in the preparation of nanochannels by covering resist patterns with silica nanobeads (Fig. 161). Closed channels are obtained after covering the polymer template structure. The channel preparation is completed by dissolving the polymer using organic liquids and a final calcination of the silica particles (D. Xia and S. R. J. Brueck [194]).

7.8.1
Nanochannel Arrays

Nanochannels may be generated either by lithographic techniques or by means of self-organized patterning. Typical nanolithographic techniques, such as excimer laser lithography, EBL, and FIBL, are used for the generation of parallel channel structures in the nanometer range.

A regular pattern of nanochannels may be obtained, for example, by the combination of FIBL with surface anodization of aluminum (Fig. 162). A thin film of PMMA is applied to enable primary lithographic patterning. The lithographic windows in the polymer film are then used for wet chemical pattern transfer into the metal. Thereby, the surface anodization is accompanied by metal oxide growth on the side walls, which inhibits lateral etching. The aluminum is converted into a porous anodic alumina film (PAAF). This film is responsible for the formation of etch holes with high aspect ratios. As a result, channel array structures are obtained in an arrangement corresponding to the lithographically generated pattern (N. W. Liu et al. [195]).

Extended arrays of nanochannels have been prepared by NIL. A gradient of channel widths was achieved by a superposition of proximity exposure with lithographic pat-

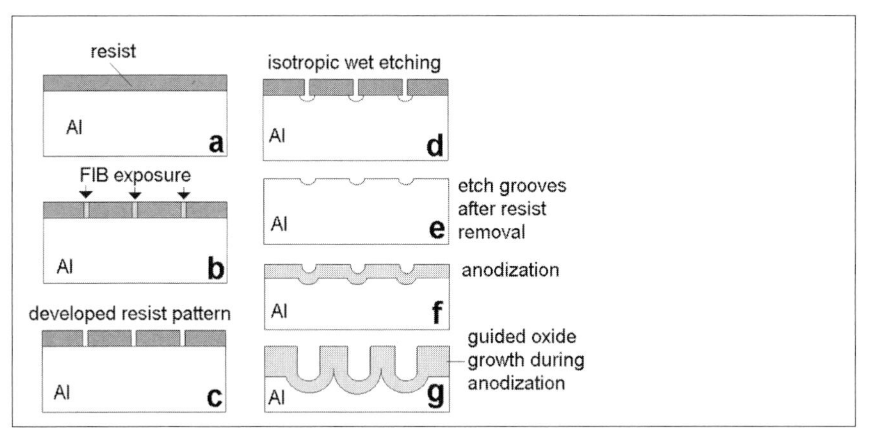

Fig. 162 Formation of dense nanochannel arrays by combination of FIBL, wet-etching, and surface anodization during directed acid etching: a) PMMA-covered aluminum, b) lithographic exposure by focussed ion beams, c) nanometer-sized openings in the PMMA film, d) initial wet-etching through the PMMA mask, e) etch grooves with nearly isotropic character after removal of the resist film, f) formation of a surface alumina film by anodization, g) preparation of nanochannels by directed wet-etching under continuous formation of side-wall oxide, thereby inhibiting lateral under-etching (N. W. Liu et al. [195])

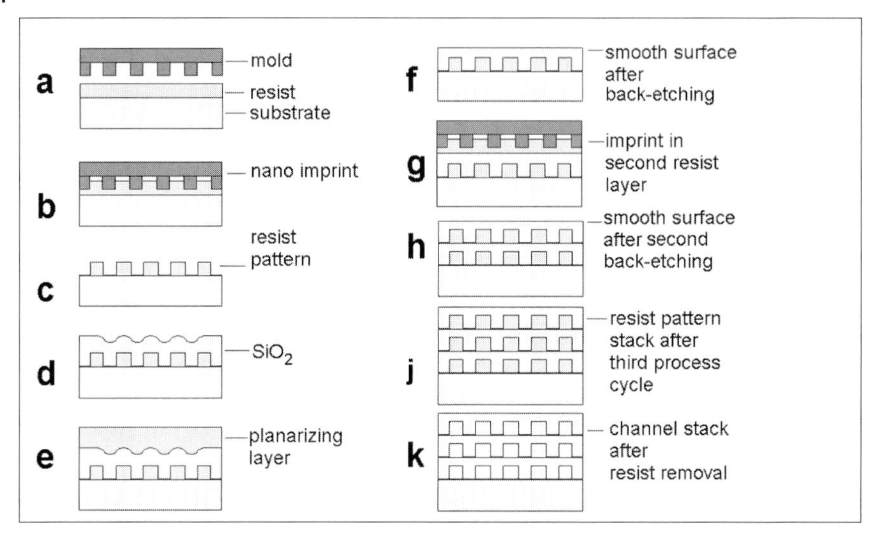

Fig. 163 Multi-level channel array technology using nano-imprinting: a) resist-covered substrate, b) imprinting of first channel structure, c) resist pattern after removal of the mold, d) CVD of SiO_2 film, e) deposition of planarizing layer, f) formation of planar SiO_2 surface by back-etching of planarizing film, g) imprinting the channel structure of the second level after a second resist deposition, h) planar surface after second SiO_2 deposition and second back-etching step, j) state after preparation of third channel plane, k) completion of nano-channel array preparation by selective etching of resist material in an oxygen plasma (R. M. Reano and S. W. Pang [198])

tern transfer during the preparation of the template (H. Cao et al. [196]). This technique was called "diffraction gradient lithography" (DGL). The gradient fluid channel structures induce the stretching of longer DNA molecules for their transport through the nanochannels. Nanochannels are also used to monitor the transport of single DNA molecules (S. M. Stavis et al. [197]).

Stacks with several levels of parallel nanochannels have been prepared by multilayer technology using polymers as sacrificial layers and planarization steps with inorganic dielectric material to cover the polymer structures by CVD (Fig. 163). First, a polymer film is patterned with the first channel structure through the use of a micro- or nanomold. The polymer pattern obtained by this imprinting procedure is then covered with SiO_2 by means of CVD. Then, a planarizing layer is applied and etched back in order to obtain a smooth, planar SiO_2 surface that is ready for the preparation of the next channel level. The organic sacrificial material can be removed after completion of channel stacking by selective plasma etching with oxygen. As a result, three-dimensional stacks of channel arrays are formed (R. M. Reano and S. W. Pang [198]).

7.8.2
Nanofluidic Electrospraying

S. Arscott and D. Troadec [199] have reported a nanofluidic device for electrospraying. The nozzle for the release of liquid ended in a final channel of length 4 μm with a cross-sectional area of 50 nm × 300 nm. The structures were generated by FIB using Ga^{+} ions. The device was operated at a flow rate of 100 pL min^{-1}. A droplet volume of about 6.5×10^{-23} m^3 (6.5×10^{-2} atL) was estimated by calculation. This size implies a frequency of droplet release of about 2.6 million droplets per second.

7.8.3 Liquid Transport in Nanotubes

The movement of molecules through nanotubes has been measured by fluorescence microscopy. DNA molecules migrate through lithographically prepared nanochannels or through carbon nanotubes if an electrical field is applied between the two open ends of the nanotubes. The propagation of a diffusion front can be visualized by time-dependent fluorescence imaging. This can be applied for surface modification by time-controlled diffusion. The technique is termed "diffusion-limited patterning", DLP (R. Karnik et al. [200]).

7.8.4 Nanofluidic Actuators for Optical Application

The chemical or electrical tuning of surface wetting or the shift of a three-phase contact line can be used to define the three-dimensional geometries of interfaces. This may be used, for example, to assess the curvature between two immiscible liquids of different refractive indices. The change of curvature of such a liquid/liquid interface allows control of the light path by local refraction. Cylindrically or spherically shaped liquids can act as liquid lenses. L. Dong et al. [201] reported a tunable liquid micro lens based on such wetting behavior.

7.8.5 Functional Molecular Devices for Nanofluidics

Chemical synthesis provides an alternative approach to nanofluidic elements besides nanolithographic patterning and related methods for the preparation of nanochannels and functional nanofluidic structures. One strategy involves the formation of supermolecular structures providing nanofluidic tool functions (R. Hernandez et al. [202]). The switching of molecular nanovalves may be realized by molecular motions of cylinder-like cryptand molecules actuated by light (photochemical switching) or by current (electrochemical switching).

8
Technical Nanosystems

8.1
What are Nanosystems?

This book emphasizes methods for fabrication on the nanometer scale. The application of nanostructure techniques usually leads to devices that have to be integrated into systems in order for them to be of any use. The term technical nanosystems is used in a wider sense to describe technical arrangements with the following properties.

1. An independent function that is usable externally.
2. A defined geometric correlation with an external technical arrangement.
3. An integration into an external functional environment by appropriate interfaces.
4. The use of at least one unit with nanometer dimensions that is essential for the functioning of the device.

In general, nanosystems include all technical arrangements that have the character of such a system, which means setups that take signals from the environment or a memory and transform them into actions, with the proviso that one or more essential elements has critical dimensions in the nanometer range. If the essential nanoelement is defined on the basis of this extended definition for nanostructures, then technical nanosystems are actually a reality. In the wider sense, all modern solid-state circuits are examples of nanotechnology: the thickness of gate oxides in solid-state transistors is about 10 nm, and their lateral dimensions are in the sub-μm range, so they fulfill the definition. Hence for many years the broad use of nanotechnical systems has been observed. However, nanotechnical systems in the narrower sense of the geometrical definition of nanostructures (two dimensions below 100 nm) are also highly developed, and some applications already exist, with more developments well on the way.

Strictly speaking, nanosystems are only systems that have outer dimensions in the nanometer range. In contrast to individual units and devices, because a system has to fulfill several functions, a whole series of such functions has to be realized, e. g., from the sensory signal detection or a data recall, through to processing of the signals and data to a nanotechnical action. Combining these system–technical requirements with engineering practicabilities (the nanosystem has to be integrated into an external coordination system) leads to systems that have to be constructed from single molecules

Nanotechnology. M. Köhler and W. Fritzsche

ISBN: 978-3-527-31871-1

or clusters that are able to orient themselves geometrically and that are integrated by self-assembly processes into an external geometry of a micro- or even macroscale environment.

While a number of examples exist of technical nanosystems in the wider sense, the development of nanosystems in the narrowest sense is limited to the study of such systems in nature and to concepts that either scale down units from the classical mechanical or microtechnical world to the nanometer range or that transfer examples in nature to technical problems.

8.2
Systems with Nanocomponents

Virtually all nanosystems developed are arrangements with one or more key components in the nanometer range, but with other areas exhibiting microtechnical or macroscopic dimensions. A typical example is scanning probe microscopes: their electronic control units and their tip holders are microtechnical devices, the table and other mechanical parts are highly precise but have dimensions in the mm to cm range and are therefore macroscopic, and only the scanning tip (or more precisely the radius) exhibits nanometer dimensions.

Processes in nanostructures show, typically, small amplitudes, low particle numbers or low energies. For a constant noise level, downscaling leads to a decrease in the signal-to-noise ratio. Furthermore, entering the nanometer range means approaching atomic dimensions and so that quantum dynamic processes, such as quantum fluctuation and uncertainty in position and impulses of single particles play a greater role. Provided systems are limited to the inclusion of just one or a few nanoelements, the requirements with respect to the reliability of this system in fulfilling certain operations are moderate. In the case of several connected nanotechnical devices, failures become an issue. This problem becomes serious when the miniaturization down to the nanometer range is used to connect a great number of nanodevices in order to achieve a high integration.

Acceptable failure probabilities for highly integrated nanodevices can be achieved by providing the systems with significant redundancy. The failure problem is important for systems that are based on the controlled transport or conversion of individual elementary particles or quanta. This applies for single-electron devices (SED), including systems that utilize the tunneling or the control of tunneling of individual electrons (single electron tunneling devices), but also cellular quantum automaton (QCA) devices. The latter use the flipping of occupation states in square arrays with 4 quantum dots, which creates elementary logic cells. For integrated SED or QCA devices, which contain for example 10^{11} individual devices and will work at 1 GHz, a redundancy of between 10 and 100 is required [1]. Nanometer dimensions for individual devices are interesting, because such a high integration density for information can be realized. High densities can be achieved, for example, by the use of electron occupation in quantum dots, in individual molecular orbitals or the orbitals of small ensembles of molecules in regular arrangements. With quantum dots, wireless storage and processing systems are possible that work with single electrical charges and allow for very high operation densities. The working principle could be cellular automatons based on addition elements [2]. Plane arrangements of molecules in monolayers, as, e. g., in LB films, are suitable for the realization of high-density storage. They could work as shifting registers, with donor and acceptor molecules or photoconductive and isolating molecules positioned relative to each other and so the electron transport is controlled perpendicularly to the molecular layers. The transport of electronic information can be realized by either pulses of an electrical field or, for photoconductive materials, by light pulses [3].

Highly complex logic devices with nanostructures fabricated by scaling of the critical dimensions of the electronic solid-state devices to below 100 nm will be mass produced in the near future. There are also developments regarding alternative electronic information processing systems with logics based on single electron tunneling processes. Devices with tunneling barriers 1–2 nm thick and electron confinements of some tens to a few hundreds of nanometers have been combined in simple circuits that work only at low temperatures (<4 K).

An ultimately nanometer-scaled system for information processing has been proposed using the shift of individual atoms in a chain of atoms adsorbed onto a solid substrate. Such atomic switch units enable logic as well as storage devices. This particular vision includes the arrangement of a computer working in the THz range with 10^7 logic elements and 10^9 storage units on an area of 0.04 mm [4].

8.3 Entire Systems with Nanometer Dimensions

Nature provides a number of examples of complete systems in the nanometer range. They are determined by functional macromolecules or their complexes, but they are not limited to molecular tasks in the sense of chemical processes. The sensory and actuator functions extend widely beyond pure chemical processes, e. g., into optical, electrical and mechanical transduction.

In recent years, solutions with nanometer dimensions have found their way into classical technical applications. In contrast to nanotechnology, these examples typically include materials with phase boundaries in the nanometer range, where the morphology of the individual phases or the particles is responsible for novel and advantageous properties. Examples of this group are nanoparticles used as dye pigments or nanofibers for construction materials. For these technical solutions, the properties of the nanostructured components and sometimes a preferred orientation are important, but not the geometric position of individual particles or individual phases in their three-dimensional environment.

There is no clear distinction between nanomaterials and real nanotechnical devices. One approach to the field of nanostructure technology is represented by systems that exhibit not only a preferred orientation, but that are also geometrically determined in at least one coordinate relative to an external reference system. An example of this approach is field emission displays with carbon nanotubes (CNTs) as light sources that are oriented perpendicular to the substrate surface in a structured monolayer. CNTs provide a significantly enhanced electron efficiency and thus also a better light efficiency compared with graphite electrodes, so that they are an interesting alternative especially for compact displays [5]. Supramolecular chemistry can synthesize nanosystems that combine several functions (cf. Section 5.4). In a simple case, compact nanoparticles or dendrimers are applied as construction modules. In addition to dendrimers consisting of identical units, hierarchies of branches can also be synthesized by forming branches of chemically different partial dendrimers. Organic groups or metal ions in complex compounds are possible cores of such dendrimers. The coupling of individual dendrimers on substrate surfaces can be realized by integration into a complete molecular monolayer of a smaller compound. For these monolayers, SAMs of alkylthiols on gold substrates are particularly suitable. The integration of alkylsulfide-substituted palladium dendrimers into a decanethiol monolayer was shown by AFM imaging [6]. The insertion of chemically active single molecules into an inert monomolecular matrix allows the generation of disperse individual molecules on solid substrates. So the integration of a substituted bishydrochinons into a decanethiol layer facilitates the chemical generation of nanolocalized Ag_4 seeds, which can be developed into locally fixed Ag nanoparticles [7].

For the functional interfacing of the molecular and supramolecular architectures as well as nanohybrid structures of molecules and nanoparticles, they have to be integrated into an environment that is accessible to the structure technologies. This integration also assists in the functional coupling between the individual functional nanoelements. Owing to the planar technical basis of microtechnology, an integration of nanoparticles and molecular nanoarchitectures in planar–technical structures is an essential step towards the fabrication of nanosystems. High-resolution lithographic methods (Section 4.3) generate structures of less than 100 nm. For typical layer thicknesses of 10 to several hundreds of nanometers, the aspect ratios vary between 0.1 and 10. These structures could provide the external environment for nanoarchitectures. However, the fine positioning, that is, locating the binding position with sub-nm precision, and the exact orientation of the nanoarchitecture in the planar technical structure remains problematic. Regarding this point, a gap of typically 2 orders of magnitude can be observed for most preparations. This gap applies to the positioning and orientation, and not to particle size and technical structure size. It cannot yet be closed by lithographic techniques.

Nature provides excellent molecular and cellular architectures that have been used by mankind for particular purposes. Proteins and carbohydrates, and also cellular and tissue structures with their properties defined in the nanometer range, play an important role in our daily lives. For future nanotechnology, biogenic macromolecules and nanosystems should not be solely taken as materials, but as individual constructive and functional elements. Hence the natural self-assembly principle could be used to provide large numbers of components.

There are two ways of utilizing the molecular self-organization of individual particles in nanostructure technology:

(A) To work at the outset with the precision of the smallest component with respect to the spatial definition. Therefore, the lithographic techniques have to be developed towards atomic resolution. The transformation from material to single-particle manipulation has to be realized by an enhanced planar technology. The self-organization principle for the generation of complex supermolecular architectures would start on the exact localized basic structures.
(B) Self-organized particles pre-synthesized in or transported through the fluid phase are then assembled on the surface. Biogenic and synthetic nano modules, also with higher complexity, are first produced in larger quantities prior to integration as individual components into technical systems.

Approach (B) represents a radical change in today's technological strategy. In microelectronic as well as in the whole microsystem technology, the miniaturized individual components are fabricated as an ensemble in the material, so that in every step of their production the spatial arrangement, the relative positioning to each other, orientation and distance are defined. Only the atoms and molecules that generate the materials are randomly distributed inside the defined spatial regions. The assemblage of prestructured devices happens later at chip level, with millimeter objects containing millionfold functional microelements. To use chemical self-organization on freely mobile particles, they have to be positioned in defined locations and functionally linked. A miniaturization and multiplication of today's macroscopic robotics technology will not be able to fulfill these tasks. It is still unclear how the massive nano assembling will occur. It is probable that assisting supermolecular structures are required that provide addressable framework structures in order to support the assembling process. They provide fine positioning and have to be anchored in lithographically structured basic structures providing coarse positioning.

Complex molecular systems consist of several thousand atoms. Such macromolecules typically exhibit diameters of more than 2 nm in their folded state. Supersystems with 10^3 macromolecular units would have an extension of at least 20 nm even for a three-dimensional arrangement. This value is accessible to today's advanced lithographic methods, and will also be standard for mass production in the not so distant future. Thus further developments will focus on the design of production-ready interfaces and construction principles for the spatial and functional coupling of lithographic structures and supermolecular structures in the medium nanometer range (around 10–30 nm).

Chemically functional support structures are required for the exact positioning and orientation of nanoarchitectures in planar solid-state structures. These support structures assist with their inner molecular structure and their functional groups, a fine-tuning of position and an exact orientation of nanoarchitectures in the planar–technical environment. Besides the immobilization of molecules from solution, also the additional integration of compact nanoparticles with their high mechanical and chemical stability and their high degree of geometrical definition is of interest. A variety of approaches have been developed for the combination of the object classes "solid surfaces", "molecules", and "nanoparticles". In addition to the chemical coupling, aspects of manipulation and orientation of nanoparticles and molecules play important roles. A highly ordered network of gold clusters (diameter 3.7 nm) with π-electron-rich organic monolayers as electrical tunneling barriers was prepared by deposition of alkylthiol-coupled Au clusters on an MoS_2 substrate prior to substitution of the aliphates by aryldiisonitrile [8]. Dielectric nanoparticles with diameters down to about 40 nm can be trapped and shifted by a focused laser beam in solution. As well as synthetic beads, biological micro- and nanoobjects in particular are accessible by this technique [9][10]. Even larger molecules can be manipulated after coupling with dielectric particles [11][12]. Long DNA molecules can be localized and stretched by this technique, so that local operations with the AFM or by enzymes can be conducted prior to microscopic control [13][14]. An arrangement with two optical tweezers has been proposed to study molecular interactions in motor protein systems, such as myosin/actin [11].

While mechanical nanosensors are well established as individual devices for the diverse variations of scanning probe microscopy, parallelization of this technique is under development in order to combine the high lateral resolution with a higher processing speed. Fast scanning probe techniques are not only interesting for microscopy, but also for data storage and even high-resolution lithography (cf. Section 4.4). Therefore, developments towards whole arrays of scanning probes are interesting for both sensory and actuator applications. The "millipede" concept combines a two-dimensional array of 32 × 32 cantilever (1024 elements) on one substrate. Four additional sensors in each corner of the chip assist in the alignment relative to an object plate, which can be moved in the *x*- and *y*- and also *z*-directions. The writing of structures, e. g., in polymeric materials, is realized by a thermo-mechanical deflection of the addressed cantilever. The readout is possible by either optical or piezo-resistive principles [15]. Using parallel AFM cantilevers, structures of dimensions down to 40 nm were fabricated, resulting in a storage density of 62.5 Gbit cm^{-2} [16]. The aim is to realize (with much higher integration densities of the transducers) high-density storage systems, which despite the relatively low write and read rate, will allow an ensemble to compete with fast optical, magnetic or magneto-optical read and write systems.

While the preparation of quantum dot arrays with electrically coupled individual points is still in its infancy, there are already concepts for the design of calculating machines on the basis of quantum points. Thus, quantum dot chains provide not only wiring but also possible storage and switching elements, including logic elements such as "AND" and "OR" or inverter [17]. Switching elements could be, for example, constructed based on crossed carbon nanotubes, which can exhibit bi-stable electrical states. Theoretical estimations utilizing this principle pointed to non-volatile memory with a density of 10^{12} bit cm^{-2} [18].

Concepts of novel digital information processing systems with key elements in the nanometer range are not limited to electrical principles. Also the application of various quantized excitation states in different band regions is being discussed for the design of new generations of computers. Quantum computers with ^{31}P doped silicon based on the nuclear spin operations have been proposed with gate dimensions of about 10 nm, and are already accessible in today's laboratories [19].

The development of miniaturized systems for energy conversion is of particular interest for diverse mobile systems and for future autonomous acting micro- and nanorobots. K. W. Lux and K. J. Rodriguez [20] have reported on fuel cell arrays with single cells of dimensions in the deep submicron range based on nanopatterned Pt electrodes and Nafion® separation membranes.

Even more so than with conventional and microtechnical systems, nanosystems should be designed as partially autonomous units with complete transduction loops with the inclusion of sensory signal input, signal processing and action. Nature demonstrates this principle of autonomous small units in cells. These natural nanosystems are always based on material transport, chemical transformations and the movement of larger groups of atoms inside molecules. An analogous task, that is, one required for transformation or transport of material, can be addressed by partly-autonomous synthetic or semisynthetic nanosystems. For purely information processing systems, the biological example is not attractive, due to the low processing speed resulting from slow partial reactions. The time requirement for the rate-limiting step of many biomolecular reaction chains lies in the millisecond range, and this notion is therefore a million-fold slower than processes in electrical solid-state circuits with cycle frequencies in the GHz range.

Molecular-based switching processes have to utilize fast processes, such as the transfer of individual small atoms between adjacent groups of atoms. In principle, these processes require only times in the order of molecular vibrations (picoseconds and slightly below), so that operation frequencies in the THz range would be possible. Because chemical reactions are always connected with the movement of atoms or groups of atoms, the rate of operations in nanosystems is limited by the fastest chemical process. Higher rates can only be achieved with elementary particles, that reach a higher speed compared with atoms but which can also be fixed.

The further development of solid-state technology (SST) will lead to electronic devices integrating more than a trillion single components on one chip. The so-called terrascale integration (TSI) may still be based on transistor technology. It is possible to estimate the parameters of circuits of these dimensions. The critical energy of switching E_{switch} is given by:

$$E_{switch} = 0.5 \times C_{out} \times V^2$$

where C_{out} is the capacitance loading of the output terminals of the circuit and V is the voltage. Switching cells with dimensions of the order of 10 nm are possible based on silicon technology. For these, gate oxides with thicknesses of about 1 nm, a gate channel length of 10 nm, and channel thicknesses of about 3 nm are required (J. D. Meindl et al. [21]).

An alternative to pure inorganic solid-state electronics could be metal-molecule-hybrid structures. This might also allow the realization of electronic chips for TSI. Small- and medium-sized conducting molecules, such as thiolated arenes, porphyrins, or fullerenes are under discussion for such devices (G. Y. Tseng and J. C. Ellenbogen [22]). Problems associated with the application of organics in highly integrated devices may stem not only from the limited long-term stability and the oxygen-sensitivity of these compounds, but also from the attainable switching times and clock frequencies.

There is a converging development in the fast signal processing of natural and technical systems. Both in the biological and the technical world, the most advanced systems use the electron as an ideal elementary tool for switching and communication processes. A computer and a brain both apply currents, potentials and localized charges. Architectures for molecular–electronic systems are therefore of special interest, even when their practical realization is still challenging to today's technology, and so this field is dominated by theoretical studies [23].

The electron also exhibits several properties that make it particularly interesting for recent as well as for future applications towards the storage and processing of information and the transfer and conversion of signals.

- It exhibits a small mass and can be accelerated even by low energies.
- The ratio of charge to mass is very large, so that an external field achieves a relatively strong acceleration compared with atoms or molecules.
- Electrons are part of all technical materials and are therefore abundant.
- Depending on the choice of material, electrons can be strongly localized or widely delocalized.
- Electrons can be material-based (electrical conductors) or material-free (vacuum) transported.
- If necessary, electrons can also be transported via secondary, assisting particles (ions), that are mobile in a condensed (electrolyte) or gas (vacuum) phase.
- For material-based electron transport (electrical conduction), the electron mobility can be adjusted over many orders of magnitude by choice of material or the local adjustment of material properties.
- Electrons are mediators of chemical bonds and can therefore directly induce the transduction between electronic signals and material transformations.
- Electrons are sufficiently small to undergo tunneling processes efficiently.
- The thickness of typical tunneling barriers for electrons is in the lower nanometer range, so that their preparation is accessible to both thin-film technological and chemical process stages.

These outstanding properties are the basis for electrons remaining the central elementary particle for signal transduction and data processing, even beyond the solid state electronic and single-quantum techniques. The delocalization of single particles and the use of ensembles of particles or electrons in the three-dimensional and chemically unstructured solid state material of today's electronics will be replaced in the future by nanotechnical units that are highly structured in their interior by binding topologies, by single particle processes and by precisely controlled localization and delocalization. The novel systems will exhibit cluster architectures for the precise control of the strong localization and transport of single electrons or other individual particles. Movements of atoms and molecular rearrangements and therefore chemical processes in particular are used for the setting up of system architectures and for long-term system optimization, whereas fast switch and transduction processes will probably apply the technically established highly mobile particles such as photons, excitons and electrons.

Tab. 9 Examples of lithographic and particle manipulations on solid substrate surfaces in the medium and lower nanometer range

Material	*Structure width (nm)*	*Aspect ratio (height/width)*	*Method*	*Reference*
Ag,Se	3	0.3	STM	Y. Utsugi 1992 [24]
Al	100	0.07	AFM/electr.	A. Boisen et al. 1998 [25]
Al	3	0.1–0.2	STM	T Mitsui et al. 1999 [26]
AlF_3	10	–	EBL	A. Murray. et al 1984 [27]
AlF_3	30	ca. 2	EBL	E. Kratschmer and M. Isaacson 1987 [28]
AlGaSb	100	0.15	AFM/electr.	S. Sasa et al. 1999 [29]
Au	40	0.25	AFM/resist	B. Klehn and U. Kunze 1999 [30]
Au	20	–	STM	R. Wiesendanger 1995 [31]
Au	5–10	0.06	STM	X. Hu et al. 1999 [32]
Au	30	0.2	STM/chem.	H. Brückl et al. 1999 [33]
Au	16	0.3	STM	X. Hu and P. von Blancken-hagen 1999 [34]
Au	3		STM	H.J. Mamin et al. 1991 [35]
Au	200	18	EBD	K.L. Lee and M. Hatzakis 1989 [36]
Au	35	13	XBL/galvanic	G. Simon et al. 1997 [37]
Au	25	15	EBL/galvanic	M. Haghiri-Gosnet et al. 1995 [38]
Au	15	about 14	IL	S. Zhang et al. [39]
	10 (gap)	–	EBL/FIBM	S. Cabrini et al. [40]
AuPd	20	0.75	e-beam/lift-off	H.G. Craighead 1984 [41]
AuPd	7	1.3	EBL/RIE	W. Chen and H. Ahmed 1993 [42]
Au/Ti	10	1	imprinting/lift-off	S.Y. Chou et al. 1997 [43]
Au/Ti	100 nm	about 0.3		NIL H. Lee et al. [44]
Bi	50 nm	0.8	EBL/RIE	S. H. Choi et al. [45]
Cd	20		STM	R.M. Silver et al. 1987 [46]
Co	25	1.2	NI/RIE	W. Wu et al. 1998 [47]
CoFeNi	30	8	EBL	X. Yang et al. [48]
	22	5	EBL	X. Yang et al. [48]
$Co_{35}Tb_{65}$	10	0.7	STM	R. Wiesendanger 1992 [49]
Co	single atom		STM	G. Meyer et al. 1998 [50]
Cr	100	–	nanopipette	A. Lewis et al. 1999 [51]
Cr	20 nm	–	NIL	M. D. Austin [52]
CrF_2	80	–	EBL	E. Kratschmer and M. Isaacson 1987 [28]
Cr_2O_3	25	–	STM	H.J. Song et al. 1994 [53]
Cu	60	3	EBL, CMP	Y. Hsu et al. 1998 [54]
Cu	single atom		STM	G. Meyer et al. 1998 [50]
Diethylhexylphthalate	single molecule		STM	J.S. Foster et al. 1988 [55]
Fe	60	0.8	STM	T. Schaub et al. 1992 [56]
Fe	93 nm	about 0.1	Laser interf.	G. Myszkiewicz et al. [57]
FeF_2	4	–	EBL	E. Kratschmer and M. Isaac-son 1987 [28]

Tab. 9 Examples of lithographic and particle manipulations on solid substrate surfaces in the medium and lower nanometer range (continued)

Material	*Structure width (nm)*	*Aspect ratio (height/ width)*	*Method*	*Reference*
FeSi	3	–	STM	F. Thibaudau et al. 1994 [58]
InAs	50	–	AFM/electr.	S. Sasa et al. 1999 [29]
GaAs	25	>2	EBL/PET	H. Craighead 1984 [41]
GaAs	55	0.5	AFM/resist	B. Klehn and U. Kunze 1999 [30]
GaAs	20	0.75	EBL	H. Craighead et al. 1983 [41]
GaAs	6 nm	–	NIL	M. D. Austin [52]
GaAs (Ga-implantation)	2–3	–	FIB	G.A.C. Jones et al. 1998 [59]
GaP	180 nm	about 12	Laser ablatn.	A. L. Roest et al. [60]
GaSb	100	0.1	AFM/electr.	S. Sasa et al. 1999 [29]
Ge/W	0.8	1	STM	R.S. Becker et al. 1987 [61]
$GeSb_2Te_4$	54 nm	about 0.3	fs-laser	Y. Lin et al. [62]
HSQ	50 nm	1	IL	T. M. Bloomstein et al. [63]
InAs	40 nm	about 50–100	EBL, chem. growth	T. Bryllert et al. [64]
InP	50 nm	0.2	IL	T. M. Bloomstein et al. [63]
Ni2Si	25.3 nm	about 1	STL	Z. Zhang et al. [65]
Nb-Oxid	70	0.05	AFM/electr.	J. Shirakashi et al. 1997 [66]
Ni	50	0.2	STM	T. Schaub et al. 1992 [56]
Ni	20	1.75	EBL	S.Y. Chou et al. 1996 [67]
Ni	75	9.3	EBL/galvanic	S.Y. Chou et al. 1996 [67]
Pd	16	0.6	EBL	S.P. Beaumont et al. 1981 [68]
Polymers (organic)				
– AZPN-114 (e-beam resist)	80	6	EBL	D. Park et al. 1998 [69]
– Calixarene-resist	15	2	EBL	J. Sone et al. 1999 [70]
– Methylstyrol	7	3	EBL	J. Sone et al. 1999 [70]
– Novolak-resist	20	2	EBL	J. Yamamoto et al. 1997 [71]
– Novolak-resist AZ 1450J	100	7	FIBL/RIE	S. Matsui et al. 1986 [72]
– PMMA	20	1.5	EBL	A.N. Broers 1981 [73]
– PMMA	15	27	EBL	G. Simon et al. 1997 [37]
– PMMA	10...40	1	AFM	G. Binnig et al. 1999 [74]
– PMMA	60	1.7	hot imprint	S.Y. Chou et al. 1995 [75]
– PMMA	25	–	hot imprint	S.Y. Chou et al. 1995 [75]
– PMMA	20	13	XBL	G. Simon et al. 1997 [37]
– P4BCMU (p-diacetylene urethane)	20		STM	C.R.K. Marrian et al. 1990 [76]
– SAL601 (e-beam resist)	20	0.05	STM	A. Archer et al. 1994 [77]
– SAL601 (e-beam resist)	70	4	XBL	L. Malmquist et al. 1997 [78]
– ZEP520	10	5	EBL	K. Kurihara et al. 1995 [79]

Tab. 9 Examples of lithographic and particle manipulations on solid substrate surfaces in the medium and lower nanometer range (continued)

Material	*Structure width (nm)*	*Aspect ratio (height/ width)*	*Method*	*Reference*
Polymer-Resist	60 nm	–	SPL	W. Srituravanich et al. [80]
– Resist	45 nm	–	PXL	Q. Leonhard et al. [81]
– Shipley UV-135	20 nm	about 8	UV, TF	H. L. Chen et al. [82]
– Resist LUVR 99071	30 nm	about 1.5	IL, immersion	M. Switkes and M. Rothschild [83]
– dye-doped PMMA	50 nm	about 15	EBL	R. R. Panepucci et al. [84]
– photoresist	2 nm	about 1	NFL (365 nm)	T. Ito et al. [85]
– polystyrene	20 nm	–	EBL	M. D. Austin et al. [52]
– PUA	80 nm	about 4	CFL	H. Yoon et al. [86]
Porphyrin	single molecule		STM	(IBM)
Pt	ca. 70	–	STM/chem.	H. Brückl et al. 1999 [33]
Pt	100	10	EBD	H.W. Koops et al. 1994 [87]
Pt	30 nm	–	ISL	Y.-K. Choi et al. [88]
$Rh_{25}Zr_{25}$	25	–	STM	U. Staufer et al. 1989 [89]
Ru	100	4.5	EBL/RIE	S. Skaberna et al. 2000 [90]
SAMs				
– Hexadecylthiol	20	0.1	STM	J. Hartwich et al. 1999 [91]
– Octadecylthiol	50	0.04	DPN	S. Hong et al. 1999 [92]
– Mercaptohexadecanoic acid	15	0.13	DPN	S. Hong et al. 1999 [92]
– $C_{22}H_{45}SH$ on Au	40	0.07	EBL	J.A.M. Sonday-Huethhorst et al. 1994 [93]
Si	100	3	STM/RIE	K. Kragler et al. 1996 [94]
Si	50	4	EBL/RIE	D.W. Carr et al. 1998 [95]
Si	55	0.5	AFM/Resist	B. Klehn and U. Kunze 1999 [30]
Si	30	1	AFM(anod./ECR)	E.S. Snow et al. 1995 [96]
Si	50	0.5	STM/TMAH-WE	N. Kramer et al. 1995 [97]
Si (poly-Si)	14	2	EBL	J. Sone et al. 1999 [70]
Si	10	7	EBL/Lift-off/ECR	T. Tada and T. Kanayama (1998) [98]
Si	12 nm	–	EBL	W. Henschel et al. [99]
Si	10 nm	–	RSL	Y.-K. Choi et al. [88a]
Si	7 nm	about 18	STL	Y.-K. Choi et al. [88b]
Si/Si–H	1–3	–	STM	J.W. Lyding et al. 1994 [100]
poly-Si	18 nm	about 1.6	STL	Z. Zhang et al. [65]
poly-Si	20 nm	10	EBL, Plasma	L. Dreeskornfeld et al. [101]
SiO_2	50	8	STM/RIE	K. Kragler et al. 1996 [94]
SiO_2	25	0.06	AFM/resist	B. Klehn and U. Kunze 1999 [30]
SiO_2	50	0.2	AFM/SAM	H. Sugimura et al. 1999 [102]
SiO_2	20 nm	about 1	EBL/RIE	M. D. Austin et al. [52]
SiO_x	100	–	STM/anod.	H. Sugimura et al. 1995 [103]
SiO_x	50	–	STM/heat	N. Li et al. 1999 [104]

Tab. 9 Examples of lithographic and particle manipulations on solid substrate surfaces in the medium and lower nanometer range (continued)

Material	*Structure width (nm)*	*Aspect ratio (height/ width)*	*Method*	*Reference*
Siloxane				
– Methylsiloxane (Accuglass 111)	40	1	AFM	S.W. Park et al. 1995 [105]
SnO_2	85 nm	–	EBL/RIE	P. Candeloro et al. [106]
Ti	75	0.13	AFM/resist	B. Klehn and U. Kunze 1999 [30]
Ti	75	0.13	AFM/electr.	R.J.M. Vullers et al. 1999 [107]
Ti	50 nm	0.8	FIB	Ch. Santschi et al. [108]
TiO_2	80	0.05	STM oxidation	K. Matsumoto 1997 [109]
W	ca. 100	–	STM	X. Hu and P. von Blancken-hagen 1999 [34]
W	20	ca. 10	XBL/RIE	G. Simon et al. 1997 [37]
WSi_x on Si	60	ca. 3	EBL/HD-Plasma	D. Tennant et al. 1997 [110]
ZnO	50 nm	about 20	VTD	H. J. Fan et al. [111]

References

Chapter 1

[1] B. C. Crandell, J. Lewis (Eds.), *Nanotechnology* (MIT Press, Cambridge, 1989).
[2] K. E. Drexler, *Nanosystems* (John Wiley & Sons, New York, **1992**).
[3] M. Cross, *Travel to the Nanoworld* (Plenum Trade, New York, **1999**).
[4] W. Henschel, T. Wahlbrink, Y. M. Georgiev et al., *J. Vac Sci. Technol. B* 21 (**2003**) 2975.

Chapter 2

[1] M. Mammen, S.-K. Choi and G. M. Whitesides, *Angew. Chem.* 110 (1998) 2908; Angew. Chem. Int. Ed. 37 (**1998**) 2754.
[2] J.-M. Lehn, *Proceedings of the 2nd Internat. Kyoto Conf. New Aspects of Org. Chem.*, Kodansha, Tokyo, **1983**.
[3] R. Sikorski and R. Peters, *Science* 279 (**1998**) 1967.
[4] A. Dove, *Nature Biotechnology* 156 (**1998**) 830.
[5] C. Mao, T. H. LaBean, J. H. Relf and N. C. Seeman, *Nature* 407 (**2000**) 493.
[6] Q. Liu, L. Wang, A. G. Frutos, A. E. Condon, R. M. Corn and L. M. Smith, *Nature* 403 (**2000**) 175.
[7] M. Ogihara and A. Ray, *Nature* 403 (**2000**) 143.

Chapter 3

[1] W. M. Moreau, *Semiconductor Lithography* (Plenum Press, New York, **1988**).
[2] A. Heuberger (Hrsg.), *Mikromechanik* (Springer, Berlin, **1991**).
[3] M. J. Bowden, in: L. F. Thompson, C. G. Willson, M. J. Bowden (Eds.), *Introduction to Microlithography* (ACS, 2nd Edition, Washington D.C., **1994**) 19.
[4] M. Madou, *Fundamentals of Microfabrication* (CRC, Boca Raton, **1997**).
[5] P. Rai-Choudhury (Ed.), *Handbook of Microlithography, Micromachining and Microfabrication, Vol. 1 Microlithography* (SPIE Press, London, **1997**).
[6] W. Menz, J. Mohr, *Mikrosystemtechnik für Ingenieure* (VCH, Weinheim, **1997**).
[7] F. Völklein, Th. Zetterer, *Einführung in die Mikrosystemtechnik* (Vieweg, Braunschweig, **2000**).
[8] Y. Nishi, R. Doering (Eds.), *Handbook of Semiconductor Manufacturing Technology* (Marcel Dekker, New York, **2000**).
[9] W. Menz, J. Mohr, O. Paul, *Mikrosystem Technology* (Wiley-VCH, Weinheim, **2001**).
[10] F. W. Kern, G. W. Gale, in: *Handbook of Semiconductor Manufacturing Technology* (Marcel Dekker, New York, **2000**).
[11] M. Schulz, in: H.-F. Hadamovsky (Ed.), *Halbleiterwerkstoffe* (Leipzig, **1972**) 97.
[12] H.-F. Hadamovsky (Ed.), *Halbleiterwerkstoffe* (Leipzig, **1972**).
[13] J. S. Logan, M. J. Hait, H. C. Jones, G. R. Firth and D. B. Thompson, *J. Vac. Sci. Technol. A 7* (**1989**) 1392.
[14] S. Maniv, C. Miner and W. D. Westwood, *J. Vac. Sci. Technol. 18* (**1981**) 195.

Nanotechnology. M. Köhler and W. Fritzsche

ISBN: 978-3-527-31871-1

[15] A. Sherman, *Chemical vapour deposition for microelectronics* (Noyes Publications, Park Ridge, **1987**).
[16] H. O. Pierson, *Handbook of Chemical Vapor Deposition* (Noyes Publications, Park Ridge, **1992**).
[17] L. Q. Xia et al., in: *Handbook of Semiconductor Manufacturing Technology* (Marcel Dekker, New York, **2000**) 287.
[18] T. F. Kuech, M. A. Tischler, in: K. A. Jackson, W. Schröder (Eds.), *Handbook of Semiconductor Technology. Vol. 2, Processing of Semiconductors* (Wiley-VCH, Weinheim, **2000**) 111.
[19] T. Suntola, *Mater. Sci. Rep.* 4 (**1989**) 261.
[20] M. Ritala and M. Leskelä, *Nanotechnology* 10 (**1999**) 19.
[21] G. Binnig, M. Despont, U. Drechsler, W. Häberle, M. Lutwyche, P. Vettiger, H. J. Mamin, B. W. Chui and T. W. Kenny, *Appl. Phys. Lett.* 74 (**1999**) 1329.
[22] M. M. Deshmukh, A. C. Ralph, M. Thomas and J. Silcox, *Appl. Phys. Lett.* 75 (**1999**) 1631.
[23] R. Lüthi, R. R. Schlittler, J. Brugger, P. Vettiger, M. E. Welland and J. K. Gimzewski, *Appl. Phys. Lett.* 75 (**1999**) 1314.
[24a] T. A. Taton, C. A. Mirkin and R. L. Letsinger, *Science* 289 (2000) 1757.
[24b] T. A. Taton, R. C. Mucic, C. A. Mirkin and R. L. Letsinger, *J. Am. Chem. Soc.* (**2000**) 6305.
[25] C. M. Niemeyer, B. Ceyhan, S. Gao, L. Chi, S. Peschel and U. Simon, *Colloid and Polymer Science* 279 (**2001**) 68.
[26a] W. Fritzsche, H. Porwol, A. Wiegand, S. Bornmann, J. M. Köhler, *Nanostructured Materials* 10 (**1998**) 89.
[26b] W. Fritzsche, K. J. Böhm, E. Unger and J. M. Köhler, *Nanotechnology* 9 (**1998**) 177.
[27] R. Leuschner, G. Pawlowski, in: K. A. Jackson, W. Schröder (Eds.), *Handbook of Semiconductor Technology, Vol. 2, Processing of Semiconductors* (Wiley-VCH, Weinheim, **2000**) 177.
[28] G. E. Fuller, in: *Handbook of Semiconductor Manufacturing Technology* (Marcel Dekker, New York, **2000**) 461.
[29] J. Vollmer, H. Hein, W. Menz and F. Walter, *Transducers 93*, Yokohama, June 1993 (**1993**) 116.
[30] P. Bley, *Interdisciplinary Science Reviews* 18 (**1993**) 267.
[31] P. Bley, J. Göttert, M. Harmening, M. Himmelhaus, W. Menz, J. Mohr, C. Müller and U. Wallrabe, *Proc. MICRO SYSTEM Technologies* 91 (**1991**) 302.
[32] Y. Xia and G. M. Whitesides, *Angew. Chem.* 110 (1998) 568; *Angew. Chem. Int. Ed.* 37 (**1998**) 550.
[33] A. Kumar and G. M. Whitesides, *Appl. Phys. Lett.* 63 (**1993**) 2002.
[34] R. J. Jackman, J. L. Wilburg, G. M. Whitesides, *Science* 269 (**1995**) 664.
[35] H. Wu, S. Brittain, J. Anderson, B. Grzybowski, S. Whitesides and G. M. Whitesides, *J. Am. Chem. Soc.* 122 (**2000**) 12691.
[36] H.-C. Scheer, H. Schulz, T. Hoffmann and C. M. Sotomayor-Torres, *J. Vac. Sci. Technol. B* 16 (**1998**) 3917.
[37] M. Köhler, *Ätzverfahren für die Mikrotechnik* (Wiley-VCH, Weinheim, **1998**).
[38] A. F. Bogenschütz, *Ätzpraxis für Halbleiter* (München, **1967**).
[39] J. A. Mucha, D. W. Hess, E. S. Aydil, in: *Introduction to Microlithography* (ACS, 2nd Edition, Washington D.C., **1994**) 377.
[40] V. A. Yunkin, I. W. Rangelow, J. A. Schäfer, D Fischer, E. Voges and S. Sloboshanin, *Microelectronic Engineering* 23 (**1994**) 361.
[41] W. Chen and H. Ahmed, *J. Vac. Sci. Technol. B* 11 (**1993**) 2519.
[42] A. Gruber, J. Gspann and H. Hoffmann, *Appl. Phys. A* 68 (**1999**) 197.
[43] J. Matsuo, N. Toyoda and I. Yamada, *J. Vac. Sci. Technol. B* 14 (**1996**) 3951.
[44] D. I. Amey, in: K. A. Jackson, W. Schröder (Eds.), *Handbook of Semiconductor Technology*, Vol. 2, Processing of Semiconductors (Wiley-VCH, Weinheim **2000**) 607.
[45] R. C. Bracken et al., in: *Handbook of Semiconductor Manufacturing Technology* (Marcel Dekker, New York, **2000**) 999.
[46] Q.-Y. Tong, T. H. Lee, U. Gosele, M. Reiche, J. Ramm and E. Beck, *J. Electrochem. Soc.* 144 (**1997**) 384.
[47] M. C. Wu, in: *2nd German-American Frontiers of Engineering* (8.–10. 4. **1999, Irvine, CA, USA**).
[48] E. Smela, O. Inganas and I. Lundstrom, *Science* 268 (**1995**) 1735.

[49] M. E. Thomas, R. H. Havemann, *Handbook of Semiconductor Manufacturing Technology* (Marcel Dekker, New York, **2000**) 287.

[50] M. A. Gallop, R. W. Barrett, W. J. Dower, S. P. A. Fodor and E. M. Gordon, *J. Med. Chem.* 37 (**1994**) 1233.

[51] E. M. Gordon, R. W. Barrett, W. J. Dower, S. P. A. Fodor and M. A. Gallop, *J. Med. Chem.* 37 (**1994**) 1385.

[52] J. M. Köhler, R. Pechmann, A. Schaper, A. Schober, T. M. Jovin, M. Thürk and A. Schwienhorst, *Microsystem Technology* 1 (**1995**) 202.

[53] G. Mayer, J. M. Köhler, *Sensors and Actuators A* 60 (**1997**) 202.

[54] G. Mayer, K. Wohlfart, A. Schober, J. M. Köhler, in: *Microsystem technology: A powerful tool for biomolecular studies* (Eds.: J. M. Köhler, T. Mejevaia and H. P. Saluz) (Basel, **1999**) 75.

[55] S. P. A. Fodor, J. L. Read, M. C. Pirrung, L. Stryer, A. T. Lu and D. Solas, *Science* 251 (**1991**) 767.

[56] S. P. A. Fodor and J. W. Jacobs, *Tibtech* 12 (**1994**) 19.

[57] S. P. A. Fodor, *Science* 277 (**1997**) 393.

[58] J. Maskos and E. M. Southern, *Nucleic Acids Research* 20 (**1992**) 1679.

[59] M. Schena, D. Shalon, R. W. Davis, P. O. Brown, *Science* 270 (**1995**) 467.

[60] M. A. Northrup, C. Gonzalez, D. Hadley, R. F. Hills, P. Landre, S. Lehew, R. Saiki, J. J. Sninsky, R. Watson and R. J. Watson, *Transducers* 95 (**1995**) 764.

[61] M. U. Kopp, A. J. de Mello and A. Manz, *Science* 280 (**1998**) 1046.

[62] S. Poser, T. Schulz, U. Dillner, V. Baier, J. M. Köhler, D. Schimkat, G. Mayer and A. Siebert, *Sensors and Actuators A* 62 (**1997**) 672.

[63] J. H. Daniel, S. Iqbal, R. B. Millington, D. F. Moore, C. R. Lowe, D. Leslie, M. A. Lee and M. J. Pearce, *Sensors and Actuators A* 71 (**1998**) 81.

[64] A. T. Woolley and R. A. Mathies, *Anal. Chem.* 67 (**1995**) 3676.

[65] A. T. Woolley, G. F. Sensabaugh and R. A. Mathies, *Anal. Chem.* 69 (**1997**) 2181.

[66] C. H. Mastrangelo, M. A. Burns and D. T. Burke, *Proceedings of the IEEE* 86 (**1998**) 1769.

[67] A. T. Woolley, D. Hadley, P. Landre, A. J. de Mello, R. A. Mathies and M. A. Northrup, *Anal. Chem.* 68 (**1996**) 4081.

[68] M. A. Burns, C. H. Mastrangelo, T. S. Sammarco, F. P. Man, J. R. Webster, B. N. Johnson, B. Foerster, D. Jones, Y. Fields and A. R. Kaiser, *Proc. Natl. Acad. Sci. USA* 93 (**1996**) 5556.

[69] F. Scheller and F. Schubert, *Biosensoren* (Berlin, **1989**).

Chapter 4

[1] J. M. Köhler, O. Ritzel, W. Brodkorb, P. Dittrich, *Exp. Techn. d. Physik* 39 (**1991**) 277.

[2] K. A. Lister, S. Thoms, D. S. MacIntyre et al., *J. Vac. Sci. Technol. B* 22 (**2004**) 3257.

[3] T. Nomura and R. Suzuki, *Nanotechnology* 3 (**1992**) 21.

[4] N. Ikawa, S. Shimada and H. Tanaka, *Nanotechnology* 3 (**1992**) 6.

[5] S. Hong, J. Zhu and C. A. Mirkin, *Science* 286 (**1999**) 523.

[6] R. D. Piner, J. Zhu, F. Xu, S. Hong and C. A. Mirkin, *Science* 283 (**1999**) 661.

[7] J. A. Stroscio, L. J. Whitman, R. A. Dragoset and R. J. Celotta, *Nanotechnology* 3 (**1992**) 133.

[8] T. Ichinokawa, H. Itzumi, C. Haginoya and H. Itoh, *Nanotechnology* 3 (**1992**) 147.

[9] S. Y. Chou, P. R. Krauss and P. J. Renstrom, *Appl. Phys. Lett.* 67 (**1995**) 3114.

[10] X. Sun, L. Zhuang, W. Zhang and S. Y. Chou, *J. Vac. Sci. Technol. B* 16 (**1998**) 3922.

[11] S. Y. Chou, P. R. Krauss, W. Zhang, L. Guo and L. Zhuang, *J. Vac. Sci. Technol. B* 15 (**1997**) 2897.

[12] E. Kim, Y. Xia, X.-M. Zhao and G. M. Whitesides, *Adv. Mater.* 9 (**1997**) 651.

[13] J. Haisma, M. Verheijen, K. van den Heuvel and J. van den Berg, *J. Vac. Sci. Technol. B* 14 (**1996**) 4124.

[14] Y. Xia and G. M. Whitesides, Angew. Chem. 110 (1998) 568; *Angew. Chem. Int. Ed.* 73 (**1998**) 550.

[15] N. L. Abbott, A. Kumar, G. M. Whitesides, *Chem Mater.* 6 (**1994**) 586.

[16a] S. Y. Chou, P. R. Krauss and L. Kong, *J. Appl. Phys.* 79 (**1996**) 6101.

[16b] S. Y. Chou, P. R. Krauss and P. J. Renstrom, *Science* 272 (**1996**) 85.
[17] B. D. Terris, H. J. Mamin, M. E. Best, J. A. Logan, D. Rugar and S. A. Rishton, *Appl. Phys. Lett.* 69 (**1996**) 4262.
[18] A. Kumar and G. M. Whitesides, *Appl. Phys. Lett.* 63 (**1993**) 2002.
[19] Y. Xia, E. Kim, X.-M. Zhao, J. A. Rogers, M. Prentiss and G. M. Whitesides, *Science* 273 (**1996**) 347.
[20] R. J. Jackman, J. L. Wilburg, G. M. Whitesides, *Science* 269 (**1995**) 664.
[21] H. Lee, S. Hong and K. Yang, *Appl. Phys. Lett.* 88 (**2006**) 143112.
[22] H. Y. Low, W. Zhao and J. Dumond, *Appl. Phys. Lett.* 89 (**2006**) 023109.
[23] D. Pisignano, G. Maruccio, E. Mele, L. Persano, F. DiBenedetto and R. Cingolani, *Appl. Phys. Lett.* 87 (**2005**) 123109.
[24] L.-R. Bao, X. Cheng, X.-D. Huang, L.-J. Guo, S.-W. Pang and A.-F. Yee, *J. Vac. Sci. Technol. B* 20 (**2002**) 2881.
[25] A. Erbe, W. Jiang, Z. Bao et al., *J. Vac. Sci. Technol. B* 23 (**2005**) 3132.
[26] C. C. Huang and K. L. Ekinci, *Appl. Phys. Lett.* 88 (**2006**) 093110.
[27] H. Yoon, T.-I. Kim, S. Choi, K.-Y. Suh, M.-J. Kim and H.-H. Lee, *Appl. Phys. Lett.* 88 (**2006**) 254104.
[28] N. Farkas, J. R. Corner, G. Zhang et al., *Appl. Phys. Lett.* 85 (**2004**) 5691.
[29] W. Zhang and S.-Y. Chou, *Appl. Phys. Lett.* 83 (**2003**) 1632.
[30] J. Tallal, K. Berton, M. Gordon and D. Peyrade, *J. Vac. Sci. Technol. B* 23 (**2005**) 2914.
[31] T. Ohtake, K. I. Nakamatsu, S. Matsui, H. Tabata and T. Kawai, *J. Vac. Sci. Technol. B* 22 (**2004**) 3275.
[32] W. R. Anderson, C. C. Bradley, J. J. McClelland and R. J. Celotta, *Phys. Rev. A* 59 (**1999**) 2476.
[33] G. Kostovski, A. Mitchell, A. Holland, E. Fardin and M. Austin, *Appl. Phys. Lett.* 88 (**2006**) 133128.
[34] Y. Lin, M. H. Hong, T. C. Chong et al., *Appl. Phys. Lett.* 89 (**2006**) 041108.
[35] H. Ito, G. M. Walraff, N. Fender et al., *J. Vac. Sci. Technol. B* 19 (**2001**) 2678.
[36] M. Toriumi, T. Yamazaki, T. Furukawa, S. Irie, S. Ishikawa and T. Itani, *J. Vac. Sci. Technol. B* 20 (**2002**) 2909.
[37] T. Itani, M. Torumi, T. Naito et al., *J. Vac. Sci. Technol. B* 19 (**2001**) 2705.
[38] G. M. Walraff, D. R. Medeiros, M. Sanchez et al., *J. Vac. Sci. Technol. B* 22 (**2004**) 3479.
[39] S.-W. Lee, J. Finders, S.-G. Woo and H.-K. Cho, *J. Vac. Sci. Technol. B* 23 (**2005**) 2875.
[40] M. Rothschild, T. M. Bloomstein, R. R. Kunz et al., *J. Vac. Sci. Technol. B* 22 (**2004**) 2877.
[41] M. Switkes and M. Rothschild, *J. Vac. Sci. Technol. B* 19 (**2001**) 2353.
[42] B. W. Smith, Y. Fan, J. Zhou et al., *J. Vac. Sci. Technol. B* 22 (**2004**) 3439.
[43] D. Chao, A. Patel, T. Barwicz, H. I. Smith and R. Menon, *J. Vac. Sci. Technol. B* 23 (**2005**) 2657.
[44] A. Frauenglass, S. Smolev, A. Biswas and S. R. J. Brueck, *J. Vac. Sci. Technol. B* 22 (**2004**) 3465.
[45] T. M. Bloomstein, P. W. Juodawlkis, R. B. Swint et al., *J. Vac. Sci. Technol. B* 23 (**2005**) 2617.
[46] N. D. Wittels, in: R. Newman (Ed.), *Fine line lithography* (NHPC, New York, **1980**) 1.
[47] J. P. Silverman, in: *Handbook of Semiconductor Manufacturing Technology* (Marcel Dekker, New York, **2000**) 543.
[48] M. Burkhardt, H. I. Smith, D. A. Antoniadis, T. P. Orlando, M. R. Melloch, K. W. Rhee and M. C. Peckerar, *J. Vac. Sci Technol. B* 12(6) (**1994**) 3611.
[49] H. Kinoshita, *J. Vac. Sci. Technol. B* 23 (**2005**) 2584.
[50] T. A. Brunner, *J. Vac. Sci. Technol. B* 21 (**2003**) 2632.
[51] M. Khan, G. Han, G. Tsvid, T. Kitayama, J. Maldonado, F. Cerrina, *J. Vac. Sci. Technol. B* 19 (**2001**) 2423.
[52] E. Toyota, T. Hori, M. Khan and F. Cerrina, *J. Vac. Sci. Technol. B* 19 (**2001**) 2428.
[53] E. Toyota and M. Washio, *J. Vac. Sci. Technol. B* 20 (**2002**) 2979.
[54] Q. Leonhard, D. Malueg, J. Wallace et al., *J. Vac. Sci. Technol. B* 23 (**2005**) 2896.
[55] L. Malmqvist, A. L. Bogdanov, L. Montelius and H. M. Hertz, *J. Vac. Sci. Technol. B* 15 (**1997**) 814.
[56] Y. Chen, G. Simon, A. M. Haghiri-Gosnet, F. Carcenac, D. Decanini, F. Rousseaux and H. Launois, *J. Vac. Sci. Technol. B* 16 (**1998**) 3521.

[57] G. Simon, A. M. Haghiri-Gosnet, J. Bourneix, D. Decanini, Y. Chen, F. Rousseaux, H. Launois and B. Vidal, *J. Vac. Sci. Technol. B* 15 (**1997**) 2489.
[58] K. Fujii, Y. Tanaka, T. Tagguchi, M. Yamabe, K. Suziki, Y. Gomei and T. Hisatsugu, *J. Vac. Sci. Technol. B* 16 (**1998**) 3504.
[59] C. W. Gwyn, R. Stulen, D. Sweeney and D. Attwood, *J. Vac. Sci. Technol. B* 16 (**1998**) 3142.
[60a] L. F. Thompson, in: L. F. Thompson, C. G. Willson, M. J. Bowden (Eds.), *Introduction to Microlithography* (ACS, 2nd Edition, Washington D.C., **1994**) 1.
[60b] L. F. Thompson, in: L. F. Thompson, C. G. Willson, M. J. Bowden (eds.), *Introduction to Microlithography* (ACS, 2nd Edition, Washington D.C., **1994**) 269.
[61] B. J. Lin and T. H. P. Chang, *J. Vac. Sci. Technol.* 16 (**1979**) 1669.
[62] J. M. Moran and D. Maydan, *J. Vac. Sci. Technol.* 16 (**1979**) 1620.
[63] C. Li and J. Richards, *IEEE Transact.* (**1980**) 412.
[64] P. R. West, P. Griffing, *SPIE Proc.* 394 (**1983**) 33.
[65] T. Ito, T. Yamada, Y. Inao, T. Yamaguchi, N. Mizutani and R. Kuroda, *Appl. Phys. Lett.* 89 (**2006**) No 033113.
[66] W. Srituravanich, S. Durant, H. Lee, C. Sun, X. Zhang, *J. Vac. Sci. Technol. B* 23 (**2005**) 2636.
[67] X. Luo and T. Ishihara, *Appl. Phys. Lett.* 84 (**2004**) 4780.
[68] N. Saitou, in: *Handbook of Semiconductor Manufacturing Technology* (Marcel Dekker, New York, **2000**) 571.
[69] R. Catanescu, J. Binder and J. Zacheja, *Proc. „MICRO SYSTEM Technologies 96"* (**1996**) 717.
[70] V. Bögli, *Dissertation*, Aachen (**1988**) 18.
[71] D. K. Ferry, M. Khoury, D. P. Pivin, K. M. Connolly, T. K. Whidden, M. N. Kosicki and D. R. Allee, *Semicond. Sci. Technol.* 11 (**1996**) 1552.
[72] A. N. Broers, *J. Electrochem. Soc.* 128 (**1981**) 166.
[73] S. Uchino, J. Yamamoto, S. Migitaka, K. Kojima, M. Hashimoto and H. Shiraishi, *J. Vac. Sci. Technol. B* 16 (**1998**) 3684.
[74] K. Kurihara, K. Iwadate, H. Namatsu, M. Nagase, H. Takenaka and K. Murase, *Jpn. J. Appl. Phys. Pt. 1*, 34 (**1995**) 6940.
[75] H. G. Craighead, R. E. Howard, L. D. Jackel and P. M. Mankiewich, *Appl. Phys. Lett.* 42 (**1983**) 38.
[76] O. Dial, C. C. Cheng and A. Scherer, *J. Vac. Sci. Technol. B* 16 (**1998**) 3887.
[77] A. M. Haghiri-Gosnet, C. Vieu, G. Simon, F. Carcenac, A. Madouri, Y. Chen, F. Rousseaux and H. Launois, *J. Vac. Sci. Technol. B* 13 (**1995**) 3066.
[78] J. Sone, J. Fujita, Y. Ochiai, S. Manako, S. Matsui, E. Nomura, T. Baba, H. Kawaura, T. Sakamoto, C. D. Chen, Y. Nakamura and J. S. Tsai, *Nanotechnology* 10 (**1999**) 135.
[79a] J. Fujita, Y. Ohnishi, Y. Ochiai, E. Nomura and S. Matsui, *J. Vac. Sci. Technol. B* 14 (**1996**) 4272.
[79b] J. Fujita, Y. Ohnishi, Y. Ochiai, S. Matsui, *Appl. Phys. Lett.* 68 (**1996**) 1297.
[80] S. Tedesco, T. Mourier, B. Dal'Zotto, A. McDougall, S. Blanc-Coquant, Y. Quéré, P. J. Paniez and B. Mortini, *J. Vac. Sci. Technol. B* 16 (**1998**) 3676.
[81] L. E. Ocola, C. J. Biddick, D. M. Tennant, W. K. Waskiewicz and A. E. Novembre, *J. Vac. Sci. Technol. B* 16 (**1998**) 3705.
[82] T. Ishii, H. Nozawa, T. Tamamura and A. Ozawa, *J. Vac. Sci. Technol. B* 15 (**1997**) 2570.
[83] E. Kratschmer and M. Isaacson, *J. Vac. Sci. Technol. B* 5 (**1987**) 369.
[84] P. M. St. John and H. G. Craighead, *J. Vac. Sci. Technol. B* 14 (**1996**) 69.
[85] M. J. Lercel, R. C. Tiberio, P. F. Chapman, H. G. Craighead, C. W. Sheen, A. N. Parikh and D. L. Allara, *J. Vac. Sci. Technol. B* 11 (**1993**) 2823.
[86] K. Ogai, Y. Kimura, R. Shimizu, J. Fujita and S. Matsui, *Appl. Phys. Lett.* 66 (**1995**) 1560.
[87] T. Nakasugi, A. Ando, R. Inanami et al., *J. Vac. Sci. Technol. B* 20 (**2002**) 2651.
[88] M. J. Van Bruggen, B. Van Someren and P. Kruit, *J. Vac. Sci. Technol. B* 23 (**2005**) 2833.
[89] T. Miura, *J. Vac. Sci. Technol. B* 20 (**2002**) 2622.
[90] S. E. Kubatkin, A. V. Danilov, A. L. Bogdanov, H. Olin and T. Claeson, *Appl. Phys. Lett.* 73 (**1998**) 3604.
[91] B. M. Siegel, *Ion-beam lithography*, in: N. G. Einspruch, R. K. Watts (Eds.), *VLSI Microstructure Science* (AP, Orlando, **1987**), 148.

[92] G. Gross, R. Kaesmaier, H. Löschner and G. Stengl, *J. Vac. Sci. Technol. B* 16 (**1998**) 3150.

[93] S. Matsui, K. Mori, K. Saigo, T. Shiokawa, K. Toyoda and S. Namba, *J. Vac. Sci. Technol. B* 4 (**1986**) 845.

[94] G. A. C. Jones, P. D. Rose and S. Brown, *J. Vac. Sci. Technol. B* 16 (**1998**) 2570.

[95] H. König, N. Mais, E. Höfling, J. P. Reithmaier, A. Forchel, M. Müssig and H. Brugger, *J. Vac. Sci. Technol. B* 16 (**1998**) 2562.

[96] B. D. Terris, L. Folks, D. Weller, J. E. E. Baglin, A. J. Kellock, H. Rothuizen and P. Vettiger, *Appl. Phys. Lett.* 75 (**1999**) 403.

[97] S. Cabrini, R. J. Barsotti, A. Carpentiero et al., *J. Vac. Sci. Technol. B* 23 (**2005**) 2806.

[98] J. H. Thywissen, K. S. Johnson, R. Younkin, N. H. Dekker, K. K. Berggren, A. P. Chu, M. Prentiss and S. A. Lee, *J. Vac. Sci. Technol. B* 15 (**1997**) 2093.

[99] B. Brezger, T. Schulze, U. Drodofsky, J. Stuhler, S. Nowak, T. Pfau and J. Mlynek, *J. Vac. Sci. Technol. B* 15 (**1997**) 2905.

[100] S. B. Hill, C. A. Haich, F. B. Dunning, G. K. Walters, J. J. McClelland, R. J. Celotta, H. G. Craighead, J. Han and D. M. Tanenbaum, *J. Vac. Sci. Technol. B* 17 (**1999**) 1087.

[101] A. Gruber, J. Gspann and H. Hoffmann, *Appl. Phys. A* 68 (**1999**) 197.

[102] Y. Ochiai, H. Watanabe, J. Fujita, M. Baba, S. Manako and S. Matsui, *Jpn. J. Appl. Phys. Pt. 1,* 32 (**1993**) 6147.

[103] F. J. Palacios, M. P. Iniguez, M. J. López and J. A. Alonso, *Phys. Rev. B* 60 (**1999**) 2908.

[104a] W. Fritzsche, H. Porwol, A. Wiegand, S. Bornmann, J. M. Köhler, *Nanostructured Materials* 10 (**1998**) 89.

[104b] W. Fritzsche, K. J. Böhm, E. Unger and J. M. Köhler, *Nanotechnology* 9 (**1998**) 177.

[105] K. I. Schiffmann, *Nanotechnology* 4 (**1993**) 163.

[106] K. L. Lee and M. Hatzakis, *J. Vac. Sci. Technol. B* 7 (**1989**) 1941.

[107] H. W. Koops, J. Kretz, M. Rudolph, M. Weber, G. Dahm and K. L. Lee, *Jpn. J. Appl. Phys. Pt. 1,* 33 (**1994**) 7099.

[108] S. Matsui and K. Mori, *J. Vac. Sci. Technol. B* 4 (**1986**) 299.

[109] S. Matsui, *Proceedings of the IEEE* 85 (**1997**) 629.

[110] S. H. M. Persson, L. Olofsson and L. Gunnarsson, *Appl. Phys. Lett.* 74 (**1999**) 2546.

[111a] N. Li, T. Yoshinobu and H. Iwasaki, *Jpn. J. Appl. Phys. Pt. 2,* 38 (**1999**) L252.

[111b] N. Li, T. Yoshinobu and H. Iwasaki, *Appl. Phys. Lett.* 74 (**1999**) 1621.

[112] K. Tang, D.-J. Fu, S. Kötter, C. R. Cantor, H. Köster, *Nucleic Acids Research* 23 (**1995**) 3126.

[113] A. Yokoo, *J. Vac. Sci. Technol. B* 21 (**2003**) 2966.

[114] M. Winzer, M. Kleiber, N. Dix and R. Wiesendanger, *Appl. Phys. A* 63 (**1996**) 617.

[115] R. Negishi, T. Hasegawa, K. Terabe, M. Aono, T. Ebihara, H. Tanaka, T. Ogawa, *Appl. Phys. Lett.* 88 (**2006**) 223111.

[116] R. Wiesendanger, *Technology of Proximal Probe Lithography* (**1995**) 162.

[117] B. Klehn and U. Kunze, *J. Appl. Phys.* 85 (**1999**) 3897.

[118] T. Junno, S. B. Carlsson, H. Q. Xu, L. Montelius and L. Samuelson, *Appl. Phys. Lett.* 72 (**1998**) 548.

[119] C. L. Cheung, H. J. Hafner, T. W. Odom, K. Kim and C. M. Lieber, *Appl. Phys. Lett.* 76 (**2000**) 3136.

[120] E. Harel, S. E. Meltzer, A. A. Requicha, M. E. Thompson and B. E. Koel, *Nano Lett.* 5 (**2005**) 2624.

[121] B. Geisler, F. Noll and N. Hampp, *Scanning* 22 (**2000**) 7.

[122] M. Abplanalp, J. Fousek and P. Gunter, *Phys. Rev. Lett.* 86 (**2001**) 5799.

[123] P. Hinterdorfer and Y. F. Dufrene, *Nat. Methods* 3 (**2006**) 347.

[124] W. Liu, L. M. Jawerth, E. A. Sparks, M. R. Falvo, R. R. Hantgan, R. Superfine, S. T. Lord and M. Guthold, *Science* 313 (**2006**) 634.

[125] N. Oyabu, O. Custance, I. Yi, Y. Sugawara and S. Morita, *Phys. Rev. Lett.* 90 (**2003**) 176102.

[126] R. Wiesendanger, *J. Vac. Sci. Technol. B* 12(2) (**1994**) 515.

[127] G. Meyer, L. Bartels and K.-H. Rieder, *Jpn. J. Appl. Phys. Pt. 1,* 37 (**1998**) 7143.

[128] C. T. Salling, *J. Vac. Sci. Technol.* 14(2) (**1996**) 1322.

[129] D. M. Eigler and E. K. Schweizer, *Nature* 344 (**1990**) 524.

[130] J. Repp, G. Meyer, S. Paavilainen, F. E. Olsson and M. Persson, *Science* 312 (**2006**) 1196.

[131] J. A. Stroscio, F. Tavazza, J. N. Crain, R. J. Celotta and A. M. Chaka, *Science* 313 (**2006**) 948.

[132] M. Alemani, L. Gross, F. Moresco, K.-H. Rieder, W. Cheng, X. Bouju, A. Gourdon and C. Joachim, *Chem. Phys. Lett.* 402 (**2005**) 180.

[133] L. Grill, F. Moresco, P. Jiang, C. Joachim, A. Gourdon and K.-H. Rieder, *Phys. Rev. B* 69 (**2004**) 035416.

[134] S.-W. Hla, K.-F. Braun, B. Wassermann and K.-H. Rieder, *Phys. Rev. Lett.* 93 (**2004**) 208302.

[135] G. Binnig, M. Despont, U. Drechsler, W. Häberle, M. Lutwyche, P. Vettiger, H. J. Mamin, B. W. Chui and T. W. Kenny, *Appl. Phys. Lett.* 74 (**1999**) 1329.

[136a] P. Vettiger, J. Brugger, M. Despont, U. Drechsler, U. Dürig, W. Häberle, M. Lutwyche, H. Rothuizen, R. Stutz, R. Widmer and G. Binnig, *Microelectron. Engn.* 46 (**1999**) 11.

[136b] P. Vettiger, M. Despont, U. Drechsler, U. Dürig, W. Häberle, M. Lutwyche, H. Ruthuizen, R. Stutz, R. Widmer and G. Binnig, F&M 107 (**1999**) 46.

[137] S. Lyuksyutov, P. B. Paramonov, S. Juhl and R. A. Vaia, *Appl. Phys. Lett.* 83 (**2003**) 4405.

[138] R. Wiesendanger, *Appl. Surf. Sci.* 54 (**1992**) 271.

[139] R. J. M. Vullers, M. Ahlskog and C. Van Haesendonck, *Appl. Surf. Sci.* 144–145 (**1999**) 584.

[140] S. Sasa, T. Ikeda, K. Anjiki and M. Inoue, *Jpn. J. Appl. Phys. Pt. 1*, 38 (**1999**) 480.

[141] S. Sasa, T. Ikeda, M. Akahori, A. Kajiuchi and M. Inoue, *Jpn. J. Appl. Phys. Pt. 1*, 38 (**1999**) 1064.

[142] A. Boisen, K. Birkenund, O. Hansen and F. Grey, *J. Vac. Sci. Technol. B* 16 (**1998**) 2977.

[143] G. Abadal, A. Boisen, Z. J. Davis, O. Hansen and F. Grey, *Appl. Phys. Lett.* 74 (**1999**) 3206.

[144] N. Farkas, G. Zhang, E. A. Evans, R. D. Ramsier and J. A. Dagata, *J. Vac. Sci. Technol. A* 21 (**2003**) 1188.

[145] S. W. Park, H. T. Soh, C. F. Quate and S.-L. Park, *Appl. Phys. Lett.* 67 (**1995**) 2415.

[146a] H. Sugimura and N. Nakagiri, *Jpn. J. Appl. Phys. Pt. 1*, 34 (**1995**) 3406.

[146b] H. Sugimura and N. Nakagiri, *Langmuir* 11 (**1995**) 3623.

[147] H. Sugimura and N. Nakagiri, *Nanotechnology* 8 (**1997**) A15.

[148] J. Hartwich, M. Sundermann, U. Kleineberg and U. Heinzmann, *Appl. Surf. Sci.* 144–145 (**1999**) 538.

[149] H. Sugimura, O. Takai and N. Nakagiri, *J. Vac. Sci. Technol. B* 17 (**1999**) 1605.

[150] K. Wilder, H. T. Soh, A. Atalar and C. F. Quate, *Rev. Sci. Instrum.* 70 (**1999**) 2822.

[151] D. Walters, M. Viani, G. T. Paloczi, T. E. Schäffer, J. P. Cleveland, M. A. Wendman, G. Gurley, V. Elings and P. K. Hansma, SPIE 3009 (**1997**) 43.

[152] T. Ando, N. Kodera, E. Takai, D. Maruyama, K. Saito and A. Toda, *Proc. Natl. Acad. Sci. USA* 98 (**2001**) 12468.

[153] M. D. Ward, *Science* 308 (**2005**) 1566.

[154] M. B. Viani, T. E. Schäffer, G. T. Paloczi, L. I. Pietrasanta, B. L. Smith, J. B. Thompson, M. Richter, M. Rief, H. E. Gaub, K. W. Plaxco, A. N. Cleland, H. G. Hansma and P. K. Hansma, *Rev. Sci. Instrum.* 70 (**1999**) 1.

[155] P. K. Hansma, G. Schitter, G. E. Fantner and C. Prater, *Science* 314 (**2006**) 601.

[156] R. Köning, *PTB-Bericht*, F-26 (**1997**) 28.

[157] M. Schwartzkopff, P. Radojkkovic, M. Enachescu, E. Hartmann and F. Koch, *J. Vac. Sci. Technol. B* 14 (**1996**) 1336.

[158] J. A. Dagata, J. Schneier, H. H. Habary, C. J. Evans, M. T. Postek and J. Bennett, *Appl. Phys. Lett.* 56 (**1990**) 2001.

[159] N. Kramer, J. Jorritsma, H. Birk and C. Schönenberger, *J. Vac. Sci. Technol. B* 13 (**1995**) 805.

[160] J. W. Lyding, G. C. Abeln, T.-C. Shen, C. Wang and J. R. Tucker, *J. Vac. Sci. Technol. B* 12(6) (**1994**) 3735.

[161] T. Mitsui, E. Hill and E. Ganz, *J. Appl. Phys.* 85 (**1999**) 522.

[162] G. C. Abeln, M. C. Hersam, D. S. Thompson, S.-T. Hwang, H. Choi, J. S. Moore and J. W. Lyding, *J. Vac. Sci. Technol. B* 16 (**1998**) 3874.

[163] J. A. Van Kan, J. L. Sanchez, T. Osipowicz and F. Watt, *Microsystem Technologies* 6 (**2000**) 82.

[164] H. J. Song, M. J. Rack, K. Abugharbieh, S. Y. Lee, V. Khan, D. K. Ferry and D. R. Allee, *J. Vac. Sci. Technol. B* 12(6) (**1994**) 3720.
[165] M. Aono, A. Kobayashi, F. Grey, H. Uchida and D.-H. Huang, *Jpn. J. Appl. Phys. Pt. 1.* 32 (**1993**) 1470.
[166] R. Otero, F. Hummelink, F. Sato, S. B. Legoas, P. Thostrup, E. Laegsgaard, I. Stensgaard, D. S. Galvao and F. Besenbacher, *Nat. Mater.* 3 (**2004**) 779.
[167a] X. Hu and P. von Blanckenhagen, *Appl. Phys. A* 68 (1999) 137.
[167b] X. Hu and P. von Blanckenhagen, *J. Vac. Sci. Technol. B* 17 (**1999**) 265.
[168] H. J. Mamin, P. H. Guethner and D. Rugar, *Phys. Rev. Lett.* 65 (**1990**) 2418.
[169] H. J. Mamin, S. Chiang, H. Birk, P. H. Guethner and D. Rugar, *J. Vac. Sci. Technol. B* 9 (**1991**) 1398.
[170] X. Hu, D. Sarid and P. von Blanckenhagen, *Nanotechnology* 10 (**1999**) 209.
[171] R. S. Becker, J. A. Golovchenko and B. S. Swartzentruber, *Nature* 325 (**1987**) 419.
[172] H. Brückl, J. Kretz, H. W. Koops and G. Reiss, *J. Vac. Sci. Technol. B* 17 (**1999**) 1350.
[173] E. E. Ehrichs, R. M. Silver and A. L. de Lozanne, *J. Vac. Sci. Technol. A* 6 (**1988**) 540.
[174] R. M. Silver, E. E. Ehrichs and A. L. de Lozanne, *Appl. Phys. Lett.* 51 (**1987**) 247.
[175] J. S. Foster, J. E. Frommer and P. C. Arnett, *Nature* 331 (**1988**) 324.
[176] F. Thibaudau, J. R. Roche and F. Salvan, *Appl. Phys. Lett.* 64 (**1994**) 523.
[177] S.-T. Yau, D. Saltz, A. Wriekat and M. H. Nayfeh, *J. Appl. Phys.* 69 (**1991**) 2970.
[178] R. Maoz, S. R. Cohen and J. Sagiv, *Adv. Mater.* 11 (**1999**) 55.
[179] C. R. K. Marrian, E. A. Dobisz and R. J. Colton, *J. Vac. Sci. Technol. A* 8 (**1990**) 3563.
[180] A. Archer, J. M. Hetrick, M. H. Nayfeh and I. Adesida, *J. Vac. Sci. Technol. B* 12(6) (**1994**) 3166.
[181] K. Kragler, E. Günther, R. Leuschner, G. Falk,, H. von Seggern and G. Saemann-Ischenko, *J. Vac. Sci. Technol. B* 14 (**1996**) 1327.
[182] M. A. McCord and R. F. W. Pease, *J. Vac. Sci. Technol. B* 4 (**1986**) 86.
[183] W. T. Müller, D. L. Klein, T. Lee, J. Clarke, P. L. McEuen and P. G. Schultz, *Science* 268 (**1995**) 272.
[184] A. Lewis, Y. Kheifetz, E. Shambrodt, A. Radko, E. Khatchatryan and C. Sukenik, *Appl. Phys. Lett.* 75 (**1999**) 2689.
[185] J. J. Davis, K. S. Coleman, K. L. Busuttil and C. B. Bagshaw, *J. Am. Chem. Soc.* 127 (**2005**) 13082.
[186a] S. S. Wong, E. Joselevich, A. T. Woolley, C. L. Cheung and C. M. Lieber, *Nature* 394 (**1998**) 52.
[186b] S. S. Wong, A. T. Woolley, E. Joselevich, C. L. Cheung and C. M. Lieber, *J. Am. Chem. Soc.* 120 (**1998**) 8557.
[187] S. Nolte, B. N. Chichkov, H. Welling, Y. Shani, K. Lieberman and H. Terkel, *Optics Lett.* 24 (**1999**) 914.
[188] R. D. Piner, J. Zhu, F. Xu, S. Hong and C. A. Mirkin, *Science* 283 (**1999**) 661.
[189] L. M. Demers, D. S. Ginger, S. J. Park, Z. Li, S. W. Chung and C. A. Mirkin, *Science* 296 (**2002**) 1836.
[190] K. B. Lee, J. H. Lim and C. A. Mirkin, *J. Am. Chem. Soc.* 125 (**2003**) 5588.
[191] J.-H. Lim and C. A. Mirkin, *Adv. Mater.* 14 (**2002**) 1474.
[192] L. Fu, X. Liu, Y. Zhang, V. P. Dravid and C. A. Mirkin, *Nano Lett.* 3 (**2003**) 757.
[193] H. Zhang, K.-B. Lee, Z. Li and C. A. Mirkin, *Nanotechnology* 14 (**2003**) 1113.
[194] K. Unal, J. Frommer and H. K. Wickramasinghe, *Appl. Phys. Lett.* 88 (**2006**) 183105.
[195] F. Chen, Q. Qing, L. Ren, Z. Wu and Z. Liu, *Appl. Nanophys. Lett.* 86 (**2005**) 123105.
[196] C. S. Ah, Y. J. Yun, J. Sung, H. J. Park, D. H. Ha, W. S. Yun, *Appl. Phys. Lett.* 88 (**2006**) 133116.
[197a] Y.-K. Choi, J. S. Lee, J. Zhu et al., *J. Vac. Sci. Technol. B* 21 (**2003**) 2951.
[197b] Y.-K. Choi, J. Zhu, J. Grunes, J. Bokor and G. A. Somorjai, *J. Phys. Chem.* 107 (**2003**) 3340.
[198] H.-L. Chen, C.-H. Chen, F.-H. Ko, T.-C. Chi, C.-T. Pan, H. C. Lin, *J. Vac. Sci. Technol. A* 20 (**2002**) 2973.
[199] K. Zhao, R. S. Averback and D. G. Cahill, *Appl. Phys. Lett.* 89 (**2006**) 053103.
[200] J. Lian, L. Wang, X. Sun, Q. Yu and R. C. Ewing, *Nano Lett.* 6 (**2006**) 1047.

Chapter 5

[1] A. Murray, M. Isaacson and I. Adesida, *Appl. Phys. Lett.* 45 **(1984)** 589.

[2] E. Kratschmer and M. Isaacson, *J. Vac. Sci. Technol. B* 5 **(1987)** 369.

[3] J. Gierak, C. Vieu, H. Launois, G. B. Asaayag and A. Septier, *Appl. Phys. Lett.* 70 **(1997)** 2049.

[4] H. J. Fan, B. Fuhrmann, R. Scholz et al., *Nanotechnology* 17 **(2006)** 231.

[5] J. S. Lee, M. S. Islam and S. Kim, *Nano Lett.* 6 **(2006)** 1487.

[6] L. Wang, D. Major, P. Paga, D. Zhang, M. G. Norton and D. N. McIlroy, *Nanotechnology* 17 **(2006)** S298.

[7] J. Viernow, D. Y. Petrovykh, F. K. Men, A. Kirakosian, J.-L. Lin and F.J. Himpsel, *Appl. Phys. Lett.* 74 **(1999)** 2125.

[8] H. Rauscher, T. A. Jung, J.-L. Lin, A. Kirakosian, F. J. Himpsel, U. Rohr and K. Müllen, *Chem. Phys. Lett.* 303 **(1999)** 363.

[9] Y. Akiyama, F. Mizukami, Y. Kiyozumi, K. Maeda, H. Izutsu and K. Sakaguchi, *Angew. Chem.* 111 (1999) 1510; *Angew. Chem. Int. Ed.* 38 **(1999)** 1420.

[10] S. P. Beaumont, P. G. Bower, T. Tamamura and C. D. W. Wilkinson, *Appl. Phys. Lett.* 38 (1981) 436.

[11] S. Y. Chou, P. R. Krauss and L. Kong, *J. Appl. Phys.* 79 **(1996)** 6101.

[12] Ch. Santschi, M. Jenke, P. Hoffmann, J. Brugger, *Nanotechnology* 17 **(2006)** 2722.

[13] S. Hashioka, M. Saito, E. Tamiya, H. Matsumura, *J. Vac. Sci. Technol. B* 21 **(2003)** 2937.

[14] T. Ishii, H. Tanaka, E. Kuramochi and T. Tamamura, *Jpn. J. Appl. Phys. Pt. 1,* 37 **(1998)** 7202.

[15] N. Hirose, H. Ohta and T. Matsui, *IEEE Transactions on Applied Superconductivity* 7 **(1997)** 2635.

[16] Y. Hsu, T. E. F. M. Standaert, G. S. Oehrlein, T. S. Kuan, E. Sayre, K. Rose, K. Y. Lee and S. M. Rossnagel, *J. Vac. Sci. Technol. B* 16 **(1998)** 3344.

[17] W. Wu, B. Cui, X. Sun, W. Zhang, L. Zhuang, L. Kong and S. Y. Chou, *J. Vac. Sci. Technol. B* 16 **(1998)** 3825.

[18] G. Myszkiewicz, J. Hohlfeld, A. J. Toonen et al., *Appl. Phys. Lett.* 85 **(2004)** 3842.

[19] O. Jessensky, F. Müller and U. Gösele, *Appl. Phys. Lett.* 72 **(1998)** 1173.

[20] M. Lee, S. Hong and D. Kim, *Appl. Phys. Lett.* 89 **(2006)** 043120.

[21] N. A. Melosh, A. Boukai, F. Diana et al., *Science* 300 **(2003)** 112.

[22] S. Y. Chou, P. R. Krauss and P. J. Renstrom, *Science* 272 **(1996)** 85.

[23] Y. Utsugi, *Nanotechnology* 3 **(1992)** 161.

[24] H. J. Mamin, P. H. Guethner and D. Rugar, *Phys. Rev. Lett.* 65 **(1990)** 2418.

[25] H. J. Mamin, S. Chiang, H. Birk, P. H. Guethner and D. Rugar, *J. Vac. Sci. Technol. B* 9 **(1991)** 1398.

[26] R. S. Becker, J. A. Golovchenko and B. S. Swartzentruber, *Nature* 325 **(1987)** 419.

[27] E. E. Ehrichs, R. M. Silver and A. L. de Lozanne, *J. Vac. Sci. Technol. A* 6 **(1988)** 540.

[28] R. M. Silver, E. E. Ehrichs and A. L. de Lozanne, *Appl. Phys. Lett.* 51 **(1987)** 247.

[29] D. Wang, L. Tsau, K. L. Wang and P. Chow, *Appl. Phys. Lett.* 67 **(1995)** 1295.

[30] O. Carny, D. E. Shalev and E. Gazit, *Nano Lett.* 6 **(2006)** 1594.

[31] T. Sakamoto, H. Kawaura, T. Baba, J. Fujita and Y. Ochiai, *J. Vac. Sci. Technol. B* 15 **(1997)** 2806.

[32] H. G. Craighead, R. E. Howard, L. D. Jackel and P. M. Mankiewich, *Appl. Phys. Lett.* 42 **(1983)** 38.

[33] P. A. Lewis, H. Ahmed and T. Sato, *J. Vac. Sci. Technol. B* 16 **(1998)** 2938.

[34] I. W. Rangelow, F. Shi, P. Hudek, P. B. Grabiec, B. Volland, E. I. Givargizov, A. N. Stepanova, L. N. Obolenskaya, E. S. Mashkova and V. A. Mollchanow, *J. Vac. Sci. Technol. B* 16 **(1998)** 3185.

[35] E. I. Givargizov, A. N. Stepanova, L. N. Obolenskaya, E. S. Mashkova, V. A. Molchanov, M. E. Givargizov and I. W. Rangelow, *Ultramicroscopy* 82 **(2000)** 57.

[36] J. A. Dagata, J. Schneier, H. H. Habary, C. J. Evans, M. T. Postek and J. Bennett, *Appl. Phys. Lett.* 56 **(1990)** 2001.

[37] M. Aono, A. Kobayashi, F. Grey, H. Uchida and D.-H. Huang, *Jpn. J. Appl. Phys. Pt. 1.* 32 **(1993)** 1470.

[38] N. Kramer, J. Jorritsma, H. Birk and C. Schönenberger, *J. Vac. Sci. Technol. B* 13 **(1995)** 805.

[39] K. Zhang, J. Lu, P.-E. Hellström, M. Östling and S.-L. Zhang, *Appl. Phys. Lett.* 88 **(2006)** 213103.

[40] W. Henschel, T. Wahlbrink, Y. M. Georgiev et al., *J. Vac Sci. Technol. B* 21 (**2003**) 2975.
[41] L. Dreeskornfeld, J. Hartwich, J. Kretz, L. Risch, W. Roesner, D. Schmitt-Landsiedel, *J. Vac. Sci. Technol. B* 20 (**2002**) 2777.
[42] A. Tilke, L. Pescini, R. H. Blick, H. Lorenz and J. P. Kotthaus, *Appl. Phys. A* 71 (**2000**) 357.
[43a] L. Guo, E. Leobandung and S. Y. Chou, *Science* 275 (**1997**) 649.
[43b] L. Guo, E. Leobandung, L. Zhuang and S. Y. Chou, *J. Vac. Sci. Technol. B* 15 (**1997**) 2840.
[44] G. A. C. Jones, P. D. Rose and S. Brown, *J. Vac. Sci. Technol. B* 16 (**1998**) 2570.
[45] D. Tennant, F. Klemens, T. Sorsch, F. Baumann, G. Timp, N. Layadi, A. Kornblit, B. J. Sapjeta, J. Rosaamilia, T. Boone, B. Weir and P. Silverman, *J. Vac. Sci. Technol. B* 15 (**1997**) 2799.
[46] M. D. Austin, W. Zhang, H. Ge, D. Wasserman, S. A. Lyon, S. Y. Chou, *Nanotechnology* 16 (**2005**) 1058.
[47] A. L. Roest, M. A. Verheijen, O. Wunnicke, S. Serafin, H. Wondergem and E. P. A. M. Bakkers, *Nanotechnology* 17 (**2006**) S271.
[48] P. Candeloro, E. Comini, C. Baratto, G. Faglia, G. Sberveglieri, R. Kumar, A. Carpentiero and E. DiFabrizio, *J. Vac. Sci. Technol. B* 23 (**2005**) 2784.
[49] D. Drouin, J. Beauvais, E. Lavallée, S. Michel, J. Mouine and R. Gauvin, *J. Vac. Sci. Technol. B* 15 (**1997**) 2269.
[50] C. Single, F. Zhou, H. Heidemeyer, F. E. Prins, D. P. Kern and E. Plies, *J. Vac. Sci. Technol. B* 16 (**1998**) 3938.
[51] M. Bockrath, D. H. Cobden, P. L. McEuen, N. G. Chopra, A. Zettl, A. Thess and R. E. Smalley, *Science* 275 (**1997**) 1922.
[52] T. W. Ebbesen and P. M. Ajayan, *Nature* 358 (**1992**) 220.
[53] W.-S. Yun, J. Kim, K.-H. Park et al., *J. Vac. Sci. Technol. A* 18 (**2000**) 1329.
[54] S. Iijima, *Nature* 354 (**1991**) 56.
[55] S. S. Wong, E. Joselevich, A. T. Woolley, C. L. Cheung and C. M. Lieber, *Nature* 394 (**1998**) 52.
[56] S. S. Wong, A. T. Woolley, E. Joselevich, C. L. Cheung and C. M. Lieber, *J. Am. Chem. Soc.* 120 (**1998**) 8557.
[57] P. G. Collins, A. Zettl, H. Bando, A. Thess and R. E. Smaley, *Science* 278 (**1997**) 100.
[58] J. Li, C. Papadopoulos, J.-M. Xu and M. Moskovits, *Appl. Phys. Lett.* 75 (**1999**) 367.
[59] M. J. Bronikowski, H. M. Manohara and B. D. Hunt, *J. Vac. Sci. Technol. A* 24 (**2006**) 1318.
[60] H. Sato, K. Hata, H. Miyake, K. Hiramatsu and Y. Saito, *J. Vac. Sci. Technol.* 23 (**2005**) 754.
[61] S. M. Vieira, K. B. K. Teo, W. I. Milne et al., *Appl. Phys. Lett.* 89 (**2006**) 022111.
[62] T. Y. Choi, D. Poulikakos, J. Tharian and U. Sennhauser, *Nano Lett.* 6 (**2006**) 1589.
[63] N. O. V. Plank, M. Ishida and R. Cheung, *J. Vac. Sci. Technol. B* 23 (**2005**) 3178.
[64] C. Klinke, J. B. Hannon, A. Afzali and P. Avouris, *Nano Lett.* 6 (**2006**) 906.
[65] A. Nojeh, A. Ural, R. F. Pease and H. Dai, *J. Vac. Sci. Technol. B* 22 (**2004**) 3421.
[66] J. Luo and J. Zhu, *Nanotechnology* 17 (**2006**) S262.
[67] Z. Ni, Q. Li, D. Zhu and J. Gong, *Appl. Phys. Lett.* 89 (**2006**) 053107.
[68] L. Jiang, Y. Kim, T. Iyoda, J. Li, K. Kitazawa, A. Fujishima and K. Hashimoto, *Adv. Mater.* 11 (**1999**) 649.
[69] J. Fujita, Y. Ohnishi, Y. Ochiai, S. Matsui, *Appl. Phys. Lett.* 68 (**1996**) 1297.
[70] J. Hartwich, M. Sundermann, U. Kleineberg and U. Heinzmann, *Appl. Surf. Sci.* 144–145 (**1999**) 538.
[71] R. C. Tiberio, H. G. Craighead, M. Lercel, T. Lau, C. W. Sheen and D. L. Allara, *Appl. Phys. Lett.* 62 (**1993**) 476.
[72] W. Geyer, V. Stadler, W. Eck, M. Zharnikov, A. Gölzhäuser and M. Grunze, *Appl. Phys. Lett.* 75 (**1999**) 2401.
[73] C. R. K. Marrian, F. K. Perkins, S. L. Brandow, T. S. Koloski, E. A. Dobisz and J. M. Calvert, *Appl. Phys. Lett.* 64 (**1994**) 390.
[74] M. A. McCord and R. F. W. Pease, *J. Vac. Sci. Technol. B* 4 (**1986**) 86.
[75] S. Manako, J. Fujita, Y. Ochiai, E. Nomura and S. Matsui, *Jpn. J. Appl. Phys. Pt. 1*, 36 (**1998**) 7773.
[76] C. R. K. Marrian, E. A. Dobisz and R. J. Colton, *J. Vac. Sci. Technol. A* 8 (**1990**) 3563.

[77] J. M. Moran and D. Maydan, *The Bell System Technical Journal* 58 **(1979)** 1027.
[78] E. Bassous, L. M. Ephrath, G. Pepper and D. J. Mikalsen, *J. Electrochem. Soc.* 130 **(1983)** 478.
[79] H. Gokan, M. Itoh and S. Esho, *J. Vac. Sci. Technol. B* 2 **(1984)** 34.
[80] T. Bjornholm, T. Hassenkam, D. R. Greve, R. D. McCullough, M. Jayaraman, S. M. Savoy, C. E. Jones and J. T. McDevitt, *Adv. Mater.* 11 **(1999)**, 1218.
[81] P. Leclere, V. Parente, J. L. Brédas, B. Francois and R. Lazzaroni, *Chem. Mater.* 10 **(1998)** 4010.
[82] C. R. Martin, *Adv. Mater.* 3 **(1991)** 457.
[83] J. N. Randall and B. L. Newell, *J. Vac. Sci. Technol. B* 12(6) **(1994)** 3631.
[84] N. C. Greenham, S. C. Moratti, D. D. C. Bradley, R. H. Friend and A. B. Holmes, *Nature* 365 **(1993)** 628.
[85] J. J. M. Halls, C. A. Walsh, N. C. Greenham, E. A. Marseglia, R. H. Friend, S. C. Moratti and A. B. Holmes, *Nature* 376 **(1995)** 498.
[86] R. H. Friend, R. W. Gymer, A. B. Holmes, J. H. Burroughes, R. N. Marks, C. Taliani, D. D. C. Bradley, D. A. Dos Santos, J. L. Brédas, M. Lögdlund and W. R. Salaneck, *Nature* 397 **(1999)** 121.
[87] H. Sirringhaus, P. J. Brown, R. H. Friend, M. M. Nielsen, K. Bechgaard, B. M. W. Langeveld-Voss, A. J. H. Spiering, R. A. J. Janssen, E. W. Meijer, P. Herwig and D. M. de Leeuw, *Nature* 401 **(1999)** 685.
[88] P. K. H. Ho, J. S. Kim, J. H. Burroughes, H. Becker, S. F. Y. Li, T. M. Brown, F. Cacialli and R. H. Friend, *Nature* 404 **(2000)** 481.
[89] F. Scheller and F. Schubert, *Biosensoren* (Berlin, **1989**).
[90] W. Göpel, *Sensors and „Smart" Molecular Nanostructures: Components for Future Information Technologies; in Sensors, vol. 8; Micro- and Nanosensor Technology* (VCH, Weinheim, **1995**) 295–336.
[91] D. Pum, M. Weinhandl, C. Hödl and U. B. Sleytr, *J. Bacteriology* 175 **(1993)** 2762.
[92] U. B. Sleytr, P. Messner, D. Pum and M. Sára, Angew. Chem. 111 **(1999)** 1098; Angew. Chem. Int. Ed. 38 (1999) 1034.
[93] A. Ulman, *Ultrathin Organic Films* (Academic Press, London, **1991**).
[94] Y. Xia and G. M. Whitesides, *Angew. Chem.* 110 **(1998)** 568; *Angew. Chem. Int. Ed.* 37 **(1998)** 550.
[95] S. P. A. Fodor, *Science* 277 **(1997)** 393.
[96] C. M. Niemeyer and D. Blohm, *Angew. Chem.* 111 **(1999)** 3039; *Angew. Chem. Int. Ed.* 38 (1999) 2865.
[97] I. W. Hamley, *Angew. Chem.* 115 **(2003)** 1730; *Angew. Chem. Int. Ed.* 42 **(2003)** 1692.
[98] D. G. Choi, S. Kim, E. Lee and S.-M. Yang, *J. Am. Chem. Soc.* 127 **(2005)** 1636.
[99] A. M. Brozell, M. A. Muha and A. N. Parikh, *Langmuir* 21 **(2005)** 11588.
[100] Y. Wang, K. Zang, S. Chua, M. S. Sander, S. Tripathy, C. F. Fonstad, *J. Phys. Chem. B* 110 **(2006)** 11081.
[101] K. W. Guarini, C. T. Black, K. R. Milkove and R. L. Sandstrom, *J. Vac. Sci. Technol. B* 19 **(2001)** 2784.
[102] J.-M. Lehn, Angew. Chem. *Int. Ed. Engl.* 27 **(1988)** 89.
[103] J.-M. Lehn, *Supramolecular Chemistry* (VCH, Weinheim, **1995**).
[104] V. Balzani, A. Credi, F. M. Raymo and J. F. Stoddart, *Angew. Chem. Interntl. Ed.* 39 **(2000)** 3349.
[105] B. C. Crandell, J. Lewis (Eds.), *Nanotechnology* (MIT Press, Cambridge, **1989**).
[106] K. E. Drexler, *Nanosystems* (John Wiley & Sons, New York, **1992**).
[107] J. T. Moore, P. D. Beale, T. A. Winningham and K. Douglas, *Appl. Phys. Lett.* 71 **(1997)** 1264.
[108] C. Huber, J. Liu, E. M. Egelseer, D. Moll, W. Knoll, U. B. Sleytr and M. Sara, *Small* 2 **(2006)** 142.
[109] Y. Huang, X. Duan and C. M. Lieber, *Small* 1 **(2005)** 142.
[110] S. K. Lee, D. S. Yun and A. M. Belcher, *Biomacromolecules* 7 **(2006)** 14.
[111] K. T. Nam, D. W. Kim, P. J. Yoo, C. Y. Chiang, N. Meethong, P. T. Hammond, Y. M. Chiang and A. M. Belcher, *Science* 312 **(2006)** 885.
[112] D. Sabbert, S. Engelbrecht and W. Junge, *Nature* 381 **(1996)** 623.
[113] W. Junge, H. Lill and S. Engelbrecht, Trends. *Biochem. Sci.* 22 **(1997)** 420.
[114] D. Stock, A. G. W. Leslie and J. E. Walker, *Science* 286 **(1999)** 1700.
[115] N. C. Seeman, *DNA and Cell Biology* 10 **(1991)** 475.

[116] N. C. Seeman, *Angew. Chem.* 110 (**1998**) 3408; *Angew. Chem. Int. Ed.* 37 (**1998**) 3220.
[117] K. B. Mullis, F. Ferré, R. A. Gibbs (Eds.), *The Polymerase Chain Reaction* (Birkhäuser, Basel, **1984**).
[118] A. Bensimon, A. Simon, A. Chiffaudel, V. Croquette, F. Heslot and D. Bensimon, *Science* 265 (**1994**) 2096.
[119] E. Braun, Y. Eichen, U. Sivan and G. Ben-Yoseph, *Nature* 391 (**1998**) 775.
[120] G. Maubach, A. Csaki, R. Seidel, M. Mertig, W. Pompe, D. Born and W. Fritzsche, *Nanotechnology* 14 (**2003**) 546.
[121] G. Maubach and W. Fritzsche, *Nano Lett.* 4 (**2004**) 607.
[122] G. Maubach, D. Born, A. Csaki and W. Fritzsche, *Small* 1 (**2005**) 619.
[123] C. M. Niemeyer, *Curr. Opinion Chem. Biol.* 4 (**2000**) 609.
[124] C. M. Niemeyer, *Appl. Phys. A* 68 (**1999**) 119.
[125] C. Hamon, T. Brandstetter and N. Windhab, *Synlett* S1 (**1999**) 940.
[126] X. J. Li, X. P. Yang, J. Qi and N. C. Seeman, *J. Am. Chem. Soc.* 118 (**1996**) 6131.
[127] E. Winfree, F. Liu, L. A. Wenzler and N. C. Seeman, *Nature* 394 (**1998**) 539.
[128] Z. Deng, S. H. Lee and C. Mao, *J. Nanosci. Nanotechnol.* 5 (**2005**) 1954.
[129] C. Lin, Y. Liu, S. Rinker and H. Yan, *ChemPhysChem* 7 (**2006**) 1641.
[130] T. H. LaBean, H. Yan, J. Kopatsch, F. R. Liu, E. Winfree, J. H. Reif and N. C. Seeman, *J. Am. Chem. Soc.* 122 (**2000**) 1848.
[131] Y. He, Y. Tian, Y. Chen, Z. Deng, A. E. Ribbe and C. Mao, *Angew. Chem. Int. Ed.* 44 (**2005**) 6694.
[132] H. Liu, Y. He, A. E. Ribbe and C. Mao, *Biomacromolecules* 6 (**2005**) 2943.
[133] H. Liu, Y. Chen, Y. He, A. E. Ribbe and C. Mao, *Angew. Chem. Int. Ed.* 45 (**2006**) 1942.
[134] J. D. Le, Y. Pinto, N. C. Seeman, K. Musier-Forsyth, T. A. Taton and R. A. Kiehl, *Nano Lett.* 4 (**2004**) 2343.
[135] K. Lund, Y. Liu, S. Lindsay and H. Yan, *J. Am. Chem. Soc.* 127 (**2005**) 17606.
[136] S. H. Park, C. Pistol, S. J. Ahn, J. H. Reif, A. R. Lebeck, C. Dwyer and T. H. LaBean, *Angew. Chem. Int. Ed.* 45 (**2006**) 735.
[137a] Y. Liu, Y. Ke and H. Yan, *J. Am. Chem. Soc.* 127 (2005) 17140.
[137b] Y. Liu, C. Lin, H. Li and H. Yan, *Angew. Chem. Int. Ed.* 44 (**2005**) 4333.
[138] A. Chworos, I. Severcan, A. Y. Koyfman, P. Weinkam, E. Oroudjev, H. G. Hansma and L. Jaeger, *Science* 306 (**2004**) 2068.
[139] E. Winfree, *J. Biomolecular Structure and Dynamics* 11 (**2000**) 263.
[140] R. D. Barish, P. W. Rothemund and E. Winfree, *Nano Lett.* 5 (**2005**) 2586.
[141a] H. Yan, T. H. LaBean, L. Feng and J. H. Reif, *Proc. Natl. Acad. Sci. USA* 100 (**2003**) 8103.
[141b] H. Yan, S. H. Park, G. Finkelstein, J. H. Reif and T. H. LaBean, *Science* 301 (**2003**) 1882.
[142] P. W. Rothemund, *Nature* 440 (**2006**) 297.
[143] S. H. Park, P. Yin, Y. Liu, J. H. Reif, T. H. LaBean and H. Yan, *Nano Lett.* 5 (**2005**) 729.
[144] J. Malo, J. C. Mitchell, C. Venien-Bryan, J. R. Harris, H. Wille, D. J. Sherratt and A. J. Turberfield, *Angew. Chem. Int. Ed.* 44 (**2005**) 3057.
[145] H. Li, S. H. Park, J. H. Reif, T. H. LaBean and H. Yan, *J. Am. Chem. Soc.* 126 (**2004**) 418.
[146] C. J. Loweth, W. B. Caldwell, X. Peng, A. P. Alivisatos and P. G. Schultz, *Angew. Chem. Int. Ed.* 38 (**1999**) 1808.
[147] Z. X. Deng, Y. Tian, S. H. Lee, A. E. Ribbe and C. D. Mao, *Angew. Chem. Int. Ed.* 44 (**2004**) 3582.
[148] J. Sharma, R. Chhabra, Y. Liu, Y. Ke and H. Yan, *Angew. Chem. Int. Ed.* 45 (**2006**) 730.
[149] F. Vögtle, *Supramolekulare Chemie* (Stuttgart, **1992**).
[150] K. A. Joliffe, P. Timmerman and D. N. Reinhoudt, *Angew. Chem.* 111 (**1999**) 983; *Angew. Chem. Int. Ed.* 38 (1999) 933.
[151] L. R. MacGillivray and J. L. Atwood, *Angew. Chem.* 111 (**1999**) 1081; *Angew. Chem. Int. Ed.* 38 (1999) 1018.
[152] G. M. Whitesides, J. P. Mathias and C. T. Seto, *Science* 254 (**1991**) 1312.
[153] A. J. Berresheim, M. Müller and K. Müllen, *Chem. Rev.* 99 (**1999**) 1747.
[154] M. Fischer and F. Vögtle, *Angew. Chem.* 111 (**1999**) 934; *Angew. Chem. Int. Ed.* 38 (**1999**) 884.

[155] L. J. Prins, P. Timmerman and D. N. Reinhoudt, *Pure & Appl. Chem.* 70 (**1998**) 1459.

[156] H.-A. Klok, K. A. Jolliffe, C. L. Schauer, I. J. Prins, J. P. Spatz, M. Möller, P. Timmerman and D. N. Reinhoudt, *J. Am. Chem. Soc.* 121 (**1999**) 7154.

[157] H. J. Choi, T. S. Lee and M. P. Suh, *Angew. Chem.* 111 (**1999**) 1490; *Angew. Chem. Int. Ed.* 38 (**1999**) 1405.

[158] L. J. Prins, J. Huskens, F. de Jong, P. Timmerman and D. N. Reinhoudt, *Nature* 398 (**1999**) 498.

[159] L. J. Prins, F. de Jong, P. Timmerman and D. N. Reinhoudt, *Nature* 408 (**2000**) 181.

[160a] A. M. Cassell, C. L. Asplund and J. M. Tour, *Angew. Chem.* 111 (**1999**) 2565; *Angew. Chem. Int. Ed.* 38 (1999) 2403.

[160b] A. M. Cassell, N. R. Franklin, T. W. Tombler, E. M. Chan, J. Han and H. Dai, *J. Am. Chem. Soc.* 121 (**1999**) 7975.

[161] W. Yang, X. Chai, L. Chi, X. Liu, Y. Cao, R. Lu, Y. Jiang, X. Tang, H. Fuchs and T. Li, *Chem. Eur. J.* 5 (**1999**) 1144.

[162] M. M. Tedesco, B. Ghebremariam, N. Sakai and S. Matile, *Angew. Chem.* 111 (**1999**) 523; *Angew. Chem. Int. Ed.* 38 (1999) 540.

[163] D. Fenske and H. Krautscheid, *Angew. Chem.* 102 (**1990**) 1513; *Angew. Chem. Int. Ed.* 29 (**1990**) 1452.

[164] G. Schön and U. Simon, *Colloid Polym. Sci.* 273 (**1995**) 101.

[165] D. Fitzmaurice, S. N. Rao, J. A. Preece, J. F. Stoddart, S. Wenger and N. Zaccheroni, *Angew. Chem.* 111 (**1999**) 1220; *Angew. Chem. Int. Ed.* 38 (**1999**) 1147.

[166] R. F. Service, *Science* 277 (**1997**) 1036.

[167] B.-H. Huisman, H. Schönherr, W. T. S. Huck, A. Friggeri, H.-J. van Manen, E. Menozzi, G. J. Vansco, F. C. J. M. van Veggel and D. N. Reinhoudt, *Angew. Chem.* 111 (**1999**) 2385; *Angew. Chem. Int. Ed.* 38 (**1999**) 1147.

[168] U.-W. Grummt, M. Geissler, T. Drechsler, H. Fuchs and R. Staub, *Angew. Chem.* 110 (**1998**) 3480; *Angew. Chem. Int. Ed.* 37 (**1998**) 3286.

[169] R. A. Reynolds, C. A. Mirkin and R. L. Letsinger, *J. Am. Chem. Soc.* 122 (**2000**) 3795.

[170a] T. A. Taton, C. A. Mirkin and R. L. Letsinger, *Science* 289 (**2000**) 1757.

[170b] T. A. Taton, R. C. Mucic, C. A. Mirkin and R. L. Letsinger, *J. Am. Chem. Soc.* 122 (**2000**) 6305.

[171] R. Möller, A. Csáki, J. M. Köhler and W. Fritzsche, *Nucleic Acids Research* 28 (**2000**).

[172] R. C. Mucic, J. J. Storhoff, C. A. Mirkin and R. L. Letsinger, *J. Am. Chem. Soc.* 120 (**1998**) 12674.

[173] G. Ladam, P. Schaad, J. C. Voegel, P. Schaaf, G. Decher and F. Cuisinier, *Langmuir* 16 (**2000**) 1249.

[174] J. A. Liddle, Y. Cui and P. Alivisatos, *J. Vac. Sci. Technol. B* 22 (**2004**) 3409.

[175] D. Xia and S. R. J. Brueck, *J. Vac. Sci. Technol. B* 22 (**2004**) 3415.

[176] J. Reichert, A. Csáki, J. M. Köhler and W. Fritzsche, *Anal. Chem.* 72 (**2000**) 6025.

[177] J. M. Köhler, A. Csáki, J. Reichert, R. Möller, W. Straube and W. Fritzsche, *Sensors and Actuators B* 76 (**2001**) 166.

[178] W. Fritzsche, K. J. Böhm, E. Unger and J. M. Köhler, *Appl. Phys. Lett.* 75 (**1999**) 2854.

[179] W. Fritzsche, J. M. Köhler, K. J. Böhm, E. Unger, T. Wagner, R. Kirsch, M. Mertig and W. Pompe, *Nanotechnology* 10 (**1999**) 331.

[180] I. Alexandre, S. Hamels, S. Dufour, J. Collet, N. Zammatteo, F. Longneville, J. L. Gala and J. Remacle, *Anal. Chem.* 295 (**2001**) 1.

[181] Y. F. Wang, D. W. Pang, Z. L. Zhang, H. Z. Zheng, J. P. Cao and J. T. Shen *J. Med. Virology* 70 (**2003**) 205.

Chapter 6

[1] M. Schubert, A. Gleichmann, M. Hemmleb, J. Albertz and J. M. Köhler, *Ultramicroscopy* 63 (**1996**) 57.

[2] D. Nyyssonen, in: N. G. Einspruch, R. K. Watts (Eds.), *VLSI Microstructure Science* (AP, Orlando, **1987**) 266.

[3] A. C. Diebold, in: *Handbook of Semiconductor Manufacturing Technology* (Marcel Dekker, New York, **2000**) 745.

[4] G. G. Barna, B. v. Eck, in: *Handbook of Semiconductor Manufacturing Technology* (Marcel Dekker, New York **2000**), 797.

[5] Ch. P. Ausschnitt, *Ion-beam lithography*, in: N. G. Einspruch, R. K. Watts (Eds.), VLSI Microstructure Science (AP, Orlando **1987**) 320.

[6] N. J. DiNardo, *Nanoscale Characterization of Surfaces and Interfaces* (VCH, Weinheim, **1994**).

[7] H. M. Marchman, *J. Vac. Sci. Technol. B* 15 (**1997**) 2155.

[8] H. M. Marchman, J. E. Griffith, J. Z. Y. Guo, J. Frackoviak and G. K. Celler, *J. Vac. Sci. Technol. B* 12 (**1994**) 3585.

[9] K. Yamazaki, A. Fujiwara, Y. Takahashi, H. Namatsu and K. Kurihara, *Jpn. J. Appl. Phys. Pt. 1*, 37 (**1998**) 6788.

[10] A. C. Chen, A. L. Flamholz, R. Rippstein, R. H. Fair, D. A. Heald and R. J. Amodeo, *J. Vac. Sci. Technol. B* 15 (**1997**) 2476.

[11] T. Miyatake, M. Hirose, T. Shoki, R. Ohkubo and K. Yamazaki, *J. Vac. Sci. Technol. B* 15 (**1997**) 2471.

[12] D. Brune, R. Hellborg, H. J. Whitlow and O. Hunderi (Eds.), *Surface Characterization* (Wiley-VCH, Weinheim, **1997**).

[13] K. Kristiansen, in: D. Brune, R. Hellborg, H. J. Whitlow, O. Hunderi (Eds.), *Surface Characterization* (Wiley-VCH, Weinheim, **1997**) 111.

[14] P. Auger, *Surf. Sci.* 48 (**1975**) 1.

[15] H. J. Dudek, in: *Angewandte Oberflächenanalyse mit SIMS, AES, und XPS* (Berlin, **1986**) 97.

[16] C.-O. A. Olsson, S. E. Hörnström, S. Hogmark, in: D. Brune, R. Hellborg, H. J. Whitlow, O. Hunderi (Eds.), *Surface Characterization* (Wiley-VCH, Weinheim, **1997**) 272.

[17] M. Grasserbauer, in: *Angewandte Oberflächenanalyse mit SIMS, AES, und XPS* (Berlin, **1986**) 1.

[18] A. R. Lodding, U. S. Södervall, in: D. Brune, R. Hellborg, H. J. Whitlow, O. Hunderi (Eds.), *Surface Characterization* (Wiley-VCH, Weinheim, **1997**) 205.

[19] I. Olefjord, in: D. Brune, R. Hellborg, H. J. Whitlow, O. Hunderi (Eds.), *Surface Characterization* (Wiley-VCH, Weinheim, **1997**) 291.

[20] M. F. Ebel, in: *Angewandte Oberflächenanalyse mit SIMS, AES, und XPS* (Berlin, **1986**) 221.

[21] M. Rief, H. Clausen-Schaumann and H. E. Gaub, *Nat. Struct. Biol.* 6 (**1999**) 346.

[22] H. Li, M. Rief, F. Oesterhelt and H. E. Gaub, *Appl. Phys. A* 68 (**1999**) 407.

[23] C. Albrecht, K. Blank, M. Lalic-Multhaler, S. Hirler, T. Mai, I. Gilbert, S. Schiftmann, T. Bayer, H. Clausen-Schaumann and H. E. Gaub, *Science* 301 (**2003**) 367.

Chapter 7

[1a] S. S. Wong, E. Joselevich, A. T. Woolley, C. L. Cheung and C. M. Lieber, *Nature* 394 (**1998**) 52.

[1b] S. S. Wong, A. T. Woolley, E. Joselevich, C. L. Cheung and C. M. Lieber, *J. Am. Chem. Soc.* 120 (**1998**) 8557.

[2] H. P. Lang et al., *Anal. Chim. Acta* 393 (**1999**) 59–65.

[3] P. C. D. Hobbs, D. W. Abraham, H. K. Wickramasinghe, *Appl. Phys. Lett.* 55 (**1989**) 2357.

[4] D. W. Abraham, F. A. McDonald, *Appl. Phys. Lett.* 56 (**1990**) 1181.

[5] R. Glöss and P. Pertsch, *F & M* 107 (**1999**) 64.

[6] P. Kim and C. M. Lieber, *Science* 286 (**1999**) 2148.

[7] D. W. Carr, L. Sekaric and H. G. Craighead, *J. Vac. Sci. Technol. B* 16 (**1998**) 3821.

[8] A. Bezryadin and C. Dekker, *J. Vac. Sci. Technol. B* 15 (**1997**) 793.

[9] W. A. Schooveld, J. Wildeman, D. Fichou, P. A. Bobbert, B. J. van Wees and T. M. Klapwijk, *Nature* 404 (**2000**) 977.

[10] K. I. Nakamatsu, M. Nagase, J.-Y. Igaki, H. Namatsu, S. Matsui, *J. Vac. Sci. Technol. B* 23 (**2005**) 2801.

[11] J. A. M. Sondag-Huethhorst, H. R. J. van Helleplutte and L. G. Fokkink, *Appl. Phys. Lett.* 64 (**1994**) 285.

[12] Y. Hsu, T. E. F. M. Standaert, G. S. Oehrlein, T. S. Kuan, E. Sayre, K. Rose, K. Y. Lee and S. M. Rossnagel, *J. Vac. Sci. Technol. B* 16 (**1998**) 3344.

[13] F. J. Himpsel, T. Jung, A. Kirakosian, J.-L. Lin, D. Y. Petrovykh, H. Rauscher and J. Viernow, *MRS Bulletin*, August (**1999**) 20.

[14] D. Rogers and H. Nejoh, *J. Vac. Sci. Technol. B* 17 (**1999**) 1323.

[15] C. Dekker and M. Ratner, *Physics World* (**2001**) 29.
[16] D. Porath, A. Bezryadin, S. de Vries and C. Dekker, *Nature* 403 (**2000**) 635.
[17] P. J. de Pablo, F. Moreno-Herrero, J. Colchero, J. Gómez Herrero, P. Herrero, A. M. Baró, P. Ordejón, J. M. Soler and E. Artacho, *Phys. Rev. Lett.* 85 (**2000**) 4992.
[18] W. Fritzsche, K. J. Böhm, E. Unger and J. M. Köhler, *Appl. Phys. Lett.* 75 (**1999**) 2854.
[19] W. Fritzsche, J. M. Köhler, K. J. Böhm, E. Unger, T. Wagner, R. Kirsch, M. Mertig and W. Pompe, *Nanotechnology* 10 (**1999**) 331.
[20] A. Richter, R. Seidel, R. Kirsch, M. Mertig, W. Pompe, J. Plaschke and H. K. Schackert, *Adr. Mat.* 12 (**2000**) 507.
[21] E. Braun, Y. Eichen, U. Sivan and B.-Y. Gdalyahu, *Nature* 391 (**1998**) 775.
[22] M. Mertig, L. Colomb Ciachi, R. Seidel, W. Pompe and A. De Vita, *Nano Lett.* 2 (**2002**) 841.
[23] K. Keren, M. Krueger, R. Gilad, G. Ben-Yoseph, U. Sivan and E. Braun, *Science* 297 (**2002**) 72.
[24] C. F. Monson and A. T. Woolley, *Nano Lett.* 3 (**2003**) 359.
[25] P. A. Smith, C. D. Nordquist, T. N. Jackson, T. S. Mayer, B. R. Martin, J. Mbindyo and T. E. Mallouk, *Appl. Phys. Lett.* 77 (**2000**) 1399.
[26] M. D. Musick, C. D. Keating, M. H. Keefe and M. J. Natan, *Chem. Mater.* 9 (**1997**) 1499.
[27] O. D. Velev and E. W. Kaler, *Langmuir* 15 (**1999**) 3693.
[28] R. Möller, A. Csáki, J. M. Köhler and W. Fritzsche, *Nucleic Acids Research* 28 (**2000**).
[29] S. J. Park (**2001**).
[30] G. Maubach, A. Csaki, R. Seidel, M. Mertig, W. Pompe, D. Born and W. Fritzsche, *Nanotechnology* 14 (**2003**) 546.
[31] C. M. Niemeyer, W. Bürger and J. Peplies, *Angew. Chem.* 110 (1998) 2391; *Angew. Chem. Int. Ed.* 37 (**1998**).
[32] C. M. Niemeyer, M. Adler, B. Pignataro, S. Lenhert, S. Gao, H. Fuchs and D. Blohm, *Nucleic Acids Research* 27 (**1999**) 4553.
[33] T. Sato and H. Ahmed, *Appl. Phys. Lett.* 70 (**1997**) 2759.
[34] M. Burghard, G. Philipp, S. Roth and K. von Klitzing, *Appl. Phys. A* 67 (**1998**) 591.
[35] K. D. Hermanson et al. (**2001**).
[36] Y. Li, F. Qian, J. Xiang and C. M. Lieber, *Mater. Today* 9 (**2006**) 18.
[37] Y. Cui, X. Duan, Y. Huang and C. M. Lieber, in *Nanowires and Nanobelts – Materials, Properties and Devices* (Ed.: Z. L. Wang), Kluwer Academic/Plenum Publishers, **2003**.
[38] Y. Wu, J. Xiang, C. Yang, W. Lu and C. M. Lieber, *Nature* 430 (**2004**) 61.
[39] L. J. Lauhon, M. S. Gudiksen, D. Wang and C. M. Lieber, *Nature* 420 (**2002**) 57.
[40] G. Zheng, W. Lu, S. Jing and C. M. Lieber, *Adv. Mater.* 16 (**2004**) 1890.
[41] R. S. Friedman, M. C. McAlpine, D. S. Ricketts, D. Ham and C. M. Lieber, *Nature* 434 (**2005**) 1085.
[42] F. Patolsky, G. Zheng and C. M. Lieber, *Anal. Chem.* 78 (**2006**) 4260.
[43] G. Zheng, F. Patolsky, Y. Cui, W. U. Wang and C. M. Lieber, *Nat. Biotechnol.* 23 (**2005**) 1294.
[44] Y. Huang, X. Duan, Y. Cui, L. J. Lauhon, K. H. Kim and C. M. Lieber, *Science* 294 (**2001**) 1313.
[45] Y. Huang, C. Y. Chiang, S. K. Lee, Y. Gao, E. L. Hu, J. De Yoreo and A. M. Belcher, *Nano Lett.* 5 (**2005**) 1429.
[46] Z. Zhong, D. Wang, Y. Cui, M. W. Bockrath and C. M. Lieber, *Science* 302 (**2003**) 1377.
[47] F. T. Edelmann, *Angew. Chem.* 111 (1999) 1473; *Angew. Chem. Int. Ed.* 38 (**1999**).
[48] T. W. Ebbesen, H. J. Lezec, H. Hiura, J. W. Bennett, H. F. Ghaemi and T. Thio, *Nature* 382 (**1996**) 54.
[49] A. Bachthold et al. (**2001**).
[50] H. W. Ch. Postma, T. Teepen, Z. Yao, M. Grifoni, C. Dekker, *Science* 293 (**2001**) 76.
[51] S. J. Tans et al. (**1997**).
[52] R. E. Martin and F. Diederich, *Angew. Chem.* 111 (**1999**) 1440; *Angew. Chem. Int. Ed.* 38 (1999).
[53] T. Bjornholm, T. Hassenkam, D. R. Greve, R. D. McCullough, M. Jayaraman, S. M. Savoy, C. E. Jones and J. T. McDevitt, *Adv. Mater.* 11 (**1999**), 1218.

[54] F. S. Schoonbeek, J. H. van Esch, B. Wegewijs, D. B. A. Rep, M. P. de Haas, T. M. Klapwijk, R. M. Kellogg and B. L. Feringa, *Angew. Chem.* 111 (1999) 1486; *Angew. Chem. Int. Ed.* 38 (**1999**).

[55] G. Zhang, P. Qi, X. Wang, Y. Lu, X. Li, R. Tu, S. Bangsaruntip, D. Mann, L. Zhang and H. Dai, *Science* 314 (**2006**) 974.

[56] Y. Y. Wei and G. Eres, *Nanotechnology* 11 (**2000**) 61.

[57] Y. Y. Wei and G. Eres, *Nanotechnology* 11 (**2000**) 61.

[58] K. Keren, R. S. Berman, E. Buchstab, U. Sivan and E. Braun, *Science* 302 (**2003**) 1380.

[59] Z. Chen, J. Appenzeller, Y. M. Lin, J. Sippel-Oakley, A. G. Rinzler, J. Tang, S. J. Wind, P. M. Solomon and P. Avouris, *Science* 311 (**2006**) 1735.

[60a] A. M. Cassell, C. L. Asplund and J. M. Tour, *Angew. Chem.* 111 (**1999**) 2565; *Angew. Chem. Int. Ed.* 38 (1999).

[60b] A. M. Cassell, N. R. Franklin, T. W. Tombler, E. M. Chan, J. Han and H. Dai, *J. Am. Chem. Soc.* 121 (**1999**) 7975.

[61] T. Lee, J. Liu, D. B. Janes, V. R. Kolagunta, J. Dicke, R. P. Andres, J. Lauterbach, M. R. Melloch, D. McInturff, J. M. Woodall and R. Reifenberger, *Appl. Phys. Lett.* 74 (**1999**) 2869.

[62] E. S. Snow, D. Park and P. M. Campbell, *Appl. Phys. Lett.* 69 (**1996**) 269.

[63] A. Tilke, L. Pescini, R. H. Blick, H. Lorenz and J. P. Kotthaus, *Appl. Phys. A* (**2000**).

[64] M. A. Reed, C. Zhou, C. J. Muller, T. P. Burgin and J. M. Tour, *Science* 278 (**1997**) 252.

[65] A. Richter, *Transducers 93*, Yokohama, June 1993 (**1993**) 310.

[66] F. Chen, Q. Qing, L. Ren, Z. Wu and Z. Liu, *Appl. Nanophys. Lett.* 86 (**2005**) 123105.

[67] M. L. Roukes, *Physica B* 263–264 (**1999**) 1.

[68] U. Simon, G. Schön and G. Schmid, *Angew. Chem.* 105 (1993) 264; *Angew. Chem. Int. Ed.* 32 (**1993**).

[69] M. F. Crommie, C. P. Lutz and D. M. Eigler, *Science* 262 (**1993**) 218.

[70] K. Matsumoto, M. Ishii and K. Segawa, *J. Vac. Sci. Technol. B* 14 (**1996**) 1331.

[71] K. Matsumoto, *Proceedings of the IEEE* 85 (**1997**) 612.

[72] E. M. Ford and H. Ahmed, *J. Vac. Sci. Technol. B* 16(6) (**1998**) 3800.

[73a] R. P. Andres, T. Bein, M. Dorogi, S. Feng, J. I. Henderson, C. P. Kubiak, W. Mahoney, R. G. Osifchin and R. Reifenberger, *Science* 272 (**1996**) 1323.

[73b] R. P. Andres, J. D. Bielefeld, J. I. Henderson, D. B. Janes, V. R. Kolagunta, C. P. Kubiak, W. Mahoney and R. G. Osifchin, *Science* 273 (**1996**) 1690.

[74] C. Zhou, M. R. Deshpande, M. A. Reed, L. Jones II and J. M. Tour, *Appl. Phys. Lett.* 71 (**1997**) 611.

[75] C. Z. Li, H.-X. He, A. Bogozi, J. S. Bunch and N. J. Tao, *Appl. Phys. Lett.* 76 (**2000**) 1333.

[76] A. Vilan, A. Shanzer and D. Cahen, *Nature* 404 (**2000**) 166.

[77] W. Zhang and S.-Y. Chou, *Appl. Phys. Lett.* 83 (**2003**) 1632.

[78] J. Collet and D. Vuillaume, *Appl. Phys. Lett.* 73 (**1998**) 2681.

[79] T. A. Fulton and G. J. Dolan, *Phys. Rev. Lett.* 59 (**1987**) 109.

[80] H. Ahmed, *J. Vac. Sci. Technol. B* 15 (**1997**) 2101.

[81] K. Blüthner, M. Götz, A. Hädicke, W. Krech, Th. Wagner, H. Mühlig, H.-J. Fuchs, U. Hübner, D. Schnelle and E.-B. Kley, *IEEE Transact. Appl. Superconductivity* 7 (**1997**) 3099.

[82] T. Koester, F. Goldschmidtboeing, B. Hadam, J. Stein, S. Altmeyer, B. Spangenberg and H. Kurz, *J. Vac. Sci. Technol. B* 16(6) (**1998**) 3804.

[83a] J. Shirakashi, K. Matsumoto, N. Miura and M. Konagai, *Jpn. J. Appl. Phys. Pt. 2*, 36 (**1997**) L2210.

[83b] J. Shirakashi, K. Matsumoto, N. Miura and M. Konagai, *Jpn. J. Appl. Phys. Pt. 2*, 36 (**1997**) L1257.

[84] S. E. Kubatkin, A. V. Danilov, A. L. Bogdanov, H. Olin and T. Claeson, *Appl. Phys. Lett.* 73 (**1998**) 3604.

[85] H. Ishikuro, T. Fujii, T. Saraya, G. Hashiguchi, T. Hiramoto and T. Ikoma, *Appl. Phys. Lett.* 68 (**1996**) 3585.

[86] A. Nakajima, T. Futatsugi, K. Kosemura, T. Fukano and N. Yokoyama, *Appl. Phys. Lett.* 70 (**1997**) 1742.

[87a] L. Guo, E. Leobandung and S. Y. Chou, *Science* 275 (**1997**) 649.

[87b] L. Guo, E. Leobandung, L. Zhuang and S. Y. Chou, *J. Vac. Sci. Technol. B* 15 (1997) 2840.

[88] A. Fujiwara, Y. Takahashi, K. Murase and M. Tabe, *Appl. Phys. Lett.* 67 (**1995**) 2957.

[89] C. Vieu, A. Pepin, J. Gierak, C. David, Y. Jin, F. Carcenac and H. Launois, *J. Vac. Sci. Technol. B* 16(6) (**1998**) 3789.

[90] K.-H. Park, J.-S. Ha, W.-S. Yun, M. Shin, K.-W. Park and E.-H. Lee, *Appl. Phys. Lett.* 71 (**1997**) 1469.

[91] M. Mejias, C. Lebreton, C. Vieu, A. Pépin, F. Carcenac, H. Launois and M. Boero, *Microelectron. Engn.* 41/42 (**1998**) 563.

[92] X. Luo, M. Tomcsanyi, A. O. Orlov, T. H. Kosel, G. L. Snider, *Appl. Phys. Lett.* 89 (**2006**) 043511.

[93] J. A. Liddle, Y. Cui and P. Alivisatos, *J. Vac. Sci. Technol. B* 22 (**2004**) 3409.

[94] X. Luo, A. L. Orlov, G. L. Snider, *J. Vac. Sci. Technol. B* 22 (**2004**) 3128.

[95] S.-J. Park, J. A. Liddle, A. Persaud, F. I. Allen and T. Schenkel, *J. Vac. Sci. Technol. B* 22 (**2004**) 3115.

[96] D. L. Klein et al. (**1997**).

[97] D. L. Klein, P. L. McEuen, J. E. Bowen Katari, R. Roth and A. P. Alivisatos, *Appl. Phys. Lett.* 68 (**1996**) 2574.

[98] M. Bockrath, D. H. Cobden, P. L. McEuen, N. G. Chopra, A. Zettl, A. Thess and R. E. Smalley, *Science* 275 (**1997**) 1922.

[99] C. Joachim and J. K. Gimzewski, *Chem. Phys. Lett.* 265 (**1997**) 353.

[100] T. Bryllert, L.-E. Wernersson, T. Löwgren, L. Samuelson, *Nanotechnology* 17 (**2006**) 227.

[101] U. Simon, *Adv. Mater.* 10 (**1999**) 1487.

[102] W. Fritzsche et al. (**2002**).

[103] S. H. M. Persson, L. Olofsson and L. Gunnarsson, *Appl. Phys. Lett.* 74 (**1999**) 2546.

[104] C. Weiss and W. Zwerger, *Europhys. Lett.* 47 (**1999**) 97.

[105] T. Morita, K. I. Nakamatsu, K. Kanda et al., *J. Vac. Sci. Technol. B* 22 (**2004**) 3137.

[106] O. Ikkala, J. Ruokolainen, R. Mäkinen, M. Torkkeli, R. Serimaa, T. Mäkelä and G. ten Brinke, *Synth. Metals* 102 (**1999**) 1498.

[107] S. J. Wind et al. (**2002**).

[108] K. Ishibashi, S. Moriyama, D. Tsuya, T. Fuse and M. Suzuki, *J. Vac. Sci. Technol. A* 24 (**2006**) 1349.

[109] K. K. Likharev, *IBM J. Res. Develop.* 32 (**1988**) 144.

[110] A. Fujiwara, Y. Takahashi, K. Yamazaki, H. Namamtsu, M. Nagase, K. Kurihara and K. Murase, *IEEE Transactions on Electron Devices* 46 (**1999**) 954.

[111] Y. A. Pashkin, Y. Nakamura and J. S. Tsai, *Appl. Phys. Lett.* 76 (**2000**) 2256.

[112] L. J. Geerligs, V. F. Anderegg, P. A. M. Holweg, J. E. Mooij, H. Pothier, D. Esteve, C. Urbina and M. H. Devoret, *Phys. Rev. Lett.* 64 (**1990**) 2691.

[113] K. Yano, T. Ishii, T. Hashimoto, T. Kobayashi, F. Murai and K. Seki, *IEEE Transactions on Electron Devices* 41 (**1994**) 1628.

[114] Z. Zhong, D. Wang, Y. Cui, M. W. Bockrath, C. M. Lieber, *Science* 302 (**2003**) 1377.

[115] J.-Y. Lee and J.-H. Cho, *Appl. Phys. Lett.* 89 (**2006**) 023124.

[116] Y. C. Tseng, K. Phoa, D. Carlton and J. Bokor, *Nano Lett.* 6 (**2006**) 1364.

[117] E. G. Emiroglu, Z. A. K. Durrani, D. G. Hasko, D. A. Williams, *J. Vac. Sci. Technol. B* 20 (**2002**) 2806.

[118] G. Bottari, D. A. Leigh and E. M. Pérez, *J. Am. Chem. Soc.* 125 (**2003**) 13360.

[119] S. Muramatsu, K. Kinabara, H. Taguchi, N. Ishii and T. Aida, *J. Am Chem. Soc.* 128 (**2006**) 3764.

[120] H.-S. Kuo, I.-S. Hwand, T.-Y. Fu, Y.-C. Lin, C.-C. Chang, T.-T. Tsong, *e-Journal of Surface Science and Nanotechnology* 4 (**2006**) 233.

[121] S. M. Vieira, K. B. K. Teo, W. I. Milne et al., *Appl. Phys. Lett.* 89 (**2006**) 022111.

[122] S. L. Gilat, A. Adronov and J. M. J. Fréchet, *Angew. Chem.* 111 (**1999**) 1519; *Angew. Chem. Int. Ed.* 38 (**1999**).

[123] G. Bauer, F. Pittner and T. Schalkhammer, *Mikrochim. Acta* 131 (**1999**) 107.

[124] L. He, M. D. Musick, S. R. Nicewarner, F. G. Salinas, S. J. Benkovic, M. J. Natan and C. D. Keating, *J. Am. Chem. Soc.* 122 (**2000**) 9071.

[125] J. Reichert, A. Csáki, J. M. Köhler and W. Fritzsche, *Anal. Chem.* (**2000**).

[126] D. Basko, G. C. La Rocca, F. Bassani and V. M. Agranovich, *Eur. Phys. J. B* 8 (**1999**) 353.

[127] J.-C. Weeber, C. Girard, J. R. Krenn, A. Dereux and J.-P. Goudonnet, *J. Appl. Phys.* 86 (**1999**) 2576.

[128] M. Bruchez Jr. et al. (**1998**).

[129] D. R. Larson, W. R. Zipfel, R. M. Williams, S. W. Clark, M. P. Bruchez, F. W. Wise and W. W. Webb, *Science* 300 (**2003**) 1434.

[130] J. R. Taylor, M. M. Fang and S. Nie, *Anal. Chem.* 72 (**2000**) 1979.

[131] D. Bimberg, N. Kirstaedter, N. N. Ledentsov, Z. I. Alferov, P. S. Kop'ev and V. M. Ustinov, *IEEE Journal of Selected Topics in Quantum Electronics* 3 (**1997**) 196.

[132] D. Bimberg, N. N. Ledentsov, M. Grundmann, N. Kirstaedter, O. G. Schmidt, M. H. Mao, V. M. Ustinov, A. Y. Egorov, A. E. Zhukov, P. S. Kop'ev, Z. I. Alferov, S. S. Ruvimov, U. Gösele and J. Heydenreich, *Jpn. J. Appl. Phys. Pt. 1,* 35 (**1996**) 1311.

[133] N. Kirstaedter, N. N. Ledentsov, M. Grundmann, D. Bimberg, V. M. Ustinov, S. S. Ruvimov, M. V. Maximov, P. S. Kop'ev, Zh. I. Alferov, *Electron. Lett.* 30 (**1994**) 1416.

[134] H. Hirayama, K. Matsunaga, M. Asada, Y. Suematusa, *Electron. Lett.* 30 (**1994**) 142–143, 14.

[135] H. Shoji, K. Mukai, N. Ohtsuka, M. Sugawara, T. Uchida, H. Ishikawa, *IEEE Photon. Technol. Lett.* 7 **1995**) 1385.

[136] A. Moritz, R. Wirth, A. Hangleiter, A. Kurtenbach, K. Eberl, *Appl. Phys. Lett.* 69 (**1996**) 212.

[137] H. Weller, *Angew. Chem.* 110 (1998) 1748; *Angew. Chem. Int. Ed.* 37 (**1998**).

[138] F. Geiger, M. Stoldt, H. Schweizer, P. Bäuerle and E. Umbach, *Adv. Mater.* 5 (1993) 922.

[139] M. Granström, M. Berggren and O. Inganäs, *Science* 267 (**1995**) 1479.

[140] D. F. Moore, *FED Journal* 5 (**1994**) 25.

[141] H.-I. Lee, S.-S. Park, D.-I. Park, S.-H. Hahm and J.-H. Lee, *J. Vac. Sci. Technol. B* 16 (**1998**) 762.

[142] W. A. de Heer, A. Ch_atelain and D. Ugarte, *Science* 270 (**1995**) 1179.

[143] A. G. Rinzler, J. H. Hafner, P. Nikolaev, L. Lou, S. G. Kim, D. Tománek, P. Nordlander, D. T. Colbert and R. E. Smalley, *Science* 269 (**1995**) 1550.

[144] H. Yamamoto, J. Wilkinson, J. P. Long, K. Bussmann, J. A. Christodoulides and Z. H. Kafafi, *Nano Lett.* 5 (**2006**) 2485.

[145] C. Gui, G. J. Veldhuis, T. M. Koster, P. V. Lambeck, J. W. Berenschot, J. G. E. Gardeniers and M. Elwenspoek, *Microsystem Technologies* 5 (**1999**) 138.

[146] R. R. Panepucci, B. H. Kim, V. R. Almeida and M. D. Jones, *J. Vac. Sci. Technol. B* 22 (**2004**) 3348.

[147] A. Kogan, S. Amasha, M. A. Kastner, *Science* 304 (**2004**) 1293.

[148] M. Khan, A. K. Sood, F. L. Deepak, C. N. R. Rao, *Nanotechnology* 17 (**2006**) S287.

[149] S. Zhang, W. Fan, B. K. Minhas, A. Frauenglass, *J. Vac. Sci. Technol. B* 22 (**2004**) 3327.

[150] M. N. Baibich, J. M. Broto, A. Fert, F. Nguyen Van Dau, F. Petroff, P. Eitenne, G. Creuzet, A. Friederich and J. Chazelas, *Phys. Rev. Lett.* 61 (**1988**) 2472.

[151] B. Dieny, V. S. Speriosu, S. Metin, S. S. P. Parkin, B. A. Gurney, P. Baumgart and D. R. Wilhoit, *J. Appl. Phys.* 69 (**1991**) 4774.

[152] L. Piraux, J. M. George, J. F. Despres, C. Leroy, E. Ferain, R. Legras, K. Ounadjela and A. Fert, *Appl. Phys. Lett.* 65 (**1994**) 2484.

[153] X. Yang, A. Eckert, K. Mountfield, H. Gentile and C. Seiler, *J. Vac. Sci. Technol. B* 21 (**2003**) 3017.

[154] L. Kong, L. Zhuang and S. Y. Chou, *IEEE Transactions on Magnetics* 33 (**1997**) 3019.

[155] T. Ohkubo, J. Kishigami, K. Yanagisawa and R. Kaneko, *IEEE Transactions on Magnetics* 27 (**1991**) 5286.

[156] H. Brückl, A. Hütten, G. Reiss, A. Becker and A. Pühler, *Statusseminar Magnetoelektronik* (**2000**).

[157] R. L. Edelstein, C. R. Tamanaha, P. E. Sheehan, M. M. Miller, D. R. Baselt, L. J. Whitman and R. J. Colton, *Biosensors and Bioelectronics* 14 (**2000**) 805.

[158] J. Schotter, P. B. Kamp, A. Becker, A. Pühler, D. Brinkmann, W. Schlepper, H. Brückl and G. Reiss, *IEEE Trans. Magnetics* 38 (**2002**) 3365.

[159] V. M. Mirsky, T. Hirsch, S. A. Piletsky and O. S. Wolfbeis, *Angew. Chem.* 111 (1999) 1179; *Angew. Chem. Int. Ed.* 38 (**1999**).

[160] J. Fritz, M. K. Baller, H. P. Lang, H. Rothuizen, P. Vettiger, E. Meyer, H.-J. Güntherodt, C. Gerber and J. K. Gimzewski, *Science* 288 (**2000**) 316.

[161] R. McKendry et al. (**2002**).

[162] T. Takeuchi and T. Matsui (**1996**).

[163] H. Shi, W.-B. Tsai, M. D. Garrison, S. Ferrari and B. D. Ratner, *Nature* 398 (**1999**) 593.

[164] V. Haguet, D. Martin, L. Marcon et al., *Appl. Phys. Lett.* 84 (**2004**) 1213.

[165] C. A. Savran, S. M. Knudsen, A. D. Ellington and S. R. Manalis, *Anal. Chem.* 76 (**2004**) 3194.

[166] P. Dutta, C. A. Tipple, N. V. Lavrik et al., *Anal. Chem.* 75 (**2006**) 2342.

[167] G. A. Campbell and R. Mutharasan, *Anal. Chem.* 78 (**2006**) 2328.

[168] S. W. Bishnoi, C. J. Rozell, C. S. Levin et al., *Nano Lett.* 6 (**2006**) 1687.

[169] S. W. Turner (**1998**).

[170] S. W. Turner, M. Cabodi and H. G. Craighead, *Phys. Rev. Lett.* 88 (**2002**) 128103.

[171] A. Meller and D. Branton (**2002**).

[172] B. Schäfer, H. Geheimhardt and K. O. Greulich, *Angew. Chem. Int. Ed.* 40 (**2001**) 4663.

[173] J. Yuqiu, C.-B. Juang, D. Keller, C. Bustamante, D. Beach, T. Houseal and E. Builes, *Nanotechnology* 3 (**1992**) 16.

[174] G. Che, S. A. Miller, E. R. Fisher and C. R. Martin, *Anal. Chem.* 71 (**1999**) 3187.

[175] Q. Pei and O. Inganäs, *J. Phys. Chem.* 96 (**1992**) 10507.

[176] S. R. Fletcher, F. Dumur, M. M. Pollar, B. L. Feringa, *Science* 310 (**2005**) 80.

[177] V. Balzani, A. Credi, M. Venturi, Molecular Devices and Machines (Wiley-VCH **2003**).

[178] J. Howard, A. J. Hunt and S. Baek, *Methods in Cell Biology* 39 (**1993**) 137.

[179] J. R. Dennis, J. Howard and V. Vogel, *Nanotechnology* 10 (**1999**) 232.

[180] L. Ionov, M. Stamm and S. Diez, *Nano Lett.* 6 (**2006**) 1982.

[181] L. Limberies and R. J. Stewart, *Nanotechnology* 11 (**2000**) 47.

[182] R. Stracke, K. J. Böhm, E. Unger, *8th Conference on Molecular Nanotechnology*, Bethesda, USA (Nov. **2000**) 5.

[183] C. T. Lin, M. T. Kao, K. Kurabayashi and E. Meyhofer, *Small* 2 (**2006**) 281.

[184] M. G. van den Heuvel, M. P. de Graaff and C. Dekker, *Science* 312 (**2006**) 910.

[185] T. M. Duncan, V. V. Bulygin, Y. Zhou, M. L. Hutcheon and R. L. Cross, *Proc. Natl. Acad. Sci. USA* 92 (**1995**) 10964.

[186] W. Junge, H. Lill and S. Engelbrecht, *Trends. Biochem. Sci.* 22 (**1997**) 420.

[187] D. Sabbert, S. Engelbrecht and W. Junge, *Proc. Natl. Acad. Sci. USA* 94 (**1997**) 4401.

[188] D. Stock, A. G. W. Leslie and J. E. Walker, *Science* 286 (**1999**) 1700.

[189] H. Noji, R. Yasuda, M. Yoshida and K. J. Kinosita, *Nature* 386 (**1997**) 299.

[190] Y. Sambongi, Y. Iko, M. Tanabe, H. Omote, A. Iwamoto-Kihara, I. Ueda, T. Yanagida, Y. Wada and M. Futai, *Science* 286 (**1999**) 1722.

[191] O. Pänke, K. Gumbiowski, W. Junge and S. Engelbrecht, *FEBS Lett.* 472 (**2000**) 34.

[192] G. Steinberg-Yfrach, J.-L. Rigaud, E. N. Durantini, A. L. Moore, D. Gust and T. A. Moore, *Nature* 392 (**1998**) 479.

[193] J. L. Pearson, D. R. S. Cumming, *J. Vac. Sci. Technol. B* 23 (**2005**) 2793.

[194] D. Xia and S. R. J. Brueck, *J. Vac. Sci. Technol. B* 23 (**2005**) 2694.

[195] N. W. Liu, A. Datta, C. Y. Liu and Y. L. Wang, *Appl. Phys. Lett.* 82 (**2003**) 1281.

[196] H. Cao, J. O. Tegenfeldt, R. H. Austin, S. Y. Chou, *Appl. Phys. Lett.* 81 (**2002**) 3058.

[197] S. M. Stavis, J. B. Edel, Y. Li et al., *Nanotechnology* 16 (**2005**) S314.

[198] R. M. Reano and S. W. Pang (**2005**).

[199] S. Arscott, D. Troadec, *Appl. Phys. Lett.* 87 (**2005**) 134101.

[200] R. Karnik, K. Castelino, Ch. Duan and A. Majumdar, *Nano Lett.* 6 (**2006**) 1735.

[201] L. Dong, A. K. Agarwal, D. J. Beebe and H. Jiang, *Nature* 442 (**2006**) 551.

[202] R. Hernandez, H.-R. Tseng, J. W. Wong, J. F. Stoddart and J. I. Zink, *J. Am. Chem. Soc.* 126 (**2004**) 3370.

Chapter 8

[1a] S. Spagocci and T. Fountain, *Manuskript* (1999).

[1b] S. Spagocci and T. Fountain, *Manuskript* (1999).

[2] S. C. Benjamin and N. F. Johnson, *Appl. Phys. Lett.* 70 (1997) 2321.

[3] E. G. Wilson, *Jpn. J. Appl. Phys. Pt. 1*, 34 (1995) 3775.

[4] Y. Wada, T. Uda, M. Lutwyche, S. Kondo and S. Heike, *J. Appl. Phys.* 74 (1993) 7321.

[5] W. B. Choi, D. S. Chung, J. H. Kang, H. Y. Kim, Y. W. Jin, I. T. Han, Y. H. Lee, J. E. Jung, N. S. Lee, G. S. Park and J. M. Kim, *Appl. Phys. Lett.* 75 (1999) 3129.

[6] B.-H. Huisman, H. Schönherr, W. T. S. Huck, A. Friggeri, H.-J. van Manen, E. Menozzi, G. J. Vansco, F. C. J. M. van Veggel and D. N. Reinhoudt, *Angew. Chem.* 111 (1999) 2385; *Angew. Chem. Int. Ed.* 38 (1999).

[7] U.-W. Grummt, M. Geissler, T. Drechsler, H. Fuchs and R. Staub, *Angew. Chem.* 110 (1998) 3480; *Angew. Chem. Int. Ed.* 37 (1998).

[8a] R. P. Andres, T. Bein, M. Dorogi, S. Feng, J. I. Henderson, C. P. Kubiak, W. Mahoney, R. G. Osifchin and R. Reifenberger, *Science* 272 (1996) 1323.

[8b] R. P. Andres, J. D. Bielefeld, J. I. Henderson, D. B. Janes, V. R. Kolagunta, C. P. Kubiak, W. Mahoney and R. G. Osifchin, *Science* 273 (1996) 1690.

[9] M. W. Berns, W. H. Wright, R. Wiegand-Streubing, *Int. Rev. Cytol.* 129 (1991) 1.

[10] S. C. Kuo, M. P. Sheetz, *Trends in Cell Biology* 2 (1992) 116.

[11] R. Simmons, J. Sleep, A. Trombetta and P. Marya, in: R. R. H. Coombs and D. W. Robinson (Eds.), *Nanotechnology in Medicine and the Biosciences*, Gordon & Breach Publishers, Australia-Canada-China-France-Germany-India-Japan-Luxembourg-Mala ysia-Netherlands-Russia-Singapore-Switzerland-Thailand-United Kingdom (1996) 231.

[12] J. Dapprich and N. Nicklaus, *Manuskript* (1998).

[13] G. V. Shivashankar and A. Libchaber, *Appl. Phys. Lett.* 71 (1997) 3727.

[14] J. Kondoh, Y. Matsui, S. Shiokawa and W. B. Wlodarski, *Transducer* 93, Yokohama, June 1993 (1993) 534.

[15] M. Despont, J. Brugger, U. Drechsler, U. Dürig, W. Häberle, M. Lutwyche, H. Rothuizen, R. Stutz, R. Widmer, G. Binnig, H. Rohrer and P. Vettiger, *Sensors and Actuators A* 80 (2000) 100.

[16a] P. Vettiger, J. Brugger, M. Despont, U. Drechsler, U. Dürig, W. Häberle, M. Lutwyche, H. Rothuizen, R. Stutz, R. Widmer and G. Binnig, *Microelectron. Engn.* 46 (1999) 11

[16b] P. Vettiger, M. Despont, U. Drechsler, U. Dürig, W. Häberle, M. Lutwyche, H. Ruthuizen, R. Stutz, R. Widmer and G. Binnig, *F&M* 107 (1999) 46.

[17] C. S. Lent, P. D. Tougaw, W. Porod and G. H. Bernstein, *Nanotechnology* 4 (1993) 49.

[18] T. Rueckes, K. Kim, E. Joselevich, G. Y. Tseng, C.-L. Cheung and C. M. Lieber, *Science* 289 (2000) 94.

[19] B. E. Kane et al. (1998).

[20] K. W. Lux and K. J. Rodriguez, *Nano Lett.* 6 (2006) 288.

[21] J. D. Meindl, Q. Chen and J. A. Davis, *Science* 293 (2001) 2044.

[22] G. Y. Tseng and J. C. Ellenbogen, *Science* 294 (2001) 1293.

[23] J. C. Ellenbogen and J. C. Love, *Proceedings of the IEEE* 88 (2000) 386.

[24] Y. Utsugi, *Nanotechnology* 3 (1992) 161.

[25] A. Boisen, K. Birkenund, O. Hansen and F. Grey, *J. Vac. Sci. Technol. B* 16 (1998) 2977.

[26] T. Mitsui, E. Hill and E. Ganz, *J. Appl. Phys.* 85 (1999) 522.

[27] A. Murray, M. Isaacson and I. Adesida, *Appl. Phys. Lett.* 45 (1984) 589.

[28] E. Kratschmer and M. Isaacson, *J. Vac. Sci. Technol. B* 5 (1987) 369.

[29a] S. Sasa, T. Ikeda, K. Anjiki and M. Inoue, *Jpn. J. Appl. Phys. Pt. 1*, 38 (1999) 480.

[29b] S. Sasa, T. Ikeda, M. Akahori, A. Kajiuchi and M. Inoue, *Jpn. J. Appl. Phys. Pt. 1*, 38 (1999) 1064.

[30] B. Klehn and U. Kunze, *J. Appl. Phys.* 85 (1999) 3897.

[31] R. Wiesendanger, *Jpn. J. Appl. Phys. Pt. 1*, 34 (1995) 3388.

[32] X. Hu, D. Sarid and P. von Blanckenhagen, *Nanotechnology* 10 (1999) 209.

[33] H. Brückl, J. Kretz, H. W. Koops and G. Reiss, *J. Vac. Sci. Technol. B* 17 (**1999**) 1350.

[34a] X. Hu and P. von Blanckenhagen, *Appl. Phys. A* 68 (**1999**) 137.

[34b] X. Hu and P. von Blanckenhagen, *J. Vac. Sci. Technol. B* 17 (**1999**) 265.

[35] H. J. Mamin, S. Chiang, H. Birk, P. H. Guethner and D. Rugar, *J. Vac. Sci. Technol. B* 9 (**1991**) 1398.

[36] K. L. Lee and M. Hatzakis, *J. Vac. Sci. Technol. B* 7 (**1989**) 1941.

[37] G. Simon, A. M. Haghiri-Gosnet, J. Bourneix, D. Decanini, Y. Chen, F. Rousseaux, H. Launois and B. Vidal, *J. Vac. Sci. Technol. B* 15 (**1997**) 2489.

[38] A. M. Haghiri-Gosnet, C. Vieu, G. Simon, F. Carcenac, A. Madouri, Y. Chen, F. Rousseaux and H. Launois, *J. Vac. Sci. Technol. B* 13 (**1995**) 3066.

[39] S. Zhang, W. Fan, B. K. Minhas, A. Frauenglass, *J. Vac. Sci. Technol. B* 22 (**2004**) 3327.

[40] S. Cabrini, R. J. Barsotti, A. Carpentiero et al., *J. Vac. Sci. Technol. B* 23 (**2005**) 2806.

[41] H. G. Craighead, *J. Appl. Phys.* 55 (**1984**) 4430.

[42] W. Chen and H. Ahmed, *J. Vac. Sci. Technol. B* 11 (**1993**) 2519.

[43] S. Y. Chou, P. R. Krauss, W. Zhang, L. Guo and L. Zhuang, *J. Vac. Sci. Technol. B* 15 (**1997**) 2897.

[44] H. Lee, S. Hong and K. Yang, *Appl. Phys. Lett.* 88 (**2006**) 143112.

[45] S. H. Choi, K. L. Wang, M. S. Leung et al., *J. Vac. Sci. Technol. B* 18 (**2000**) 1326.

[46] R. M. Silver, E. E. Ehrichs and A. L. de Lozanne, *Appl. Phys. Lett.* 51 (**1987**) 247.

[47] W. Wu, B. Cui, X. Sun, W. Zhang, L. Zhuang, L. Kong and S. Y. Chou, *J. Vac. Sci. Technol. B* 16 (**1998**) 3825.

[48] X. Yang, A. Eckert, K. Mountfield, H. Gentile and C. Seiler, *J. Vac. Sci. Technol. B* 21 (**2003**) 3017.

[49] R. Wiesendanger, *Appl. Surf. Sci.* 54 (**1992**) 271.

[50] G. Meyer, L. Bartels and K.-H. Rieder, *Jpn. J. Appl. Phys. Pt. 1*, 37 (**1998**) 7143.

[51] A. Lewis, Y. Kheifetz, E. Shambrodt, A. Radko, E. Khatchatryan and C. Sukenik, *Appl. Phys. Lett.* 75 (**1999**) 2689.

[52] M. D. Austin, W. Zhang, H. Ge, D. Wasserman, S. A. Lyon, S. Y. Chou, *Nanotechnology* 16 (**2005**) 1058.

[53] H. J. Song, M. J. Rack, K. Abugharbieh, S. Y. Lee, V. Khan, D. K. Ferry and D. R. Allee, *J. Vac. Sci. Technol. B* 12(6) (**1994**) 3720.

[54] Y. Hsu, T. E. F. M. Standaert, G. S. Oehrlein, T. S. Kuan, E. Sayre, K. Rose, K. Y. Lee and S. M. Rossnagel, *J. Vac. Sci. Technol. B* 16 (**1998**) 3344.

[55] J. S. Foster, J. E. Frommer and P. C. Arnett, *Nature* 331 (**1988**) 324.

[56] T. Schaub, R. Wiesendanger and H.-J. Güntherodt, *Nanotechnology* 3 (**1992**) 77.

[57] G. Myszkiewicz, J. Hohlfeld, A. J. Toonen et al., *Appl. Phys. Lett.* 85 (**2004**) 3842.

[58] F. Thibaudau, J. R. Roche and F. Salvan, *Appl. Phys. Lett.* 64 (**1994**) 523.

[59] G. A. C. Jones, P. D. Rose and S. Brown, *J. Vac. Sci. Technol. B* 16 (**1998**) 2570.

[60] A. L. Roest, M. A. Verheijen, O. Wunnicke, S. Serafin, H. Wondergem and E. P. A. M. Bakkers, *Nanotechnology* 17 (**2006**) S271.

[61] R. S. Becker, J. A. Golovchenko and B. S. Swartzentruber, *Nature* 325 (**1987**) 419.

[62] Y. Lin, M. H. Hong, T. C. Chong et al., *Appl. Phys. Lett.* 89 (**2006**) 041108.

[63] T. M. Bloomstein, P. W. Juodawlkis, R. B. Swint et al., *J. Vac. Sci. Technol. B* 23 (**2005**) 2617.

[64] T. Bryllert, L.-E. Wernersson, T. Löwgren, L. Samuelson, *Nanotechnology* 17 (**2006**) 227.

[65] Z. Zhang, J. Lu, P. E. Hellstrom, M. Ostling and S. L. Zhang, *Appl. Phys. Lett.* 88 (**2006**) No 213103

[66a] J. Shirakashi, K. Matsumoto, N. Miura and M. Konagai, *Jpn. J. Appl. Phys. Pt. 2*, 36 (**1997**) L2210.

[66b] J. Shirakashi, K. Matsumoto, N. Miura and M. Konagai, *Jpn. J. Appl. Phys. Pt. 2*, 36 (**1997**) L1257.

[67a] S. Y. Chou, P. R. Krauss and L. Kong, *J. Appl. Phys.* 79 (**1996**) 6101.

[67b] S. Y. Chou, P. R. Krauss and P. J. Renstrom, *Science* 272 (**1996**) 85.

[68] S. P. Beaumont, P. G. Bower, T. Tamamura and C. D. W. Wilkinson, *Appl. Phys. Lett.* 38 (**1981**) 436.

[69] D. Park, C. R. K. Marrian, D. Gammon, R. Bass, P. Isaacson and E. Snow, *J. Vac. Sci. Technol. B* 16 (**1998**) 3891.

[70] J. Sone, J. Fujita, Y. Ochiai, S. Manako, S. Matsui, E. Nomura, T. Baba, H. Kawaura, T. Sakamoto, C. D. Chen, Y. Nakamura and J. S. Tsai, *Nanotechnology* 10 (**1999**) 135.

[71] J. Yamamoto, S. Uchino, H. Ohta, T. Yoshimura and F. Murai, *J. Vac. Sci. Technol. B* 15 (**1997**) 2868.

[72] S. Matsui, K. Mori, K. Saigo, T. Shiokawa, K. Toyoda and S. Namba, *J. Vac. Sci. Technol. B* 4 (**1986**) 845.

[73] A. N. Broers, *J. Electrochem. Soc.* 128 (**1981**) 166.

[74] G. Binnig, M. Despont, U. Drechsler, W. Häberle, M. Lutwyche, P. Vettiger, H. J. Mamin, B. W. Chui and T. W. Kenny, *Appl. Phys. Lett.* 74 (**1999**) 1329.

[75] S. Y. Chou, P. R. Krauss and P. J. Renstrom, *Appl. Phys. Lett.* 67 (**1995**) 3114.

[76] C. R. K. Marrian, E. A. Dobisz and R. J. Colton, *J. Vac. Sci. Technol. A* 8 (**1990**) 3563.

[77] A. Archer, J. M. Hetrick, M. H. Nayfeh and I. Adesida, *J. Vac. Sci. Technol. B* 12(6) (**1994**) 3166.

[78] L. Malmqvist, A. L. Bogdanov, L. Montelius and H. M. Hertz, *J. Vac. Sci. Technol. B* 15 (**1997**) 814.

[79] K. Kurihara, K. Iwadate, H. Namatsu, M. Nagase, H. Takenaka and K. Murase, *Jpn. J. Appl. Phys. Pt. 1*, 34 (**1995**) 6940.

[80] W. Srituravanich, S. Durant, H. Lee, C. Sun, X. Zhang, *J. Vac. Sci. Technol. B* 23 (**2005**) 2636.

[81] Q. Leonhard, D. Malueg, J. Wallace et al., *J. Vac. Sci. Technol. B* 23 (**2005**) 2896.

[82] H.-L. Chen, C.-H. Chen, F.-H. Ko, T.-C. Chi, C.-T. Pan, H. C. Lin, *J. Vac. Sci. Technol. A* 20 (**2002**) 2973.

[83] M. Switkes and M. Rothschild, *J. Vac. Sci. Technol. B* 19 (**2001**) 2353.

[84] R. R. Panepucci, B. H. Kim, V. R. Almeida and M. D. Jones, *J. Vac. Sci. Technol. B* 22 (**2004**) 3348.

[85] T. Ito, T. Yamada, Y. Inao, T. Yamaguchi, N. Mizutani and R. Kuroda, *Appl. Phys. Lett.* 89 (**2006**) No 033113.

[86] H. Yoon, T.-I. Kim, S. Choi, K.-Y. Suh, M.-J. Kim and H.-H. Lee, *Appl. Phys. Lett.* 88 (**2006**) 254104.

[87] H. W. Koops, J. Kretz, M. Rudolph, M. Weber, G. Dahm and K. L. Lee, *Jpn. J. Appl. Phys. Pt. 1*, 33 (**1994**) 7099.

[88a] Y.-K. Choi, J. S. Lee, J. Zhu et al., *J. Vac. Sci. Technol. B* 21 (**2003**) 2951.

[88b] Y.-K. Choi, J. Zhu, J. Grunes, J. Bokor and G. A. Somorjai, *J. Phys. Chem.* 107 (**2003**) 3340.

[89] U. Staufer, L. Scandella and R. Wiesendanger, *Zeitschrift für Physik B – Condensed Matter* 77 (**1989**) 281.

[90] S. Skaberna, M. Versen, B. Klehn, U. Kunze, D. Reuter and A. D. Wiek, *Ultramicroscopy* 82 (**2000**) 153.

[91] J. Hartwich, M. Sundermann, U. Kleineberg and U. Heinzmann, *Appl. Surf. Sci.* 144–145 (**1999**) 538.

[92] S. Hong, J. Zhu and C. A. Mirkin, *Science* 286 (**1999**) 523.

[93] J. A. M. Sondag-Huethhorst, H. R. J. van Helleplutte and L. G. Fokkink, *Appl. Phys. Lett.* 64 (**1994**) 285.

[94] K. Kragler, E. Günther, R. Leuschner, G. Falk„ H. von Seggern and G. Saemann-Ischenko, *J. Vac. Sci. Technol. B* 14 (**1996**) 1327.

[95] D. W. Carr, L. Sekaric and H. G. Craighead, *J. Vac. Sci. Technol. B* 16 (**1998**) 3821.

[96] E. S. Snow, W. H. Juan, S. W. Pang and P. M. Campbell, *Appl. Phys. Lett.* 66 (**1995**) 1729.

[97] N. Kramer, J. Jorritsma, H. Birk and C. Schönenberger, J. Vac. Sci. Technol. B 13 (**1995**) 805.

[98] T. Tada and T. Kanayama, *J. Vac. Sci. Technol. B* 16 (**1998**) 3934.

[99] W. Henschel, T. Wahlbrink, Y. M. Georgiev et al., *J. Vac Sci. Technol. B* 21 (**2003**) 2975.

[100] J. W. Lyding, G. C. Abeln, T.-C. Shen, C. Wang and J. R. Tucker, *J. Vac. Sci. Technol. B* 12(6) (**1994**) 3735.

[101] L. Dreeskornfeld, J. Hartwich, J. Kretz, L. Risch, W. Roesner, D. Schmitt-Landsiedel, *J. Vac. Sci. Technol. B* 20 (**2002**) 2777.

[102] H. Sugimura, O. Takai and N. Nakagiri, *J. Vac. Sci. Technol. B* 17 (**1999**) 1605.

[103a] H. Sugimura and N. Nakagiri, *Jpn. J. Appl. Phys. Pt. 1*, 34 (**1995**) 3406.

[103b] H. Sugimura and N. Nakagiri, *Langmuir* 11 (**1995**) 3623.

[104a] N. Li, T. Yoshinobu and H. Iwasaki, *Jpn. J. Appl. Phys. Pt. 2,* 38 (**1999**) L252.

[104b] N. Li, T. Yoshinobu and H. Iwasaki, *Appl. Phys. Lett.* 74 (**1999**) 1621.

[105] S. W. Park, H. T. Soh, C. F. Quate and S.-L. Park, *Appl. Phys. Lett.* 67 (**1995**) 2415.

[106] P. Candeloro, E. Comini, C. Baratto, G. Faglia, G. Sberveglieri, R. Kumar, A. Carpentiero and E. DiFabrizio, *J. Vac. Sci. Technol. B* 23 (**2005**) 2784.

[107] R. J. M. Vullers, M. Ahlskog and C. Van Haesendonck, *Appl. Surf. Sci.* 144–145 (**1999**) 584.

[108] Ch. Santschi, M. Jenke, P. Hoffmann, J. Brugger, *Nanotechnology* 17 (**2006**) 2722.

[109] K. Matsumoto, *Proceedings of the IEEE* 85 (**1997**) 612.

[110] D. Tennant, F. Klemens, T. Sorsch, F. Baumann, G. Timp, N. Layadi, A. Kornblit, B. J. Sapjeta, J. Rosaamilia, T. Boone, B. Weir and P. Silverman, *J. Vac. Sci. Technol. B* 15 (**1997**) 2799.

[111] H. J. Fan, B. Fuhrmann, R. Scholz et al., *Nanotechnology* 17 (**2006**) 231.

Index

Nanotechnology. M. Köhler and W. Fritzsche

ISBN: 978-3-527-31871-1

F

G

K

L

M

N

Q

R